Sustainable Agriculture Reviews

Volume 45

Other Publications by Dr. Eric Lichtfouse

Books
Scientific Writing for Impact Factor Journals
https://www.novapublishers.com/catalog/product_info.php?products_id=42242

Environmental Chemistry
http://www.springer.com/978-3-540-22860-8

Sustainable Agriculture
Volume 1: http://www.springer.com/978-90-481-2665-1
Volume 2: http://www.springer.com/978-94-007-0393-3

Book series
Environmental Chemistry for a Sustainable World
http://www.springer.com/series/11480

Sustainable Agriculture Reviews
http://www.springer.com/series/8380

Journal
Environmental Chemistry Letters
http://www.springer.com/10311

Sustainable agriculture is a rapidly growing field aiming at producing food and energy in a sustainable way for humans and their children. Sustainable agriculture is a discipline that addresses current issues such as climate change, increasing food and fuel prices, poor-nation starvation, rich-nation obesity, water pollution, soil erosion, fertility loss, pest control, and biodiversity depletion.

Novel, environmentally-friendly solutions are proposed based on integrated knowledge from sciences as diverse as agronomy, soil science, molecular biology, chemistry, toxicology, ecology, economy, and social sciences. Indeed, sustainable agriculture decipher mechanisms of processes that occur from the molecular level to the farming system to the global level at time scales ranging from seconds to centuries. For that, scientists use the system approach that involves studying components and interactions of a whole system to address scientific, economic and social issues. In that respect, sustainable agriculture is not a classical, narrow science. Instead of solving problems using the classical painkiller approach that treats only negative impacts, sustainable agriculture treats problem sources.

Because most actual society issues are now intertwined, global, and fast-developing, sustainable agriculture will bring solutions to build a safer world. This book series gathers review articles that analyze current agricultural issues and knowledge, then propose alternative solutions. It will therefore help all scientists, decision-makers, professors, farmers and politicians who wish to build a safe agriculture, energy and food system for future generations.

More information about this series at http://www.springer.com/series/8380

Praveen Guleria • Vineet Kumar
Eric Lichtfouse
Editors

Sustainable Agriculture Reviews 45

Legume Agriculture and Biotechnology Vol 1

Springer

Editors
Praveen Guleria
Plant Biotechnology and Genetic Engineering Lab, Department of Biotechnology
DAV University
Jalandhar, Punjab, India

Vineet Kumar
Department of Biotechnology, School of Bioengineering and Biosciences
Lovely Professional University
Phagwara, India

Eric Lichtfouse
CNRS, IRD, INRAE, Coll France, CEREGE
Aix-Marseille University
Aix-en-Provence, France

ISSN 2210-4410 ISSN 2210-4429 (electronic)
Sustainable Agriculture Reviews
ISBN 978-3-030-53016-7 ISBN 978-3-030-53017-4 (eBook)
https://doi.org/10.1007/978-3-030-53017-4

This Springer imprint is published by the registered company Springer Nature Switzerland AG.
The registered company address is: Gewerbestrasse 11, 6330 Cham, Switzerland

Preface

On January 30, 1988, my twenty-seventh birthday, I became a strict vegetarian. I developed a passion for health and nutrition. My diet consists of fruits, vegetables, grains, nuts and legumes only, and has for the past 15 years now. – Dexter Scott King.

Legumes are an important constituent of vegetarian diet across the globe. Legumes have been considered of great significance in counteracting malnutrition problems because of their high protein content and easy starch digestibility. Recent years have witnessed advancement in legume cultivation research and biotechnology owing to their potential health benefits. The bioactive and antioxidant-rich

nature of legumes is responsible for health benefits associated with them. Legumes contain micro- and macronutrients of direct human benefits. Further, legumes are a rich source of polyphenolic phytochemicals, phenolic acids, flavonoids and tannins. These polyphenols have a significant role in several physiological and metabolic processes. They are an important source of natural dietary antioxidants that act as free radical scavengers, reducing agents, chelating agents of pro-oxidant metals and quenchers of the formation of singlet oxygen, which allow them to protect cells against oxidative damage. Majority of polyphenols are present in legume seeds, primarily in the seed coat. Further, the accumulation of polyphenols and antioxidant bioactive compounds has been regulated by various seed-processing methods including dehulling, germination, roasting and pressure boiling. These methods significantly increase the antioxidant potential of legume seeds. Research on legume crops has gained specific interest due to its unexploited potential to eradicate protein energy malnutrition. Targeting the accumulation of legume bioactive compounds using various biotechnological approaches and exploring their probable therapeutic potential in context to various human diseases is of utmost need and importance.

This book, entitled *Legume Agriculture and Biotechnology*, published in the series Sustainable Agriculture Reviews, is written by 10 international contributors from 5 countries. The chapters review bioactive compounds and their applications, conventional breeding, and biotechnology-based methods for legume sustainability and nutritional enhancement.

The first chapter by Lahiri et al. summarises an introductory aspect of legumes towards biotechnology with focus on nitrogen fixation and their bioactive composition. The second chapter by Tiwari et al. provides an overview of the applications of bioactive peptides derived from legumes in food and healthcare industries. In Chap. 3, Budhwar and Chakraborty describe the comparative analysis of nutrient as well as anti-nutrient components of legumes along with their nutraceutical properties. Singh et al. discuss in Chap. 4 the antioxidant components of legumes and the impact of processing ways on antioxidant compounds as well as their activities. Lavudi and Suthari describe legume seed galactomannans and their multipurpose applications in Chap. 5, and Guaadaoui et al. discuss the medicinal application and nutraceuticals properties of the diverse bioactive compounds of legumes in Chap. 6. Feki et al. in Chap. 7 provide an overview of underlying molecular mechanism behind the symbiotic association between legumes and rhizobia, and Muthini et al. in Chap. 8 describe the benefits of rhizobia and other plant growth–promoting microorganisms for sustainable improvement of legume production in small agroecosystems. In Chap. 9, Khare and Chauhan provide new insight in the development and application of new plant breeding techniques to benefit human health and maintain cost effectiveness. The final chapter by Mangena summarises challenges faced during soybean genetic improvement, the vulnerability against drought stress and approaches to improve soybean growth and productivity.

The editors are thankful to all the authors who contributed to this book for their efforts in producing timely and high-quality chapters. The creation of this book would not have been possible without the assistance of several colleagues and friends. They have helped by choosing contributors, reviewing chapters and in many other ways. Finally, we would like to thank the staff of Springer Nature for copy-editing this book in a highly professional manner.

Jalandhar, Punjab, India — Praveen Guleria

Phagwara, India — Vineet Kumar

Aix-en-Provence France — Eric Lichtfouse

Contents

About the Editors

Praveen Guleria is Assistant Professor in the Department of Biotechnology at DAV University, Jalandhar, Punjab, India. She has worked in the areas of Plant Biotechnology, Plant Metabolic Engineering, and Plant Stress Biology at CSIR – Institute of Himalayan Bioresource Technology, Palampur, HP, India.

Her research interests include plant stress biology, plant small RNA biology, plant epigenomics, and nanotoxicity. She has published several research articles in various peer-reviewed journals. She also serves as the editorial board member and reviewer for certain international peer-reviewed journals. She has been awarded the SERB Start Up Grant by DST, GOI. She has also been awarded the prestigious "Bharat Gaurav Award" by the India International Friendship Society, New Delhi. She has also received various awards like CSIR/ICMR – junior research fellowship, CSIR – senior research fellowship, and state level merit scholarship awards.

Vineet Kumar is Assistant Professor in the Department of Biotechnology, Lovely Professional University, Jalandhar, Punjab, India. He has worked in different areas of biotechnology and nanotechnology in various institutes and universities in India, namely, Panjab University Chandigarh; CSIR –Institute of Microbial Technology, Chandigarh, India; and CSIR – Institute of Himalayan Bioresource Technology and Himachal Pradesh University. He has published many articles in these areas featuring in peer-reviewed journals. He also serves as editorial board member and reviewer for international peer-reviewed journals. He has received various awards like Dr DSK-postdoctoral fellowship, senior research fellowship, and best poster awards.

Eric Lichtfouse is a biogeochemist at Aix-Marseille University who has invented carbon-13 dating, a molecular-level method allowing to study the dynamics of organic compounds in temporal pools of complex environmental media. He is Chief Editor of the journal *Environmental Chemistry Letters*, and the book series Sustainable Agriculture Reviews and Environmental Chemistry for a Sustainable World. He is the author of the book *Scientific Writing for Impact Factor Journals*, which includes an innovative writing tool: the Micro-Article.

Contributors

Richard Awino Department of Biochemistry, Microbiology and Biotechnology, Kenyatta University, Nairobi, Kenya

Ritwik Banerjee Department of Biotechnology, University of Engineering & Management, Kolkata, India

Faiçal Brini Biotechnology and Plant Improvement Laboratory, Center of Biotechnology of Sfax (CBS), University of Sfax, Sfax, Tunisia

Savita Budhwar Department of Nutrition Biology, Central University of Haryana, Mahendergarh, Jant-Pali, Haryana, India

Manali Chakraborty Department of Nutrition Biology, Central University of Haryana, Mahendergarh, Jant-Pali, Haryana, India

Pallavi Singh Chauhan Department of Life Sciences, I.T.M. University, Gwalior, Madhya Pradesh, India

Anjani Devi Chintagunta Department of Biotechnology, Vignan's Foundation for Science, Technology and Research, Guntur, Andhra Pradesh, India

Vijaya R. Dirisala Department of Biotechnology, Vignan's Foundation for Science, Technology and Research, Guntur, Andhra Pradesh, India

Bandita Dutta Department of Biotechnology, Maulana Abul Kalam Azad University of Technology, Kolkata, West Bengal, India

Meryem Elyadini Laboratory of Biochemistry, Environment and Agrifood, Department of Biology, Faculty of Sciences and technology – Mohammedia, Hassan the Second University, Mohammedia, Morocco

Kaouthar Feki Laboratory of Legumes, Center of Biotechnology of Borj-Cédria, Hammam Lif, Tunisia

Sayantani Garai Department of Biotechnology, University of Engineering & Management, Kolkata, India

Abdelkarim Guaadaoui Laboratory of Physiology, Genetic and Ethnopharmacology (LPGE), Department of Biology, Faculty of Sciences – Oujda (FSO), Mohammed the First University (UMP), Oujda, Morocco

Abdellah Hamal Laboratory of Physiology, Genetic and Ethnopharmacology (LPGE), Department of Biology, Faculty of Sciences – Oujda (FSO), Mohammed the First University (UMP), Oujda, Morocco

Abdul A. Jalloh Department of Biochemistry, Microbiology and Biotechnology, Kenyatta University, Nairobi, Kenya

Amrita Jasu Department of Biotechnology, Maulana Abul Kalam Azad University of Technology, Kolkata, West Bengal, India

Amarbir Kaur Department of Food Science and Technology, Guru Nanak Dev University, Amritsar, Punjab, India

Amritpal Kaur Department of Food Science and Technology, Guru Nanak Dev University, Amritsar, Punjab, India

Ekta Khare Department of Pharmacy, I.T.M. University, Gwalior, Madhya Pradesh, India

Kibet Charles Kirui Department of Biochemistry, Microbiology and Biotechnology, Kenyatta University, Nairobi, Kenya

Kipkorir Koech Department of Biochemistry, Microbiology and Biotechnology, Kenyatta University, Nairobi, Kenya

N. S. Sampath Kumar Department of Biotechnology, Vignan's Foundation for Science, Technology and Research, Guntur, Andhra Pradesh, India

Dibyajit Lahiri Department of Biotechnology, University of Engineering & Management, Kolkata, India

Harikrishna Naik Lavudi Department of Plant Sciences, University of Hyderabad, Hyderabad, Telangana, India

Phetole Mangena Department of Biodiversity, School of Molecular and Life Sciences, Faculty of Science and Agriculture, University of Limpopo, Sovenga, South Africa

Haythem Mhadhbi Laboratory of Legumes, Center of Biotechnology of Borj-Cédria, Hammam Lif, Tunisia

Moncef Mrabet Laboratory of Legumes, Center of Biotechnology of Borj-Cédria, Hammam Lif, Tunisia

Dipro Mukherjee Department of Biotechnology, University of Engineering & Management, Kolkata, India

Morris Muthini Department of Biochemistry, Microbiology and Biotechnology, Kenyatta University, Nairobi, Kenya

Moupriya Nag Department of Biotechnology, University of Engineering & Management, Kolkata, India

Ezekiel Mugendi Njeru Department of Biochemistry, Microbiology and Biotechnology, Kenyatta University, Nairobi, Kenya

Rina Rani Ray Department of Biotechnology, Maulana Abul Kalam Azad University of Technology, Kolkata, West Bengal, India

Balwinder Singh Department of Biotechnology, Khalsa College, Amritsar, Punjab, India

Jatinder Pal Singh Department of Food Science and Technology, Guru Nanak Dev University, Amritsar, Punjab, India

Narpinder Singh Department of Food Science and Technology, Guru Nanak Dev University, Amritsar, Punjab, India

Sateesh Suthari Department of Plant Sciences, University of Hyderabad, Hyderabad, Telangana, India
Department of Botany, Vaagdevi Degree & PG College, Warangal, Telangana, India

Pragya Tiwari Molecular Metabolic Engineering Laboratory, Department of Biotechnology, Yeungnam University, Gyeongsan, Gyeongbuk, Republic of Korea

Chapter 1
An Introduction to Legume Biotechnology

Dibyajit Lahiri, Moupriya Nag, Amrita Jasu, Bandita Dutta, Ritwik Banerjee, Dipro Mukherjee, Sayantani Garai, and Rina Rani Ray

Abstract Legumes are the group of predominant nitrogen fixers that act as a source of wide variety of secondary metabolites that act as a line of defence against various bacterial, fungal, parasitic and predatory species. The symbiotic bacterial species establishes a syntopic interaction by quorum sensing and develop a biofilm that help is building up of the symbiotic relation with the plant. This association not only helps in the mode of nitrogen fixation but also helps in increasing the bioactive contents along with various types of abiotic components found in soil. These bioactive compounds have wide therapeutic potential, as significantly used as anti-aging, antimicrobial, antibiofilm, antioxidant, antidiabetic, anti-inflammatory and cardioprotective agents. The compounds also act as antiangiogenic agent and prevent the proliferation of disease causing pathogens. Hence, legumes not only help in sustainable development of plant and agriculture, but also possess significant therapeutic potential to promote human health. The present article thus reviews the contribution of legumes towards nitrogen fixation to facilitate plant growth and discusses their bioactive compounds having beneficial role in regulating human health.

Keywords Legumes · Secondary metabolites · Bioactive · Therapeutic · Nitrogen fixation · Antimicrobial · Antioxidant · Stress

Abbreviations

AMF	Arbuscular mycorrhizal fungi
CHS	Chalcone synthase
IFR	Isoflavone reductase
IRS	Induced response system

Authors Dibyajit Lahiri, Moupriya Nag, Amrita Jasu have equally contributed to this chapter.

D. Lahiri · M. Nag · R. Banerjee · D. Mukherjee · S. Garai
Department of Biotechnology, University of Engineering & Management, Kolkata, India

A. Jasu · B. Dutta · R. R. Ray (✉)
Department of Biotechnology, Maulana Abul Kalam Azad University of Technology, Kolkata, West Bengal, India

P. Guleria et al. (eds.), *Sustainable Agriculture Reviews 45*, Sustainable Agriculture Reviews 45, https://doi.org/10.1007/978-3-030-53017-4_1

PAL	Phenylalanine ammonia lyase
PAPPP	Proline-associated pentose phosphate pathway
ROS	Reactive oxygen species
SAR	System acquired response

1.1 Introduction

There is a chance of severe food scarcity in coming era as the population increases at a geometric progression and the food productivity increases at arithmetic progression. Researches forecast that the population of the world will reach almost to 9.6 billion by 2050. Therefore there is a substantial increase in dietary intake of legumes as replacement of cereal grains (Venn and Mann 2004) and the legumes serve as one of the important sources of food to a large population. They not only act as the richest source of proteins but also are the reservoirs of large number of bioactive compounds that in turn act as important food supplement for the human population and livestock. They also contain a large number of trace elements like magnesium, potassium, calcium, iron, zinc, copper and manganese (Stagnari et al. 2017) which are of immense importance for human metabolism. A significant quantity of vitamins, minerals, carbohydrates, fibres and energy (Frankel 1996) can be obtained from legumes.

Legumes in association with the endophytic and epiphytic bacteria like *Rhizobium leguminosarum* or *Clostridium acetobutylicum*, developing biofilm on their root surface help in maintaining as well as enhancing the nitrogenous adequacy of the soil. As a result, they are used as an important component during crop rotation in agriculture. Common examples of legumes include pulses, beans, lentils, peas, soya beans, peanuts, tamarind, etc. (Harborne 1973; Mackeen et al. 1997).

The major active compounds present in legumes comprise of polyphenols and flavonoids which impart significant antioxidant potential to them. These phytocompounds scavenge the free oxygen species and protect the cells from damage (Frankel 1996). Hence natural antioxidant phytochemicals extracted from legumes play an indispensable role for human health targeting various health disorders like cardiovascular disease, tumorous growth or ischemic diseases (Ito et al. 1983). Other active compounds including glycosides, tannins, saponins and alkaloids play a significant role in maintaining human health (Flight Clifton 2006; Huda-Faujan et al. 2009). It is well documented that these active ingredients actually help in protecting plants against various biotic and abiotic stresses (Ngoci et al. 2011).

The present chapter predominantly focuses on various types of bioactive compounds present in legumes, factors influencing their biosynthesis and their significant applications (Fig. 1.1).

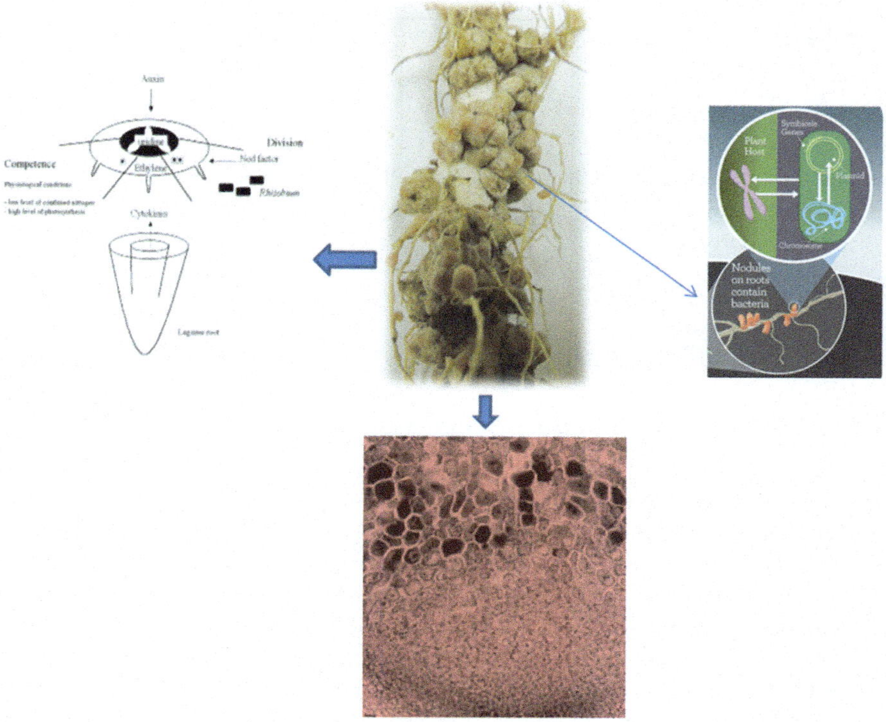

Fig. 1.1 Mutualistic relation between legumes and nitrogen fixing bacteria

1.1.1 Plant Legumes: History

Leguminosae is one of the important member of plant family which nearly comprises of 800 different types of genera and about 20,000 species and is the third largest family of flowering plants. Legumes or beans were one of the first domesticated plants which appeared before 6000 B.C. Now-a-days, there has been a high demand for nutritive foods, not only in the scientific community but also in the economic and social aspect of living. According to The United Nations, 2016 has been declared as the International Year of Pulses, recognizing pulse cultivation beneficial for the sustainable nitrogen enrichment of agricultural soil and their nutritive impact on human health.

The American 2015 Dietary Guidelines henceforth recommended pulses' consumption as "Sustainable Diets" and a high intake of legumes (beans), is linked to significantly lower risks of heart disease, high blood pressure, stroke, and type 2 diabetes, hyperlipidaemia and causes weight management (Polak et al. 2015).

With diversified dietary habits worldwide, a balanced diet chart must be maintained to reach optimum nutritional value, as like:

1. Utilization of dietary component for the purpose of providing energy;
2. Biochemical food components to maintain the regulatory functions.

Therefore, it is very important to have a set of balanced nutrients within the diet (Nishida et al. 2004). The FAO-WHO Group of Experts established the following dietary recommendations:

- 60% carbohydrates
- 3% fibre
- protein at an amount so that total calorie intake must be below 15%
- lipids not more than 25%; comprising 75% of plant foods and 25% derived from animal sources

1.2 Roles of Legumes in Agriculture

1.2.1 Rhizobium-*Legume Symbiosis*

The nitrogen starvation in the surrounding environment results in the development of symbiotic association between the rhizobial microbiota and the legumes (Jensen et al. 2012). This interaction by cellular signalling results in the development of root nodules which provides a predominant site for the conversion of the bacterial species to nitrogen fixing bacteriods. The mutualistic relation is established by the taking up of the nitrogenous compounds fixed by the bacteriods and the utilization of assimilated carbon from the plants by the bacterial species (MarÃti and Kondorosi 2014). The two predominant groups of bacterial cells that participate in the mechanism of nitrogen fixation are classified into two categories, namely α-proteobacteria and β-proteobacteria, collectively termed as *Rhizobia* (Chen et al. 2003; MacLean et al. 2007).

Wide range of gram positive *Rhizobiaceae* like bacterial species is known that help in establishing a mutualistic relationship with legumes for nitrogen fixation. These legumes develop a rhizome which are deep enough and allows to establish strong symbiotic interaction with the *Rhizobiaceae* (Gram-negative bacteria). It has a potent nutritive value for livestock farming systems, the use of nitrogenous fertilizers are reduced maximally due to the use of the forage legumes. This symbiotic relationship reduce atmospheric Nitrogen to Ammonia which later gets converted into amino acids for the synthesis of proteins in plants.

Nodule formation can be on various plant parts including stems and roots and they can be indeterminate with apical meristematic growth (Martin and Liras 1989). The process of conversion of atmospheric nitrogen to useful nitrogenous compounds is usually carried out by rhizobium bacteroids in which several parts of both partner organisms are involved. This mechanism is enunciated by the use leghaemoglobin, which helps in removing the oxygen from the symbiosomes by the nitrogenase enzyme, predominantly available within the bacteroids having microaerophilic environment. The pink colour in the nodule region provides the distinct expression on the above mentioned circumstances, which ultimately indicates an effective establishment of symbiotic relationship (Elexsona et al. 2014).

1.2.2 Symbiotic Nitrogen Fixation

Atmospheric Nitrogen, although considered to be an inert gas, is one of the essential elements that are required for survival. But most of the organisms are capable of utilizing ammonia (NH_3), rather than the inert N_2 to synthesize nucleic acids, amino acids, proteins and other nitrogen-containing components necessary for life.

Hence, the mechanism of fixation of atmospheric nitrogen is of utter importance to convert inert N_2 into biologically useful NH_3 (the only form that is utilized by all organisms) which is naturally mediated only by Nitrogen fixing *Rhizobiaceae* (α-Proteobacteria) (Sørensen and Sessitsch 2007). Nitrogen fixing bacteria is beneficial for when other plants has to release nitrogen to the environment after their death, few bacteria lives clubbed with the plant as in legumes and a few other plants where they live in small bulgings on the roots forming nodules. Within these nodules, upon nitrogen fixation they produce NH_3 that is henceforth absorbed by the plant. Thus, nitrogen fixation by legumes stands out as an important relationship between a plant and the bacteria.

There are many forms of biological nitrogen fixation in nature; with the help of lichens, blue-green algae and various soil bacteria in association with the plants. The natural ecosystem gets enriched by the significant quantities of NH_3 due to the amount of nitrogen fixed, which can be in the range of 25–75 lb of nitrogen/acre/year in a natural ecosystem to several hundred pounds in a cropping system (Burton 1972; Guldan et al. 1996; Frankow-Lindberg and Dahlin 2013).

Formation of root nodules marks the initiation of nitrogen fixation. The rhizospheric bacterium invades the root, multiplies within its cortex cells and gets supplied with all the necessary nutrients for its multiplication. Within a week or two following the infection, small nodules start to get visible with naked eye. When nodules are young, usually white or grey inside, are unable to fix nitrogen. As they gradually grow in size, they turn pink or reddish in colour, indicating the initiation of nitrogen fixation. The appearance of pink or red colour is due to the leghemoglobin (as similar to the hemoglobin present in blood) that controls by restricting the oxygen flow to the bacteria, creating an anaerobic environment.

1.2.3 Leguminous Flavonoid and Bacterial Nod (Nodulation) Gene Expression During Nitrogen Fixation

Flavonoid compounds are secreted during nitrogen limiting conditions into the rhizosphere. These flavonoids induce the expression of nod genes by activating the NodD proteins, LysR type transcription regulators (Long 1996). The synthesis of Nod factor or the nod gene expression initiates after the binding of NodD protein into the nod boxes (a conserved DNA motif) present in the upstream of the Nod operons (Fisher et al. 1998; Rostas et al. 1986).

Nod factor is an essential lipo-chito-oligosaccharide signal for the initiation of symbiotic relationship. This factor has first adapted themselves to recognize different leguminous flavonoids and the symbiotic relationship takes place only after establishing the recognition specificity. Therefore those flavonoids are called "infection flavonoids" due to their high accumulation in the infection site and increased biosynthesis after the infection by the pre recognized rhizobia. The first flavonoid identified was luteolin which has broad range of rhizobial strain specificity (Liu and Murray 2016).

Thereafter Nod factors directly bind to the LysM domain of the Nod factor receptors (e.g. NFR1, NFR2 etc.) and facilitate the activation of downstream signalling (Limpens et al. 2003). This Nod factor recognition specifically determines host range in the species level. Hence the leguminous bioactive compounds such as flavonoids play a crucial role in rhizobium-legume symbiosis.

1.3 Bioactive Compounds Present Within Legumes

Legumes are widely grown for the large number of beneficial compounds being present within them. The nutritional contents vary from proteins, fibres, minerals, starch and vitamins. It also contains a large amount of phytocompounds. These major group of secondary metabolites comprises of alkaloids, peptides, cyanogens, polyketides, simple phenolics like flavonoids (which mainly contain phytoestrogens and catechins), sterols, polyphenols, saponins, phytates and terpenoid (Tables 1.1 and 1.2) (Harborne and Turner 1984; Hegnauer and Hegnauer 1994; Kinghorn and Balandrin 1984; Seigler 1998; Southon 1994; Wink 1993b; Veitch 2010). The legumes also play an important role as nutraceuticals as they contain bioactive peptides which have an important role in human welfare. The bioactive compounds present in legumes possess antioxidant, hypocholesterolemic, antithrombotic and antioxidant activities. It has been analysed that flavonoids obtained from these legumes has anticancer activities. Legumes also contain large quantities of hydrophilic phytochemicals like ascorbic acid, polyphenols and phenolic acids which help in the redemption of cancer risk and also act as immunostimulants. This diversification also encircles in modifying lipophilic compounds like tocopherols and carotenoids which help in preventing cardiovascular diseases. The advancement in the field of ethnopharmacology in last few decades has provided us with the concept of large number of active molecules which has a sincere role in inhibiting the microbial proliferations.

Research is still going on to elucidate human health regulatory bioactive compounds present within the legumes (Hegnauer and Hegnauer 1994; Southon 1994). Majority of the identified bioactive compounds has pharmacological and toxicological activities (Wink et al. 1998; reviewed in Teuscher and Lindequist 2010). Many of the alkaloids are either neurotoxins or neuromodulators (reviewed in Wink 1992, 1993a, 2000, 2007; Wink and Schimmer 2010). Plants specifically use these compounds for protection against herbivores as well as various microbial infections (McLean 1970; Hartmann and Witte 1995; Wink and Schimmer 2010).

Table 1.1 Major bioactive compounds in legumes and associated health impacts

Legumes	Bioactive compounds	Health benefits	References
Common bean (*P. vulgaris* L.)	Tannins, anthocyanins, phenolic compounds	It can act against the liver injury in animals, breast, colon and prostrate cancer	Ganesan and Xu (2017)
Lentils (*Lensculinaris*)	Phenolic compounds	Benefits against degenerative disorder, type II diabetes, coronary heart disease and obesity	Ganesan and Xu (2017)
Soybean (*Glycinemax* (L.) Merr. Fabaceae)	Polyphenols such as isoflavone, tocopherols, anthocyanin and ascorbic acid	High antioxidant and antiradical activities, helps in preventing various types of cardiovascular diseases, cancer and osteoporosis	
Lima beans (*Phaseoluslunatus* L.)	Phytochemicals, polyphenols	Radical scavenging activities, prevent heart disease, inflammation, arthritis, and immune system decline	Ganesan and Xu (2017)
Lentils (*Lens culinaris* Medik.)	Phenolic acids, flavanols, flavonols, soy saponins, phytic acid, condensed tannins	Anticancer, angiotensin enzyme that reduces the risk of cardiovascular diseases, coronary heart disease, type II diabetes, and obesity	Ganesan and Xu (2017)
Peanut (*Arachishypogaea* L.)	Anthocyanins, proanthrocyanidines and resveratrol	Reduces weight gain, cardiovascular diseases, Alzheimer's disease, and cancer	Ganesan and Xu (2017)
Peas (*Pisumsativum* L.)	Caffeic, vanillic, p-coumaric, ferulic and sinapic acids, quercetin, kaempherol, procyanidin B2 and B3	Prevent heart disease, anti-inflammation, antioxidants	Ganesan and Xu (2017)
Adzuki beans (*Vigna angularis*)	Flavanoids, Tocopherols, and Vitamin E	Inhibit pancreatic Lipase activity and thus decrease Triglyceride concentrations, exert antioxidant activity	Ganesan and Xu (2017)
Mung beans (*Vigna radiata* L.)	Phenolic acid, Flavanoids, polyphenols and Tannins	Source of essential amino acids, Modify Glucose and Lipid metabolism, exhibits hypoglycemic and hypolipidemic effects, anticancer, anti-melanogenesis, hepatoprotective, and with immunomodulatory activities	Ganesan and Xu (2017)

1.3.1 Alkaloids

Alkaloids are group of plant secondary metabolites that protect them against microbial invasions. This class of secondary metabolites comprises of different phytoactive compounds like indo-quinolones, quinolones, agenalsine, indolizidine and many more. These have potent antimicrobial properties and also act as inhibitors of

Table 1.2 Major bioactive compounds in legumes and their sources

Bioactive compounds	Antioxidant class	Plants	Health benefits	References
Anthocyanidins, flavonols, flavones, flavan-3-ols, isoflavonoids, neoflavonoids	Flavonoids	Black bean, red kidney bean, red peanut, soybean	Possible neuroprotective agents, significant against Alzheimer's, protection against cancer, heart disease, asthma	Xu et al. (2007)
Caffeic acid, ferulic acid, procatechuic, gallic acids, syringic acids, *p*-hydroxybenzoic, *p*-coumaric, sinapic acids, vanillic acid	Phenolic acids	Bean, black bean, cowpea, lentil, lima bean, red kidney	Prevent cellular damage, promote anti-inflammatory conditions, protection against free radicals	Xu et al. (2007)
Condensed tannins, hydrolysable tannins	Tannins	Chick pea, green pea, common bean, yellow pea, lentil, soybean	Improved immune response, and blood sugar balance	Scalbert et al. (2005)
Phytic acids or Phytate	Phytic acid	Most of the *Vigna spp.*	Prevent the formation of cavities, plaque and tartar in the teeth, has hypoglycemic effects, reduce the formation of kidney stones.	Barahuie et al. (2017)
GABA	γ-Aminobutyric acid (GABA)	Adzuki beans, mung beans	Effective for the treatment of sleeplessness, depression, autonomic disorders	Ganesan and Xu (2017)

biofilms. Alkaloid like 1,3,4-oxadiazole helps in inhibiting the quorum sensing precursor molecule like 2-heptyl-4-quinolone (HHQ), the virulence effect of some organisms and the gene responsible for swarming motility can be inhibited in the presence of 7-hydroxyindole. Many alkaloids within the legumes act as neuromodulators and possess neurotoxic activities to repel the herbivorous animals. Alkaloid like squalamine has a potent antibacterial effect and it kills the bacterial cells by destroying the cell membranes. The squalamine binds to the receptor present on the exposed cell surface by penetrating through the lipopolysaccharide layer of the cell membrane resulting in the inhibition of cell membrane synthesis (Salmi et al. 2008). Electrostatic force of attraction helps in holding the squalamine with the receptors present on the bacterial cell surface. The cell membrane comprises of large quantities of free negatively charged ions that induce strong interactions with the drug like compounds (Cushnie et al. 2014). Alkaloids like isoquinolone, obtained from legume plant species like *Thalictrum minus* acts as a ligand for DNA and has the ability to get attached both with double stranded and single stranded DNA molecule. It has the ability to break down the highly conserved bacterial proteins Ftsz. Ftsz-GFPase results in the formation of Z-ring (Boberek et al. 2010). Another mode of action of these alkaloids like isoquinoline is to inhibit the synthesis of nucleic acid by inhibiting the activities of topoisomerase I and II, and on the other hand substituted quinolines like methyl quinolones prevents the bacterial cell proliferation by hindering oxygen uptake by the cells (Tominaga et al. 2002).

1.3.2 Terpenoids

These are the group of secondary metabolites that contains a 5-C isoprene units. Most of the active compounds present in this group comprises of cyclic structure that differs from one another in possessing functional groups within their ring (Harborne 1973). Classification of terpenoids is generally based on the number of isoprene units present into hemi-terpenoids, monoterpenoids, sesquiterpenoids, diterpenoids, triterpenoids, tetra terpenoids and carotenoids. Their therapeutic properties like anti-bacterial, anti-fungal, anti-viral, anti-protozoan, anti-allergens and anticancer effects are quite pronounced and are strongly reported (Mackeen et al. 1997). Presence of terpenoids is reported from many living organisms throughout nature, especially in plants where it is considered to be the major content of the natural products, in marine animals and fungi. Terpenoids support the basic functions of plants including growth, development and wound repair. In medicinal plants, different types of terpenoids are found like oleanolic acid, ursolic acid, carvacrol etc. Terpenoids are also of commercial importance due to their use in food and cosmetics industries as flavouring agents and fragrances. Ursolic acid, a secondary metabolite that is generally present in leaves, stem bark and fruit peels is a pentacyclic terpenoid (isomer of oleanolic acid) which has a wide range of pharmaceutical implications. Ursolic acid has pharmacological activities like anti-fertility effects, antibacterial activity against *Pneumococci, Staphylococci and Streptococcus mutans,* and antihelmintic activity (Ali et al. 1996). Oleanolic acid is a pentacyclic triterpenoid and its derivatives possess several therapeutic effects like antioxidant, anticancer, anti-inflammatory and hepatoprotective effects. Oleanolic acid safeguards the plants by formation of barrier against water loss and pathogens. Also, other therapeutic properties of oleanolic acid include antiviral activity, antibacterial property against *S. aureus, S. mutans, B. subtilis, M. tuberculosis, P. aeruginosa* etc., and antiprotozoal activity (Abdullah et al. 2013). Carvacrol, an essential aromatic compound found in plants like thyme, oregano, pepperwort, wild bergamot is a phenolic monoterpenoid that is capable of offering pleasant smell and taste. Some of their uses also include antibacterial and antifungal activity. Moreover it is reported in treating fever, pneumonia, nail fungus, cholecystitis and cholangitis (Abdullah et al. 2013).

1.3.3 Coumarins

Coumarins, isolated from Tonka beans, are a group of colourless and crystalline phytochemical, polyphenolic compounds. They are either free or glucose combined oxygen heterocycles. Their pharmacological, biochemical and therapeutic applications are mainly due to the substitution pattern of their pyrone classes. Coumarins are widely divided into four clusters namely simple coumarins, pyrone substituted coumarins, pyranocoumarins, and furanocoumarins. These coumarins are found abundantly in fruits, roots, stems, leaves and also in Umbelliferone and in *Rutaceae* family. Their presence is also traced in selective microorganisms like Streptomycin, aflatoxin from *Aspergillus sp*. etc. Their importance in the use as medicine for treatment of various clinical conditions alongside significant effect on physiological, anti-tumor and bacteriostatic activity has been reported. Furthermore, their potential therapeutic applications in treating chronic infections, oedema and anticoagulant, anticancer, anti-inflammation, analgesic, antidiabetic, anti-neurodegenerative activities are also significant. Esculetin, another simple coumarins has antitussive aspects. Furthermore, it's antibacterial, anti-tumor, anti-inflammatory, antioxidative and neuroprotective properties has made Esculetin as one of the most promising leads for various chemists to develop variety of drugs (Yaacob 1987).

1.3.4 Polyphenols

These are secondary metabolites widely distributed in various plant parts of cereals and beverages that are involved in defence against pathogenic aggregation or UV radiation (Yaacob 1987). The main classes of polyphenols which are conjugated forms with one or more sugar residues linked to hydroxyl groups, include phenolic acids, lignans, flavonoids and stilbenes. They are the key contributors in the color, odor, bitterness, and oxidative stability in food. Phenylalanine, a common intermediate in more than 8000 identified polyphenolic compounds, have therapeutic uses due to their properties like antidiabetic, anti-aging, cardioprotective, treatment from infections, asthma, hyper-tension, anticancer and other properties. They mainly consist of p-coumaric, caffeic acid, ferulic acid, and sinapic acids. Studies show that in polyphenol rich foods, polyphenolic compounds are responsible for treatment of chronic human diseases, increase antioxidant capacity and are known to have therapeutic vitality (Yaacob 1987).

General subclass of polyphenols includes: flavanols, flavanones, flavones, isoflavones and anthocyanins. Polyphenols include bioactive compounds like curcumin, rosmarinic acid, gingerol, shagaol etc. Curcumin (diferuloylmethane) is a polyphenol compound isolated from ground rhizomes of the plant (*Curcuma longa*) L. that has been extensively used in ayurvedic medicine over centuries as a nontoxic agent, with a variety of therapeutic properties like analgesic, anti-inflammatory, antioxidant, antiseptic, antifungal, antibacterial and antiplatelet activity (Qader et al. 2012).

Rosmarinic acid is an ester of caffeic acid that occurs mostly in perilla, sweet basil and rosemary (Yaacob 1987). Understanding the contribution towards human health and nutrition, shogaols and gingerols stands out as the major pungent compounds present in ginger (*Zingiber officinale*) (Baharum et al. 2010). Shogaols act as important biomarkers for the quality control of numerous ginger-containing products, due to their effectiveness in the treatment of inflammation, anticancer, dyspepsia, nausea, vomiting, etc. (Costerton 2007).

1.3.5 Sulphur-Containing Bioactive Compounds

The legumes are comprised of sulphur containing active compounds showing potent antibacterial properties like isothiocyanates, ajoene and allicin. They have proved their effectiveness against a wide range of microbial species like *Enterococcus faecalis, S. mutans, Prevotella nigrescens, Aggregatibacter actinomycetemcomitans* and *Clostridium perfringens.* It has been observed that the effectiveness of these bioactive compounds vary greatly from the source where they have been isolated. Isothiocyanates have even shown its effectiveness against Methicillin resistant *Staphylococcus aureus* species (Dias et al. 2012). Research has shown that allicin synergistically reacts with omeprazole to bring about its impact upon the target microbial species. Thus the synergistic effects of these bioactive compounds are showing a better path of replacing antibiotics.

1.3.6 Bioactive Compounds of Edible Legumes

Isoflavones stand out to be the most important and vividly studied bioactive compound of legumes, which together with procyanidins and phenolic acids constitute the vital phenolic compounds present in seeds. Estradiol-17 beta molecules are the group of isoflavones that has equivalent effect as that of oestrogen which has similar affectivity as that of commercially available drug molecules (Sorensen et al. 2005). They can also act effectively against various cancerous cells specially the endothelial and breast cancer (Agbafor and Nwachukwu 2011). Soy food (legume product) lowers the risk of prostate and colorectal cancer. Relatively higher content of anthocyanin and proanthocyanidins or condensed tannins are present as phenolic compounds with antioxidant potential in dark coloured seeds rather than those of pale coloured seeds. Legume seeds also contain different carotenoids, being lutein, the predominant one in economically important grain legumes, such as chickpea, lentil, soybean, peanut, cowpea, pea, faba bean, and lupin, followed by zeaxanthin and β-carotene. γ-Tocopherol is found as the most abundant isoform in lentils, soybean,, broad bean, and some lupin species (Wan Hassan 2006; Bunawan et al. 2011; Almey et al. 2010; Uyub et al. 2010) (Table 1.3).

Table 1.3 Major Secondary metabolites from legumes and their mechanisms of action

Secondary metabolites	Origin of occurrence	Activity	Examples from Fabaceae	Refer ences
β-Carboline alkaloids	*Petalostyles labicheoides*, *Acacia complanata*, *Burkea africana*, *Prosopis nigra*, *Desmodium gangeticum*	Serotonin receptor agonist DNA intercalation; mutagenic	Harman, harmalan, tetrahydroharman, leptocladine	Ganesan and Xu (2017)
Indolizidine alkaloids	Astragaleae; *Castanospermum*	Inhibits endoplasmic hydrolases	Swainsonine, castanospermine	Ganesan and Xu (2017)
Piperidine alkaloids	*Genistoid clade*	Causes malformations in embryos	Ammodendrine	Ganesan and Xu (2017)
Pyridine alkaloids	In all subfamilies; abundant in IRLC clade and *Phaseoleaesens. lat.*	Antimicrobial, bactericidal	Trigonelline	Ganesan and Xu (2017)
Pyrrolizidine alkaloids (PA)	*Crotalaria*; *Lotononis*	Mutagenic and carcinogenic as they modulate several neuroreceptors, like 5HT2, mACh, GABA, D2 and α2.	Monocrotaline, senecionine	Ganesan and Xu (2017)
Quinolizidine alkaloids (QA)	Genistoid clade; *Ormosia* clade; *Sophora secundiflora*; *Calia*, *Bolusanthus*	Modulate nAChR and mAChR; Na + channel blocker, neurotoxic in nature	Sparteine, lupanine, anagyrine, cytisine, matrine, lupinine	Ganesan and Xu (2017)
Antimicrobial peptides (AMP)	*Adenanthera* spp.	Potent antimicrobial	ApDef1	Ganesan and Xu (2017)
Amylase inhibitor	*Delonix regia*, *Vigna* sp.	Control the postprandial increase of blood glucose	Antidiabatic	Ganesan and Xu (2017)
Protease inhibitors	Several Fabaceae	Inhibits trypsin mainly in herbivores	Trypsin inhibitors and Anticarcinogenic	Ganesan and Xu (2017)
Oligo saccharide	*Vigna sp.*	Reduce colon cancer by reducing composition of mucin in colonic cancer	Prebiotic	Ganesan and Xu (2017)
Lectins	*Abrus precatorius*, *Robinia*	Inhibitors of ribosomal protein biosynthesis	Abrin, robin	Ganesan and Xu (2017)

(continued)

Table 1.3 (continued)

Secondary metabolites	Origin of occurrence	Activity	Examples from Fabaceae	References
Tannins	Mostly trees	Antimicrobial and anti-herbivore activities	Mostly catechin type	Ganesan and Xu (2017)
Phytate	*Vigna* sp.	Hypocholesterolaemic effect, antic arcinogenic and also protect DNA damage	Inositol hexaphosphate, or IP6	Ganesan and Xu (2017)
Phenyl ethylamines	Mainly *Acacia* spp.; Caesalpininoideae	Psychoactive	*N*-Methyl phenylethylamine;	Ganesan and Xu (2017)
Polyamines	Mainly Phaseoleae	Growth regulator	Spermine, spermidine	Ganesan and Xu (2017)
Tryptamines	Mimosoideae	Agonist towards Serotonin receptor; hallucinogenic	*N,N*-Dimethyltryptamine; bufotenin; *N*-methyltryptamine	Ganesan and Xu (2017)
Tyramines	Mimosoideae, Desmodieae	Psychoactive; inhibit insect feeding	*N*-Methyltyramine; hordenine; *N*-methylmescaline	Ganesan and Xu (2017)
Simple phenols	Widely distributed	Antioxidants; antimicrobial	Vanillin, syringic acid, ferulic acid, gentisic acid, gallic acid, p-hydroxy benzaldehyde	Ganesan and Xu (2017)
Flavonoids	Widely distributed, in all tribes	Antimicrobial and anti-herbivore activities; antioxidants	Quercetin, kaempferol, etc.	Ganesan and Xu (2017)
Protein	*Vigna* sp.	Anti-HIV activity, help against fungal pathogenesis	Angularin, delandin and unguilin	Ganesan and Xu (2017)

1.4 Availability of Trace Elements Within the Legumes

Legumes contain a large amount of trace elements like copper, iron, zinc and manganese. These trace elements not only maintain the health of the consumers but also provide nutritional supplements to the body. These trace elements also act as immune boosters and help in the maintenance of gastrointestinal mucosal integrity. The predominant element found within the leguminous seeds is iron. The amino acids like cysteine, fructose and citric acids help in increasing greater uptake of iron from the soil whereas compounds like phytate, oxalate and polyphenols act as inhibitors to the absorption of calcium from soil (Sandberg 2002). Commonly available amino acids like cysteine and histidine are promoters for the absorption of zinc from the soil whereas phytate, ethylenediaminetetraacetic acid, oxalate and fibres inhibit the bioavailability of zinc from the soil.

1.5 Effect of Biotic Stress on the Yield of Bioactive Compounds of Legumes

Symbiotic relations between the soil-born bacteria (rhizobia) and leguminous plants helps enhance the quality and productivity of agricultural crops thereby enhancing the animal/human nutrition and health (Graham and Vance 2003). While staying in the roots of the plant, these bacteria helps the plants in nitrogen fixation and protects against various microbial pathogens (Chakraborty et al. 2003). These symbiotic interactions have strongly driven the investigation towards a number of biotic (bacteria, fungi, viruses, insects, nematodes) stress factors thereby affecting the yield of agricultural crops.

1.5.1 Role of Microorganisms in Increasing the Bioactive Compound Content of Legumes

1.5.1.1 Arbuscular Mycorrhizae Fungi (AMF)

Arbuscular mycorrhizal fungi (AMF) shares symbiotic relation with plants and are responsible for activating different biosynthetic pathways like shikimic and malonic pathways as explained by Gor et al. (2011). Alteration in these pathways lead to alteration of the production of phenolic, terpenes and alkaloids in plants thereby affecting directly or indirectly the quality of food, medicinal plants or spices (Bunawan et al. 2011; Almey et al. 2010). For example, a significant increase in β-caryophyllene and pinene in coriander was observed by Kapoor et al. (2002) when inoculated with *G. macrocarpum*. Secondary metabolites such as allicin, was found to increase in garlic plants when inoculated with *G. fasciculatum*. Cultivars inoculated with AMF such as basil and Purple petra were found to have elevated concentration of anthocyanins. A significant rise in concentrations of carotenoids and total phenolic compounds were observed in tomato inoculated with AMF (*Glomus sp.*). In addition to fungi like AMF, soil microbiome can also act synergistically to increase some secondary metabolites. In one study, it was observed that the use of fungal consortium causes significant rise in carotenes and xanthophylls of *Capsicum annuum L* (Mena-Violante et al. 2006). In addition to the production of secondary metabolites as defence mechanism, plants also produces enzymes related to metabolic pathways. For example, inoculating *M. truncatula* and *M. sativa* roots with AMF (Glomus versiforme) resulted in rise in concentration of chalcone synthase (*CHS*), phenylalanine ammonia lyase (*PAL*) and isoflavone reductase (*IFR*) transcript levels. In another study, inoculation of barley roots with AMF resulted in increased production of endogenous jasmonic acid (Hause et al. 2002).

1.5.1.2 Rhizobial Inoculation Alters Yield of Phytocompounds of Legumes

Biofertilization with rhizobacteria is shown to provide similar legume yields than those obtained with the application of harmful chemical fertilizers. For example, seeds of medicinal legumes, such as *Psoralea corylifolia L.* (Fabaceae), are a rich source of flavones and coumarin derivatives like psoralen are used as medicines against skin disorders like vitiligo i.e., hypopigmented lesions of the skin (Nello et al. 2010; Zeng et al. 2013; Terry 2004). Inoculation with rhizobacteria like *Ensifer meliloti* and *Rhizobium leguminosarum* have been shown to enhance the psoralen content in the seeds of this legume (Kaur and Kapoor 2002). Considering the economic importance of soybean worldwide, it was observed that the inoculation of *Glycine max* with *Bradyrhizobium japonicum sv glycinearum*, increases soybean yield and nutrient content (Adaskaveg et al. 2011). Moreover, they are found to possess higher antioxidant activity and higher total fatty acids content than those from controlled ones (Sandoval-Chávez et al. 2015; Morales et al. 2010). Similarly, inoculation of chickpea (*Cicer arietinum*) with *Mesorhizobium*, is found to increase the content of flavonoids in the seeds Violante et al. (2006). In another study, inoculation of faba bean and *Arachis hypogaea* with *Rhizobium leguminosarum symbiovar-viciae* lead to a significant rise in antioxidant potential of shoots along with the content of total flavonoids, phenols, proteins and tannins (Mulas et al. 2011; Araujo et al. 2015).

Thus, biotic stress involving *rhizobial* inoculation is found to produce significant changes in the legume bioactive compounds. Hence, meticulous selection of the inoculated strains of rhizobacteria can help increase the quality, yield, efficacy of legume crops for potential benefits for human health.

1.5.1.3 Microorganisms Involved During Harvest and Post-harvest Seasons

Natural disease resistance of plants generally decreases during harvest and postharvest seasons owing to the reduced accumulation of antimicrobial compounds (Prabha et al. 2013). Therefore, a preferred strategy for enhancing the disease resistance will be inducing/acquiring natural protection of plant tissues as compared to harmful chemical. The first step of induced natural protection involves infection by microorganisms, such as bacteria, moulds, and yeasts (Couto et al. 2011) causing a defence response by controlled cell death, known as "hypersensitive response" thus activating the reactive oxygen species (ROS) generation pathway and producing secondary metabolites (Silva et al. 2013). A secondary defence response, for example, "induced response system" (IRS) or "system acquired response" (SAR) can also be activated in plants which protects them in different time period of initial attack by synthesizing metabolites like jasmonic acid, ethylene and salicylic acid. These compounds also help the plants to immunize and recognize during similar pathogen invasion in future. Sometimes, nonpathogenic microorganisms are applied to induce the response system (Farfour et al. 2015). For example, in avocado fruits, significant increase in epicatechin was observed when inoculated with *Colletotrichum magna* (nonpathogenic); Sweet cherry when inoculated with *Monilinia fructicola* alone or in combination with *Cryptococcus laurentii* as antagonist microorganisms resulted in a significant rise in catalase and superoxide dismutase. In a study it was observed that oranges upon inoculation with *P. digitatum* resulted in higher production of phenylalanine ammonia lyase PAL enzyme and antioxidant compounds, as compared to non inoculated oranges (Peix et al. 2014; Bejarano et al. 2014).

1.6 Effect of Abiotic Stress on the Yield of Bioactive Compounds of Legumes

1.6.1 Salinity

A gap in knowledge exists regarding the efficient production of legume bioactives under abiotic stress conditions like salinity. Being one of the critical abiotic stresses in agriculture, salinization of food crops will lead to impairment of plant cell physiological processes. Elevated levels of salt in the agricultural land will lead to hyperionic and hyperosmotic effects leading to membrane disintegration, essential nutrient imbalance, accumulation of reactive oxygen species (ROS) in cells and cellular toxicity causing death of plant (Hasegawa et al. 2000). This gives rise to exponential growth in global agricultural commodity yield and has detrimental impacts on global food economy. Thus, maintenance of important food crops and preservation of food quality during both the pre- and post-harvest crop stages becomes challenging. Based on this understanding novel food innovation strategies can be developed to protect food crops in the field against abiotic stresses like salinity while simultaneously producing a crop of higher nutritional value for the consumer. For example, stimulation of bioactives such as phenolics and their subsequent biosynthesis via up-regulation of proline-associated pentose phosphate pathway (PAPPP), by using elicitors as seed and foliar treatments helps to improve abiotic stress resilience and bioactive profiles in black beans (Orwat 2016). Some salt-tolerant cultivars e.g. *Salvia sp.* exhibited an increase in correlation among antioxidant activity and phenolic compounds under saline conditions although they showed 61% reduction in growth (Droby et al. 2002). In another study, the activities of superoxide dismutase, ascorbate peroxidase, catalase and peroxidase were found to increase in Quinoa plants during salt stress conditions (Ballester et al. 2013). UV-C irradiation and salts stress increases total phenolic compounds, antioxidant, flavonoids and antiradical activities when applied on lettuce seeds (Ouhibi et al. 2014). Figure 1.2 represents a general scheme of salinity stress tolerance in plants.

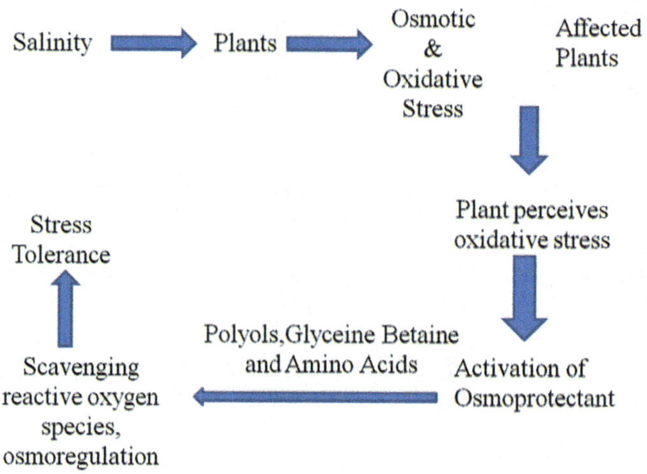

Fig. 1.2 Salinity stress tolerance in plants. (Modified from Singh et al. 2010)

1.6.2 Metal Ions

Metal ions influence expression of secondary metabolites and thereby change the nutritional value of agricultural plants. Presence of nitrogen in plants like *Matricaria chamomilla*, induces some phenolic metabolites that alters their physiological adaptations. Metal ions such as Calcium (Ca^{2+}) increases the accumulation of secondary metabolites thereby preventing plants from other abiotic stressed conditions. In a study, application of Ca^{2+} ions altered metabolism of broccoli sprouts by enhanced expression of enzymes like myrosinase and sulfotransferase and increased the antioxidant activity by formation of isothiocyanates. Biofortification with metal ions such as Selenium yielded higher amounts of phenolic compounds with enhanced antioxidant activity. Though silicon is not a true metal but its presence has been observed to enhance plant's eco-physiological adaptations during stress conditions. For example, morphological traits of the root of Indian mustard is improved in presence of silicon (Yu et al. 2008).

1.6.3 Water

Excess free radicals produced during abiotic stress conditions like water deficiency can cause functional impairment of DNA, proteins and other biomolecules. First tolerance mechanism of plants comprises of reduced stomatal conductance in drought prone agricultural lands along with efficient water usage with minimum losses. Drought conditions in plants lead to oxidative damage and ROS generation during lipid membrane peroxidation. Secondary metabolites such as carotenoids and antioxidant enzymes are produced as by-products of ROS-neutralizing pathways to maintain homeostasis. Differential exposure to water treatment leads to variations in the content of total phenols, flavonoids, and anthocyanins in cumin leaves. Drought tolerance in pomegranate fruits is best observed by increase in peel redness, vitamin C content, and total antioxidant capacity followed by decrease in cold shock injuries during storage.

In another study, low irrigation regime leads to increased concentration of total flavonols thus enhancing the antioxidant activity in blueberries (Lee et al. 2013). A study by reported that water stress is associated with rise in β-carotene, flavonoid and phenolic contents. This resulted into enhanced antioxidant activity. Machado et al. (2018) evaluated long-term regulated deficit irrigation for olives and observed that polyphenol content, phenylalanine ammonia lyase (PAL) and antioxidant activity increased significantly thus implying less water generates high values of bioactive compounds and antioxidant capacities.

1.6.4 Temperature

Plants have excellent adaptability to temperature stress as they are a rich source of endogenous antioxidants owing to their extreme survival capability during massive ultraviolet radiation exposures (Draelos 2009). Thus, the products developed from plant sources have very good safety records in the marketplace resulting in the growing demand for herbal formulations. Among these herbal components, flavonoids and phenolic acids are the most important bioactive compounds in plants that have found wide application in pharmaceuticals and cosmeceuticals as an alternative to synthetic antioxidants (Mota and Pinto 2012). Temperature stress is found to overexpress bioactive compounds such as caryophyllene oxide, β-caryophyllene, β-bisabolene, viridiflorol and manool in *Salvia lavandulifolia* plants. Prolonged exposure to elevated temperatures have also been found to induce enhanced production of secondary metabolites such as **α**-pinene, menthone, thymol, 1,8-cineole, and **α**-terpinene among *Labiatae* family (Franz and Novak 2010) This suggests that abiotic factors like temperature plays a significant role in the secondary metabolite expression for chemical phenotyping and elucidate responses of genotypes.

1.7 Health Benefit of Legumes in Agriculture

1.7.1 Legume and Its Importance to Mankind

Legumes are protein-rich plants belonging to the family Fabaceae or Leguminosae. Legumes, the second most important crop in agriculture (after cereals) are primarily grown for human consumption and livestock forage. However, they are known to have immensely critical roles in natural ecosystems, agroforestry and as colonizers of innumerable N_2 fixing symbionts. The plants are known to possess not only excellent nutritive aspects and numerous health benefits but also has been cited to have several beneficial properties like antioxidant and antibacterial effects, medicinal and therapeutic properties like anti-diabetic, anti-cancer, anti-inflammatory, anti-osteoporotic, anti-nephritic and estrogenic effects (Graham and Vance 2003; Smýkal et al. 2014; Sprent et al. 2010). Scientists have stated that these extra-nutritive qualities could be a result of the nitrogen fixing environment. The economic importance of leguminous crops upsurges to numerous folds owing to the current scenario of excessive energy consumption and soaring prices of nitrogen rich fertilisers. The plants also play a major role in restoration of the soil fertility due to their extensive use in crop rotation (systematic planting of different crops in a particular order to maintain nutrients in the soil) and intercropping.

The main classes among the commercially farmed legumes are grain, forage, blooms, pharmaceuticals, green manure, and timber species. The seeds of grain legumes including species like pea (*Pisumsativum*), soybean (*Glycine max*), beans, lentils, etc., commonly called pulses, are used for human consumption (for having high nutritional and protein values) or industrial processes (production of oils). Forage legumes are used in mainly for grazing purposes or stock feed. Indigo for industry is obtained by the cultivation of *Indigo feratinctoria*. A widely used histological stain, Hematoxylin, is produced by the flowering tree in the legume family, *Haematoxylon campechianum*. Several parts of the plants along with their metabolites are known to be pharmaceutically significant and are consumed in a regular basis for the enhancement of health and immunity, and also used as traditional medicines. Hence, legumes are often referred to as dietary supplements. According to various nutritionists, they also contain fibers, minerals, and vitamins like A, C, and E. The primary organic and mineral constituents include proteins, carbohydrates, fats, nitrogen, calcium, magnesium, zinc, potassium, and iron. Recent researches have shed light on the presence of different organic constituents like flavonoids, flavanols, flavanones, saponins, alkaloids, glucosides, rotenoids, tannins, chalcones, alkaloids, trypsin inhibitors, etc., with flavonoids are the largest and most important class. The legume specific free amino acid, canavanine, is another constituent of legumes.

Legumes are also known to have protective effects and significance in preventing abnormal conditions like insomnia, nervousness, and stress and inhibiting ailments like ulcers, diarrhea, bronchitis, and rheumatic pain. Even regulation of energy metabolism and treatment of metabolic syndromes like kwashiorkor, hyperglycemia, and hypercholesterolemic conditions can be remedied by including legumes in regular diet.

1.7.2 Significance of Bioactive Compounds from Legumes Having Health Benefits

Legumes are rich source of micronutrients, dietary fibre, proteins and essential bioactive phytochemicals. The use of the seeds from legumes in daily diet provides protection against chronic diseases due to the availability of large amount of bioactive compounds that causes degradation of proteins and availability of minerals (López et al. 2016). The seed coat of legumes are an abundant reserve for antioxidant chemicals like flavonoids, phenolic acids, lignans, and tannins. Common bean (*P. vulgaris L.*) can act effectively against liver injury in animal models, as well as breast, colon, and prostate cancer proliferation. For example, a contributing factor in the lower incidences of colon cancer registered in Latin American countries as compared with other countries was the higher consumption of common bean (Fernandez et al. 2008). The phenolic compounds present in Lentils (*Lens culinaris*) can act effectively against disorders and degenerative diseases (Shepherd et al. 1995). Polyphenols such as isoflavone and anthocyanin are a significant component of soybean seeds and have high antioxidant and antiradical activities. Consuming peanuts on a daily basis reduces the risks of weight gain, cardiovascular diseases, Alzheimer's disease, and cancer due to the presence of anthocyanins, proanthocyanidins and resveratrol. Anthocyanins are also found in black beans (i.e., delphinidin, petunidin, and malvidin), pinto beans (i.e., kaempferol), and pink beans (i.e., quercetin and kaempferol), although their physiological effects are yet to be revealed.

The various compounds are known to have significant roles in decrease of oxygen concentration, intercepting singlet oxygen, bind metal ion catalysts, and prevention of 1st chain initiation by scavenging initial radicals. Hydrolysable tannins, phenolic acids, and flavonoids has anti-mutagenic and anti-carcinogenic effects as they as protective component to DNA against the free radicals. Studies showed that the mutualistic relation and diets rich in antioxidants lowers the chances of the degenerative disorders that develop due to the presence of the reactive oxygen and nitrogen species (Naczk and Shahidi 2006), such as cancer, heart disease, arthritis, and immune system decline (Gordon 1996).

1.7.3 Prebiotic Effects of Legume

The prebiotic group of food has a beneficial effect by developing the growth of beneficial bacterial species. Humans lack the enzymes capable of digesting the α-galactosides (like raffinose) in beans. Such non-digestible a-galactosidase have recently been hypothesised to have prebiotic properties, similar to those ascribed to inulin and other fructo-oligosaccharides of cereals. Some oligosaccharides have functional effects, such as improvement of glucose control and modulation of lipid metabolism which tends to be similar to those of soluble dietary fibres. Moreover, possible enhancing effects of mineral (calcium, magnesium, iron) absorption by non-digestible carbohydrate has been reported (Swennen et al. 2006). Prebiotic properties of legume seeds with such non-digestible oligosaccharides (raffinose, stachyose, verbascose) need to be more intricately assessed in further studies.

Moreover, other than soybean many other pulses have been explored as good and healthy prebiotics, such as peanut, lupin, pigeon pea, Bambara groundnut, green gram and mung bean.

Legume seed proteins, especially those from soya bean, have notably demonstrated to have powerful bioactive compounds, which exert cholesterol-reducing properties (FDA 1999). Moreover, bioactive peptides from soy digestion have been found to bring about its impact on the immune system and on the gastrointestinal tract.

Apart from being an excellent source of protein, legumes tend to be also rich in trace elements (magnesium, potassium, calcium, iron, zinc, copper and manganese) and minerals. Beans being a good source of magnesium and potassium, might help to lower blood pressure and thereby decrease the risk of cardiovascular diseases. Legumes contain several B-vitamins. They have low content of total and saturated fats and are cholesterol-free. Beans, like *Pliaseolus* sp. L., are a major source of soluble fibre, and it is this fibre fraction that helps in lowering cholesterol levels as well as regulates blood glucose levels. Several research findings have revealed the real risk of adverse health effects from the so-called antinutritional factors present in the seed. Such risk has been limited mainly to the high content of non-proteinaceous heat-stable compounds (such as tannins and phytic acid) for susceptible subjects (i.e. vicine and convicine for people affected by favism).

1.8 Conclusion

Legumes are rich in essential nutrients and bioactive compounds such as polyphenols, alkaloids, flavonoids etc. It has been well established that leguminous plants improve the agricultural yield as well as bioactive components present in leguminous plants has high efficacy in maintaining healthy physiological condition in human beings. The symbiotic relationship between legumes and rhizobia efficiently improve the soil fertility and thus the agricultural outcome. Hence it can minimize the green house effect caused due to the use of chemical fertilizers. On the other hand the leguminous bioactive compounds pose a broad array of health effects on humans. Although considerable research is being performed on legume bioactivities that may potentially protect from the risk of cancers and from cardiovascular disease, the results of epidemiological studies do not yet provide any conclusive conclusion. Moreover, rhizobial inoculation for increasing the bioactive molecules within the leguminous plants is gaining more attention. So legume derived bioactive compounds has plenty of advantageous effects. Therefore more research is needed to identify more bioactive compounds and also to improve their efficacy for implementation in agricultural and therapeutic purposes.

References

Abdullah MZ, Mavaddat MH, Mohd AJ (2013) Antioxidant and anticancer activities of *Polygonum minus*, *Alpinia galanga* and *Etlingeraelatior*. Open Conf Proc J 4:203. https://doi.org/10.2174/2210289201304010202

Adaskaveg JE, Förster H, Wade ML (2011) Effectiveness of kasugamycin against Erwinia amylovora and its potential use for managing fire blight of pear. Plant Dis 95:448–454. https://doi.org/10.1094/PDIS-09-10-0679

Agbafor KN, Nwachukwu N (2011) Phytochemical analysis and antioxidant property of leaf extract of vitex doniana and mucunapruriens. Biochem Res Int 2011(2011):459839. https://doi.org/10.1155/2011/459839

Ali AM, Mackeen MM, El-Sharkawy SH, Abdul Hamid J, Ismail NH, Ahmad F et al (1996) Antiviral and cytotoxic activities of some plants used in Malaysian indigenous medicine. Pertanika J Trop Agric Sci 19(2/3):129–136

Almey AAA, Khan CAJ, Zahir IS, Suleiman KM, Aisyah MR, Rahim KK (2010) Total phenolic content and primary antioxidant activity of methanolic and ethanolic extracts of aromatic plants leaves. Int Food Res J 17:1077–1084

Aparicio-Fernández X, Reynoso-Camacho R, Castaño-Tostado E, García-Gasca T, de Mejía EG, Guzmán-Maldonado SH, Elizondo G, Yousef GG, Lila MA, Loarca-Pina G (2008) Antiradical capacity and induction of apoptosis on HeLa cells by a Phaseolus vulgaris extract. Plant Foods Hum Nutr 63(1):35–40. https://doi.org/10.1007/s11130-007-0066-4

Araujo J, Díaz-Alcántara C-A, Velázquez E, Urbano B, González-Andrés F (2015) Bradyrhizobium yuanmingense related strains form nitrogen-fixing symbiosis with Cajanuscajan L. in Dominican Republic and are efficient biofertilizers to replace N fertilization. Sci Hortic 192:421–428. https://doi.org/10.1016/j.scienta.2015.06.009

Aurora Sandoval-Chávez R, Álvar Martínez-Peniche R, Hernández-Iturriaga M, Teixidó-Espasa N, Usall-Rodié J, Viñas-Almenar I, Torres-Sanchis R (2015) Mechanisms of resistance in postharvest fruit-pathogen interaction. Rev Chapingo Ser Hortic XXI:185–198. https://doi.org/10.5154/r.rchsh.2014.11.050

Baharum SN, Bunawan H, Ghani MA, Mustapha WAW, Noor NM (2010) Analysis of the chemical composition of the essential oil of *Polygonum minus* Huds. using two-dimensional gas chromatography-time-of-flight mass spectrometry (GC-TOF MS). Molecules 15:7006–7015. https://doi.org/10.3390/molecules15107006

Ballester A-R, Teresa Lafuente M, González-Candelas L (2013) Citrus phenylpropanoids and defence against pathogens. Part II: gene expression and metabolite accumulation in the response of fruits to Penicillium digitatum infection. Food Chem 136(1):285–291. https://doi.org/10.1016/j.foodchem.2012.08.006

Barahuie F, Dorniani D, Saifullah B, Gothai S, Hussein MZ, Pandurangan AK, Arulselvan P, Norhaizan ME (2017) Sustained release of anticancer agent phytic acid from its chitosan-coated magnetic nanoparticles for drug-delivery system. Int J Nanomedicine 12:2361–2372

Bejarano A, Ramírez-Bahena M-H, Velázquez E, Peix A (2014) Vigna unguiculata is nodulated in Spain by endosymbionts of Genisteae legumes and by a new symbiovar (vignae) of the genus Bradyrhizobium. Syst Appl Microbiol 37(7):533–540. https://doi.org/10.1016/j.syapm.2014.04.003

Boberek JM, Stach J, Good L (2010) Genetic evidence for inhibition of bacterial division protein FtsZ by Berberine. PLoS One 5(10):e13745. https://doi.org/10.1371/journal.pone.0013745

Bunawan H, Talip N, Noor NM (2011) Foliar anatomy and micromorphology of Polygonum minus Huds and their taxonomic implications. Aust J Crop Sci 5:123–127

Burton JC (1972) Nodulation and symbiotic nitrogen fixation. In: Hanson CH (ed) Alfalfa science and technology. Monograph 15. American Society of Agronomy, Madison, pp 229–246

Castellanos-Morales V, Villegas J, Wendelin S, Vierheilig H, Eder R, Cárdenas-Navarro R (2010) Root colonisation by the arbuscular mycorrhizal fungus Glomus intraradices alters the quality of strawberry fruits (Fragaria x ananassa Duch.) at different nitrogen levels. J Sci Food Agric 90(11):1774–1782. https://doi.org/10.1002/jsfa.3998

Nello C, Maurizio C, Piero P, Luca M, Cristiana S, Manuela G (2010) Globe artichoke as a functional food. Mediterr J Nutr Metab 3:197–201. https://doi.org/10.1007/s12349-010-0021-z

Chakraborty U, Sarkar B, Chakraborty BN (2003) Protection of soybean rot by Bradyrhizobium japonicum and Trichoderma harzianum associated with changes in enzyme activities and phytoalexin production. J Mycol Plant Pathol 33:21–25. https://doi.org/10.3923/ajbs.2019.114.121

Chen WM, Moulin L, Bontemps C, Vandamme P, Béna G, BoivinMasson C (2003) Legume symbiotic nitrogen fixation by β-proteobacteria is widespread in nature. J Bacteriol 185:7266–7272. https://doi.org/10.1128/JB.185.24.7266-7272.2003

Costerton JW (2007) The biofilm primer. Springer, Berlin/Heidelberg

Couto C, Silva LR, Valentão P, Velázquez E, Peix A, Andrade PB (2011) Effects induced by the nodulation with Bradyrhizobium japonicum on *Glycine max* (soybean) metabolism and antioxidant potential. Food Chem 127(4):1487–1495. https://doi.org/10.1016/j.foodchem.2011.01.135

Cushnie TP, Cushnie B, Lamb AJ (2014) Alkaloids: an overview of their antibacterial, antibiotic-enhancing and antivirulence activities. Int J Antimicrob Agents 44(5):377–386. https://doi.org/10.1016/j.ijantimicag.2014.06.001

Dias GP, Cavegn N, Nix A, do Nascimento Bevilaqua MC, Stangl D, Zainuddin MSA, Thuret S (2012) The role of dietary polyphenols on adult hippocampal neurogenesis: molecular mechanisms and behavioural effects on depression and anxiety. Oxidative Med Cell Longev 2012:1–18. https://doi.org/10.1155/2012/541971

Draelos ZD (2009) Cosmeceuticals: undefined, unclassified, and unregulated. Clin Dermatol 27(5):431–434. https://doi.org/10.1016/j.clindermatol.2009.05.005

Droby S, Vinokur V, Weiss B, Cohen L, Daus A, Goldschmidt EE, Porat R (2002) Induction of resistance to penicillium digitatumin grapefruit by the yeast biocontrol agent candida oleophila. Phytopathology 92(4):393–399. https://doi.org/10.1094/phyto.2002.92.4.393

Elexsona N, AfsahHejria L, Rukayadia Y, Soopnaa P, Leea HY, Tuan Zainazorc TC, Nor Ainya M, Na kaguchib Y, Mitsuakib N, Sona R (2014) Effect of detergents as antibacterial agents on biofilm of antibiotics-resistant *Vibrio parahaemolyticus* isolates. Food Control 35(1):378–385

Farfour SA, Al-Saman MA, Hamouda RA (2015) Potential activity of some biofertilizer agents on antioxidant and phytochemical constituents of fava bean plant. Glob Adv Res J Agric Sci 4:26–32

Fisher RF, Egelhoff TT, Mulligan JT, Long SR (1998) Specific binding of proteins from *Rhizobium meliloti* cell-free extracts containing Nod D to DNA sequences upstream of inducible nodulation genes. Genes Dev 2:282–293. https://doi.org/10.1101/gad.2.3.282

Flight I, Clifton P (2006) Cereal grains and legumes in the prevention of coronary heart disease and stroke: a review of the literature. Eur J Clin Nutr 60(10):1145–1159. https://doi.org/10.1038/sj.ejcn.1602435

Food and Drug Administration (FDA) (1999) Final rule. Fed Regist 64:57699–57733

Frankel EN (1996) Antioxidants in lipid foods and their impact on food quality. Food Chem 57(1):51–55. https://doi.org/10.1016/0308-8146(96)00067-2

Frankow-Lindberg BE, Dahlin AS (2013) N2 fixation, N transfer, and yield in grassland communities including a deep-rooted legume or non-legume species. Plant Soil 370:567–581

Franz C, Novak J (2010) Source of essential oils. In: Baser KHC, Buchbauer G (eds) Handbook of essential oils. Science, technology and application. Taylor & Francis Group, New York, pp 40–81

Ganesan K, Xu B (2017) Polyphenol-rich lentils and their health promoting effects. Int J Mol Sci 18(11):2390. https://doi.org/10.3390/ijms18112390

Gor MC, Ismail I, Mustapha WAW, Zainal Z, Noor NM, Othman R (2011) Identification of cDNAs for jasmonic acid-responsive genes in *Polygonum minus* roots by suppression subtractive hybridization. Acta Physiol Plant 33:283–294

Gordon MH (1996) Dietary antioxidants in disease prevention. Nat Prod Rep 13(4):265–273. https://doi.org/10.1039/np9961300265

Graham PH, Vance CP (2003) Legumes: importance and constraints to greater use. Plant Physiol 131(3):872–877. https://doi.org/10.1104/pp.01700

Guldan SJ, Martin CA, Cueto-Wong J, Steiner RL (1996) Interseeding legumes into Chile: legume productivity and effect on Chile yield. HortScience 31:1126–1128

Harborne JB (1973) Phytochemical methods. Chapman and Hall Ltd., London, pp 49–188

Harborne JB, Turner BL (1984) Plant chemosystematics. Academic, London

Hartmann T, Witte L (1995) Chemistry, biology and chemoecology of the pyrrolizidine alkaloids. In: Pellitier SW (ed) Alkaloids: chemical and biological perspectives, vol 9. Pergamon, Oxford, pp 155–233

Hasegawa PM, Bressan RA, Zhu JK, Bohnert HJ (2000) Plant cellular and molecular responses to high salinity. Annu Rev Plant Physiol Plant Mol Biol 51:463–499. https://doi.org/10.1146/annurev.arplant.51.1.463

Hause B, Maier W, Miersch O, Kramell R, Strack D (2002) Induction of jasmonate biosynthesis in arbuscular mycorrhizal barley roots. Plant Physiol 130(3):1213–1220. https://doi.org/10.1104/pp.006007

Hegnauer R, Hegnauer M (1994) Chemotaxonomie der Pflanzen, vol 11a. Birkhäuser, Basle

Huda-Faujan N, Noriham A, Norrakiah AS, Babji AS (2009) Antioxidant activity of plants methanolic extracts containing phenolic compounds. Afr J Biotechnol 8(3):484–489

Ito N, Fukushima S, Hasegawa A, Shibata M, Ogiso T (1983) Carcinogenicity of butylated hydroxy anisole in F344 rats. J Natl Cancer Inst 70:343–347

Jensen ES, Peoples MB, Boddey RM, Gresshoff PM, Hauggaard-Nielsen H, Alves BJ, Morrison MJ (2012) Legumes for mitigation of climate change and the provision of feedstock for biofuels and biorefineries. A review. Agron Sustain Dev 32:329–364

Kapoor R, Giri B, Mukerji KG (2002) Glomus macrocarpum: a potential bioinoculant to improve essential oil quality and concentration in dill (Anethum graveolens L.) and carum (Trachyspermum ammi (Linn.) Sprague). World J Microbiol Biotechnol 18:459–463. https://doi.org/10.1023/A:1015522100497

Kaur C, Kapoor HC (2002) Anti-oxidant activity and total phenolic content of some Asian vegetables. Int J Food Sci Technol 37(2):153–161. https://doi.org/10.1046/j.1365-2621.2002.00552.x

Kinghorn AD, Balandrin MF (1984) Quinolizidine alkaloids of the Leguminosae: structural types, analysis, chemotaxonomy and biological activities. In: Pelletier WS (ed) Alkaloids: chemical and biological perspectives. Wiley, New York, pp 105–148

Lee MH, Cho EJ, Wi SG, Bae H, Kim JE, Cho J-Y et al (2013) Divergences in morphological changes and antioxidant responses in salt-tolerant and salt-sensitive rice seedlings after salt stress. Plant Physiol Biochem 70:325–335. https://doi.org/10.1016/j.plaphy.2013.05.047

Limpens E, Franken C, Smit P, Willemse J, Bisseling T, Geurts R (2003) LysM domain receptor kinase regulating rhizobial Nod factor- induced infection. Science 302:630–633. https://doi.org/10.1126/science.1090074

Liu CW, Murray JD (2016) The role of flavonoids in nodulation host range specificity: anupdate. Plan Theory 5:E33. https://doi.org/10.3390/plants5030033

Long SR (1996) Rhizobium symbiosis: nod factors in perspective. Plant Cell 8:1885–1898. https://doi.org/10.1105/tpc.8.10.1885

López-Cortez M d S, Rosales-Martínez P, Arellano-Cárdenas S, Cornejo-Mazón M (2016) Antioxidants properties and effect of processing methods on bioactive compounds of legumes. In: Goyal AK (ed) Grain legumes. InTech, Rijeka, Croatia. https://doi.org/10.5772/63757

Machado RMA, Alves-Pereira I, Ferreira Heliyon RMA (2018) Plant growth, phytochemical accumulation and antioxidant activity of substrate-grown spinach. Heliyon 4(8):e00751. https://doi.org/10.1016/j.heliyon.2018.e00751

Mackeen MM, Ali AM, El-Sharkawy SH, Manap MY, Salleh KM, Lajis NH, Kawazu K (1997) Antimicrobial and cytotoxic properties of some Malaysian traditional vegetables (Ulam). Int J Pharmacogn 35(3):174–178. https://doi.org/10.1076/phbi.35.3.174.13294

MacLean AM, Finan TM, Sadowsky MJ (2007) Genomes of the symbiotic nitrogen-fixing bacteria of legumes. Plant Physiol 144:615–622. https://doi.org/10.1104/pp.107.101634

MarÃti G, Kondorosi Ã (2014) Nitrogen-fixing rhizobium-legume symbiosis: are polyploidy and host peptide-governed symbiont differentiation general principles of endosymbiosis? Front Microbiol 5:326. https://doi.org/10.3389/fmicb.2014.00326

Martin MF, Liras P (1989) Organization and expression of genes involved in the biosynthesis of antibiotics and other secondary metabolites. Annu Rev Microbiol 43:173–206. https://doi.org/10.1146/annurev.mi.43.100189.001133

McLean E (1970) The toxic actions of pyrrolizidine (Senecio) alkaloids. Pharmacol Rev 22:430–483

Mena-Violante HG, Ocampo-Jiménez O, Dendooven L, Martínez-Soto G, González-Castañeda J, Davies FT Jr, Olalde-Portugal V (2006) Arbuscular mycorrhizal fungi enhance fruit growth and quality of Chile ancho (Capsicum annuum L. cv San Luis) plants exposed to drought. Mycorrhiza 16(4):261–267. https://doi.org/10.1007/s00572-006-0043-z

Mota I, Pinto PR (2012) Extraction of polyphenolic compounds from Eucalyptus globulus bark: process optimization and screening for biological activity. Ind Eng Chem Res 51:6991–7000. https://doi.org/10.1021/ie300103z

Mulas D, García-Fraile P, Carro L (2011) Distribution and efficiency of rhizobium leguminosarum strains nodulating Phaseolus vulgaris in Northern Spanish soils: selection of native strains that replace conventional N fertilization. Soil Biol Biochem 43:2283–2293

Naczk M, Shahidi F (2006) Phenolics in cereal, fruits and vegetables: occurrence, extraction and analysis. J Pharm Biomed Anal 41(5):1523–1542. https://doi.org/10.1016/j.jpba.2006.04.002

Ngoci SN, Mwendia CM, Mwaniki CG (2011) Phytochemical and cytotoxicity testing of Indigoferalupatana Baker F. J Anim Plant Sci 11(1):1364–1373

Nishida C, Uauy R, Kumanyika S, Shetty P (2004) The Joint WHO/FAO Expert Consultation on diet, nutrition and the prevention of chronic diseases: process, product and policy implications. Public Health Nutr 7(1A):245–250. https://doi.org/10.1079/PHN2003592

Orwat J (2016) Phenolic antioxidant-linked bioactive enrichment in black beans (*phaseolus vulgaris* l.) to screen for health benefits and enhancement of salinity resilience

Ouhibi C, Attia H, Rebah F, Msilini N, Chebbi M, Aarrouf J, Urban L, Lachaal M (2014) Salt stress mitigation by seed priming with UV-C in lettuce plants: growth, antioxidant activity and phenolic compounds. Plant Physiol Biochem 83:126. https://doi.org/10.1016/j.plaphy.2014.07.019

Peix A, Ramírez-Bahena MH, Velázquez E, Bedmar EJ (2014) Bacterial associations with legumes. Crit Rev Plant Sci 34(1–3):17–42. https://doi.org/10.1080/07352689.2014.897899

Polak R, Phillips EM, Campbell A (2015) Legumes: health benefits and culinary approaches to increase intake. Clin Diab 33(4):198–205. https://doi.org/10.2337/diaclin.33.4.198

Prabha C, Maheshwari DK, Bajpai VK (2013) Diverse role of fast growing rhizobia in growth promotion and enhancement of psoralen content in Psoraleacorylifolia L. Phcog Mag [serial online] [cited 2019 Sep 11]; 9(Suppl S1):57–65. Available from: http://www.phcog.com/text.asp?2013/9/36/57/117870

Qader SW, Abdulla MA, Chua LS, Sirat HM, Hamdan S (2012) Pharmacological mechanisms underlying gastroprotective activities of the fractions obtained from *Polygonum minus* in Sprague Dawley rats. Int J Mol Sci 13:1481–1496. https://doi.org/10.3390/ijms13021481

Rostas K, Kondorosi E, Horvath B, Simoncsits A, Kondorosi A (1986) Conservation of extended promoter regions of nodulation genes in *Rhizobium*. Proc Natl Acad Sci U S A 83:1757–1761. https://doi.org/10.1073/pnas.83.6.1757

Salmi C, Loncle C, Vidal N, Letourneux Y, Fantini J, Maresca M et al (2008) Squalamine: an appropriate strategy against the emergence of multidrug resistant gram-negative bacteria? PLoS One 3(7):e2765. https://doi.org/10.1371/journal.pone.0002765

Sandberg A-S (2002) Bioavailability of minerals in legumes. Br J Nutr 88(S3):281. https://doi.org/10.1079/bjn/2002718

Scalbert A, Manach C, Morand C, Remesy C, Jimenez L (2005) Dietary polyphenols and the prevention of diseases. Crit Rev Food Sci Nutr 45:287–306

Seigler DS (1998) Plant secondary metabolism. Kluwer, Norwell

Shepherd J, Cobbe SM, Ford I, Isles CG, Lorimer AR, Macfarlane PW, Mckillop JH, Packard CJ (1995) Prevention of coronary heart disease with pravastatin in men with hypercholesterolemia. N Engl J Med 333(20):1301–1308. https://doi.org/10.1056/NEJM199511163332001
Silva LR, Pereira MJ, Azevedo J, Mulas R, Velazquez E, González-Andrés F et al (2013) Inoculation with Bradyrhizobium japonicum enhances the organic and fatty acids content of soybean (Glycine max (L.) Merrill) seeds. Food Chem 141(4):3636–3648. https://doi.org/10.1016/j.foodchem.2013.06.045
Singh AL, Hariprasanna K, Chaudhari V, Gor HK, Chikani BM (2010) Identification of groundnut (Arachis hypogaea L.) cultivars tolerant of soil salinity. J Plant Nutr 33:1761–1776. https://doi.org/10.1080/01904167.2010.503779
Smýkal P, Coyne CJ, Ambrose MJ, Maxted N, Schaefer H, Blair MW et al (2014) Legume crops phylogeny and genetic diversity for science and breeding. Crit Rev Plant Sci 34(1–3):43–104. https://doi.org/10.1080/07352689.2014.897904
Sorensen J, Sessitsch A (2007) Plant-associated bacteria – lifestyle and molecular interactions. In: Van Elsas JD, Jansson JD, Trevors JT (eds) Modern soil microbiology, 2nd edn. CRC Press, Boca Raton, pp 211–236
Sorensen SJ, Bailey M, Hansen LH, Kroer N, Wuertz S (2005) Studying plasmid horizontal transfer in situ: a critical review. Nat Rev Microbiol 3:700–710
Southon IW (1994) Phytochemical dictionary of the leguminosae. Chapman & Hall, London
Sprent JI, Odee DW, Dakora FD (2010) African legumes: a vital but under-utilized resource. J Exp Bot 61(5):1257–1265. https://doi.org/10.1093/jxb/erp342
Stagnari F, Maggio A, Galieni A, Pisante M (2017) Multiple benefits of legumes for agriculture sustainability: an overview. Chem Biol Technol Agric 4(1):1. https://doi.org/10.1186/s40538-016-0085-1
Swennen K et al (2006) Crit Rev Food Sci Nutr 46:459–471. *9* GRAINSLEGUMES No. 48 – January 2007
Terry L (2004) Elicitors of induced disease resistance in postharvest horticultural crops: a brief review. Postharvest Biol Technol 32(1):1–13. https://doi.org/10.1016/j.postharvbio.2003.09.016
Teuscher E, Lindequist U (2010) BiogeneGifte. Biologie, Chemie, Pharmakologie, Toxikologie. Wissenschaftliche Verlagsgesellschaft, Stuttgart
Tominaga K, Higuchi K, Hamasaki N, Hamaguchi M, Takashima T, Tanigawa T, Watanabe T, Fujiwara Y, Tezuka Y, Nagaoka T, Kadota S, Ishii E, Kobayashi K, Arakawa T (2002) In vivo action of novel alkyl methyl quinolone alkaloids against Heli-cobacterpylori. J Antimicrob Chemother 50(4):547–552
Uyub AM, Nwachukwu IN, Azlan AA, Fariza SS (2010) In-vitroantibacterial activity and cytotoxicity of selected medicinal plant extracts from Penang Island Malaysia on metronidazole-resistant Helicobacter pylori and some pathogenic bacteria. Ethnobot Res Appl 8:95–106
Veitch NC (2010) Flavonoid chemistry of the leguminosae. In: Santos-Buelga C, Escribano-Baillon MT, Lattanzio V (eds) Recent advances in polyphenol research. Wiley-Blackwell, Chichester, pp 23–58
Venn BJ, Mann JI (2004) Cereal grains, legumes and diabetes. Eur J Clin Nutr 58(11):1443–1461. https://doi.org/10.1038/sj.ejcn.1601995
Wan Hassan WE (2006) Healing herbs of Malaysia. Federal Land Development Authority (FELDA), Kuala Lumpur
Wink M (1992) The role of quinolizidine alkaloids in plant insect interactions. In: Bernays EA (ed) Insect–plant interactions, vol IV. CRC Press, Boca Raton, pp 133–169
Wink M (1993a) Allelochemical properties and the raison d'être of alkaloids. In: Cordell G (ed) The alkaloids, vol 43. Academic, Orlando, pp 1–118
Wink M (1993b) Quinolizidine alkaloids. In: Waterman PG (ed) Methods in plant biochemistry. Academic, London, pp 197–239
Wink M (2000) Interference of alkaloids with neuroreceptors and ion channels. In: Atta-Ur-Rahman (ed) Bioactive natural products. Elsevier, New York, pp 1–127

Wink M (2007) Molecular modes of action of cytotoxic alkaloids – from DNA intercalation, spindle poisoning, topoisomerase inhibition to apoptosis and multiple drug resistance. In: Cordell G (ed) The alkaloids, vol 64. Academic, San Diego, pp 1–48

Wink M, Schimmer O (2010) Molecular modes of action of defensive secondary metabolites. In: Wink M (ed) Functions and biotechnology of plant secondary metabolites. Annual plant reviews, vol 39. Wiley-Blackwell, Oxford, pp 21–161

Wink M, Schmeller T, Latz-Brüning B (1998) Modes of action of allelochemical alkaloids: interaction with neuroreceptors, DNA and other molecular targets. J Chem Ecol 24:1881–1937

Xu J, Yang Q, Qian X, Samuelsson J, Janson J-C (2007) Assessment of 4-nitro-1,8-naphthalic anhydride reductase activity in homogenates of bakers' yeast by reversed-phase high-performance liquid chromatography. J Chromatogr B 847(2):82–87. https://doi.org/10.1016/j.jchromb.2006.09.040

Yaacob KB (1987) Kesom oil-a natural source of aliphatic aldehydes. Perfum Flavor 12:27–30

Yu T, Zhang H, Li X, Zheng X (2008) Biocontrol of Botrytis cinerea in apple fruit by Cryptococcus laurentii and indole-3-acetic acid. Biol Control 46:171–177. https://doi.org/10.1016/j.biocontrol.2008.04.008

Zeng X, Lin X, Hou SX (2013) The Osa-containing SWI/SNF chromatin-remodeling complex regulates stem cell commitment in the adult Drosophila intestine. Development 140(17):3532–3540. https://doi.org/10.1242/dev.096891

Chapter 2
Legume Derived Bioactive Peptides

Pragya Tiwari, Anjani Devi Chintagunta, Vijaya R. Dirisala, and N. S. Sampath Kumar

Abstract Legumes constitute the key component of a balanced diet, a good source of nutrition for population across the globe. The bioactive compounds from legumes have been shown to demonstrate therapeutic properties against diseases like cancer, cardiovascular diseases and diabetes, highlighting its significance in agricultural biotechnology. Considering the benefits and potential therapeutic significance of bioactive constituents in legumes, the cultivation and consumption of legumes has witnessed tremendous upsurge in recent years. The recent perspective in research on bioactive compounds emphasize on understanding their role in food and nutrition. Moreover, the health promoting properties of the bioactive peptides from legumes suggest their potential application as "functional foods" and nutraceutical agents.

The legumes are excellent food source for bioactive compounds with profound applications in food, nutraceutical and health industries. Nutritional values of legume based food products such as tofu, natto, peanut butter and miso were discussed thoroughly by highlighting the significance of their consumption on regular basis. Further, significance of bioactive peptides derived from legumes was elaborated in the light of various properties *viz*., antimicrobial activity, anticancer activity, anti-diabetic activity and their impact against cardiovascular disease.

Keywords Amino acids · Antimicrobial activity · Anticancer activity · Bioactive peptides · Functional food · Healthcare · Legume biotechnology · Nutritional properties · Sustainable agriculture · Therapeutics

P. Tiwari (✉)
Molecular Metabolic Engineering Laboratory, Department of Biotechnology, Yeungnam University, Gyeongsan, Gyeongbuk, Republic of Korea

A. D. Chintagunta · V. R. Dirisala · N. S. Sampath Kumar (✉)
Department of Biotechnology, Vignan's Foundation for Science, Technology and Research, Guntur, Andhra Pradesh, India

P. Guleria et al. (eds.), *Sustainable Agriculture Reviews 45*, Sustainable Agriculture Reviews 45, https://doi.org/10.1007/978-3-030-53017-4_2

Abbreviations

ACE	Angiotensin converting enzyme
ADP	Adenosine 5′diphosphate
AHA	Acreage harvested area
AICRPs	All India Coordinated Research Projects
AMPs	Antimicrobial peptides
BBI	Bowman-Birk inhibitors
CVD	Cardiovascular disease
ECLT	Euglobulin clot lysis time
FAO	Food and Agriculture Organization
FMOC	Fluorenylmethyloxycarbonyl
ICRISAT	International Crops Research Institute for the Semi-Arid Tropics
IFN-β	Interferon beta
NABARD	National Bank for Agriculture and Rural Development
NCIPM	National Centre for Integrated Pest Management
PATT	Partial thromboplastin time
TNFα	Tumor Necrosis factor alpha
TVP	Textured vegetable protein
WHO	World Health Organization

2.1 Introduction

Agriculture forms the backbone of subsistence for millions worldwide, providing source of nutrition and generating employment for people. The cultivation of food crops and its manipulations (both conventional plant breeding techniques and genetic engineering), has witnessed a tremendous upsurge in recent years, owing to increasing interest and research on “bioactive compounds” from natural sources. On a worldwide level, plant-based nutrition has become a widely studied area among scientific community, considering the nutritive benefits of food as well as its medicinal importance (World Health Organization 2003; Singh et al. 2019). According to guidelines of World Health Organization, a balanced diet is important for human health, for which they suggested the following diet recommendations: it should consists of 75% plant based food and 25% ingredients from animal source comprising of 15% of protein, 60% carbohydrates and 25% lipids (Nishida et al. 2004). On epidemiological levels, studies have suggested an integral advantage of plant based therapeutics in reducing the risk of chronic diseases (Dillard and German 2000; Rochfort and Panozzo 2007), besides providing a balanced source of nutrition, for millions across the globe. The plants constitute an essential component of food, and different plant parts are used, based on the plant species and its nutritive contents, respectively.

2.2 Legume Cultivation and Agriculture: An Overview

The legume family, scientifically nomenclatured as "Leguminosae or Fabaceae", constitute plants comprising of seed or fruit (also known as pulse), eaten as food component. Legumes are cultivated worldwide and is the third largest family with about 751 genera, comprising 19,000 plant species (Christenhusz and Byng 2016; Stevens 2008). Recent statistics suggests that India was the largest producer of pulses, accounting for 23% of world total production (953.0 thousand tonnes). Other major pulses production nations are Poland (311.8 thousand tonnes), United Kingdom (280.0 thousand tonnes) and Mozambique (South Africa) (213.6 thousand tonnes) (UN food and Agriculture Organization 2018). The year 2016, was declared "The International year of pulses" by the United Nations General Assembly (United Nations 2015). Moreover, The Food and Agriculture Organization of the United Nations together with its constituent bodies aims to increase public awareness about nutritional benefits of pulses for sustainable food production, increase global production of pulses, enhance better utilization of pulse proteins and efficiently deal with trade of pulses, globally (United Nations 2015; International Year of Pulses 2016).

The plant species with nutritive bioactive food components namely proteins, starch, vitamins, fibres, among others, demonstrates the health benefits of legumes as "functional foods". Legumes constitute the best plant-based source of dietary proteins, palatable and have excellent nutritional properties (Messina 1999). Legume seeds are good source of protein for human and animal consumption, exhibiting adaptation to diverse climatic conditions (Muzquiz et al. 2012). Legumes constitute a unique fruit type-a dry fruit, a pod, which develops from a simple carpel and dehisces on two sides. The legume seed comprises of a rich source of bioactive compounds namely glycosides, proteins, alkaloids amongst others which exhibits health benefits and nutraceutical properties (Kamran and Reddy 2018). The pulses are rich source of protein and fibres and contain variable content of micro-nutrients, therefore are preferred in sustainable diets (Chaudhary et al. 2018). The pulses/grains as food ingredients offer multiple health benefits namely as gluten-free food, source of protein and carbohydrates with low fats and high fibre content, together with vitamins and minerals (Mudryj et al. 2014). Legumes are cultivated through agriculture, mainly for human consumption, as green manure for enhancing soil fertility and for livestock forage. The popular legumes comprises of *Syzygium aromaticum*, *Glycine max*, *Arachis hypogaea*, *Phaseolus vulgaris*, *Pisum sativum*, and *Tamarindus indica*, to name a few. Another significance of legume cultivation highlights the presence of nitrogen-fixing bacteria in root nodules, thus, exhibits an important role in crop rotation. Table 2.1 Bioactive component from Legumes and their physiological effect on human health.

According to United Nations Food and Agriculture Organization (FAO), "pulse" is a term generally used for legume crops, predominately for the dry seed (Pulse Canada 2016) and excludes green peas and seeds, seeds with oil content (soybeans

Table 2.1 Bioactive component from Legumes and their physiological effects on human health

Legumes	Bioactive peptides	Physiological effects	References
Cicer arietinum, Lens culinaris, Lupinus, Pisum sativum, Vicia faba	Oligosaccharides (raffinose, stachyose, verbascose and ajugose), Ciceritol	Alpha-galactosides act as probiotics	Schley and Field (2002) and Muzquiz et al. (2012)
Phaseolus vulgaris, Lens esculenta, C. arietinum, P. sativum, V. faba	Phytic acid, myo-inositol-(1,2,3,4,5,6) hexakisphosphate, Phytases	Phytic acid shows reduced bioavailability and toxicity of heavy metals in diet Myo-inositol controls hypercholesterolemia and atherosclerosis Phytases reduce phytate content in human food	Rimbach and Pallauf (1997), Burbano et al. (1999), Champ (2002) and Greiner and Konietzny (2006)
Legumes	Enzyme inhibitors (alpha amylases and pancreatic proteases) Lectins	Protein antinutrient	Lajolo et al. (2004) and Pusztai et al. (2004)
Glycine max	Protease inhibitors (Bowman-Birk inhibitor and Kunitz family)	Inhibit Trypsin and Chymotrypsin, Anticarcinogenic	Srinivasan et al. (2005) and Clemente et al. (2004)
Dolichos biflorus, Phaseolus lunatus, Phaseolus aureus, C. arietinum, P. sativum	Alpha amylase inhibitors	Insecticidal	Le Berre-Anton et al. (1997) and Muzquiz et al. (2012)
G. max, Lupinus, Lens culinaris, C. arietinum	Saponins	Prevent lipid peroxidation of DNA and proteins	Muzquiz et al. (2012)

and peanuts) and forage seeds (alfalfa), respectively. However, there is no clear distinction and the categories are sometimes commonly used to denote pulses.

Highlighting the increased awareness among the scientific community and people about the health promoting effect of legumes in sustainable diets, the present chapter discusses the increased cultivation of legume crops in agriculture. Moreover, the global scenario in legume cultivation through agriculture has been discussed in detail. Additionally, the chapter highlights the application of legume bioactive compounds in food, nutraceuticals and health care. Thus, the chapter extensively discusses the present trends in legume cultivation and the role of bioactive molecules from legumes in addressing nutritional problems for millions, worldwide.

2.3 Prospects and Challenges in Legume Cultivation: An Indian Perspective

To address the global food shortage and malnutrition in developing countries, cultivation of leguminous crops has been adopted as a major agricultural practice. Research work focusing on biofortification of legumes employing genomic technologies and elucidation of whole genomes, constitutes a key area of studies. The high protein content in legumes is due to the presence of nitrogen-fixing bacteria in root nodules of leguminous plants (Kouris-Blazos and Belski 2016). The health promoting effect of bioactive peptides in legumes can be attributed to different functions namely antioxidant activities, lowering of blood pressure, antimicrobial properties, among others (Mejia and Dia 2010; Moller et al. 2008; Zambrowicz et al. 2013).

In recent years, the Indian subcontinent has witnessed a sharp increase in the cultivation of legumes through agricultural practices. India has been a major exporter, accounting for approximately 23% of global legume production (UN food and Agriculture Organization 2018). Additionally, India is also the largest consumer (since we largely depend on vegetarian diet), using about 27% of world consumption of pulses in the world. Around 20% of land area is under cultivation for pulses, which constitutes about 7–10% of the total food grain production in the country (NABARD rural pulse). In India, the top five pulses producing states are as follows: Madhya Pradesh (20.3%), Maharashtra (13.8%), Rajasthan (16.4%), Uttar Pradesh (9.5%) and Karnataka (9.3%), respectively (Singh et al. 2015).

Pulses, the main constituents of dietary proteins, are cultivated on 23.47 million hectares of land and contribute to 18.45 million metric tonnes (Annual Group Meet of AICRP on MULLaRP & ARID Legumes 2019). Among the pulses, Gram comprises of the largest grown pulse (40% of the total production), followed by Pigeon pea (15 to 20%), Black gram and Mung bean at around 8–10% of the total production of pulses. However, certain limitations exist in legume cultivation, due to low yield of legumes and shortage of pulses, subject to increasing prices in the market. To address the concern, "All-India Coordinated Projects" were set up in 1960s for the improvement of oilseeds (groundnut) and grain legumes, respectively. Research on legumes gained further momentum when the International Crops Research Institute for the Semi-Arid Tropics (ICRISAT) was established in India. Research investigations on legumes have led to the development of improved genotypes, which were better suited for multiple cropping systems (Rao 1980). In the year 2013–2014, the total pulse production in the country was 19.5 metric tonnes, while India imported 2.5–3.5 metric tonnes every year (Singh et al. 2015). The increase in import statistics suggested an alarming concern, leading to huge economic expenditures. The recent statistics have suggested that the domestic requirement would rise by 26.50 metric tonnes by 2050, therefore proper measures should be undertaken to increase the production of pulses (Singh et al. 2015).

Several sequential and intercropping techniques were adopted to promote cultivation of high yielding varieties, in different agro-climatic conditions. Furthermore, genetic manipulations techniques to create disease free and pest resistant varieties of pulses were adopted (Singh et al. 2015). Several methods are being employed to increase the cultivation and production of pulses namely: Intercropping and sequential cropping, seed multiplication methods, efficient agronomic methods, use of plant growth regulators, usage of better pulse production techniques, among others. Table 2.2 discusses the various techniques and resources employed to improve pulse cultivation and production in India.

Table 2.2 Techniques and resources employed to improve pulse cultivation and production in India

S. no.	Technique employed	Plant(s) grown	Significance	References
1.	Sequential cropping	Oilseed, pulses, Pigeon pea, Chickpea, Mung bean	Improve pulse production by sequential cropping	Singh et al. (2015)
2.	Intercropping system	Cereal+ Pulse (Barley/Wheat + Lentil/Gram)	Enhance production by Intercropping	Singh et al. (2015)
		Pulse +pulse (Gram +Lentil Field pea)		
		Sugar cane + Pulse (Moong/ Urd)		
3.	Seed replacement/ Multiplication strategy	Pulses	New seed varieties developed for higher yields	Singh et al. (2012)
4.	Use of plant growth regulators	Legumes	To improve growth and development	Singh and Bhatt (2013)
5.	Efficient pest surveillance and management practices	Chickpea, Cotton	National centre for integrated pest management (NCIPM) guides people about pest monitoring and control	Anonymous (2013)
6.	Nutrient management in soil	Pulses	Use of sulphur containing fertilizers and zinc sulphate to renew nutrient content in soil	Singh et al. (2013)
7.	Mechanization in pulses	Legumes	Mechanization methods namely deep ploughing, line sowing, etc. reduce cost and improve resources	Singh et al. (2014)
8.	Post harvest maintenance	Pulse grains	Maintenance of processing and marketing of pulses	Anonymous (2013)

2.4 The Global Trends in Legume Cultivation Through Agriculture

Legumes play a crucial role as human food, animal feed and act as an ideal crop to reduce poverty, hunger and improve the human health and augment the ecosystem resilience. Since 1980, globally, there is a continuous increase in the area under cultivation as well as the yield of legumes such as bambara bean, chickpea, groundnut, soybeans, lupin, vetch, lentil, pigeon pea etc. The area under cultivation of soybean has reached to 123.55 million hectares by the year 2017 and the production status of America is the highest (85.6%). In South Asia and Latin America the area under soybean has been expanded remarkably and production has been increased at the rate of 4.3% per annum (Table 2.3). Though there was an increase in yield of soybean and faba bean, the increase was relatively slow. The production of pigeonpea and lentil has been increased at a rate of 2.0% and 2.2%.

Upon considering the global yield of the selected legumes from 1974 to 2017, groundnut, faba bean and soybean showed improved yield that range from 0.95 to 2.85 t ha^{-1} (Table 2.4). The yield of pigeonpea, chickpea, dry bean and lentil ranges from 0.53 to 1.15 t ha^{-1}. Cowpea, has relatively low yield (0.59 ton/hectares), but its yield is increasing annually at an average rate of 2.9%. The cowpea yield in developed countries is far higher than the average cowpea yields in developing regions. Among the developed countries, one-third of the cowpea yield was contributed by Latin America and Asia whereas Africa made major contribution by cropping cowpea in marginal areas (Table 2.4). The production status of chickpea (85%), dry bean (44.7%), Faba bean (44.5%), French bean (88%), groundnut (65.1%), Lentil (54.4%), Pigeon pea (85%) is highest in Asia whereas production status of Bambara bean is highest (100%) in Africa. Besides, maximum production status of lupin (76.8%) and vetch (54.2%) was noted in Oceania and Europe respectively (Table 2.4, www.fao.org). The crop yields of several developing regions are very low but, the yield of soybean, dry bean, chickpea, pigeonpea and lentil of some developing regions have surpassed the developed countries.

A substantial diversity in production tendency of legumes has been observed globally (Table 2.4). Over a last few decades, a declination trend in legume production has been noted in Europe, whereas, an increasing trend was noticed in the remaining parts of the world like Australia, Canada etc. Low distribution of legume cultivation is due to inadequate supply chain and market, reduced yield, high susceptibility to abiotic and biotic stress conditions. The legume cultivation also depends upon the farmer's choice as well as the policymakers who provide effective strategies for its cultivation.

Table 2.3 Worldwide statistics of area and yield of various legumes from 1974 to 2017

	1974		1984		1994		2004		2014		2017	
	Area	Yield	Area	Yield	Area	Yield	Area	Yield	Area	Yield	Area	Yield
Bambara bean (*Vigna subterranea*)	0.05	0.67	0.05	0.66	0.09	0.64	0.12	0.65	0.37	0.77	0.55	0.80
Chickpea (*Cicer arietinum*)	10.6	0.56	9.85	0.67	9.96	0.71	10.5	0.8	14.8	0.96	14.56	1.00
Cowpea (*Vigna unguiculata*)	4.70	0.35	3.66	0.31	7.35	0.38	9.18	0.45	12.52	0.45	12.58	0.59
Dry bean (*Phaseolus vulgaris*)	23.9	0.53	26.3	0.6	26.7	0.65	27.3	0.67	30.14	0.83	36.46	0.86
Faba bean (*Vicia faba*)	3.98	1.07	3.32	1.29	2.48	1.45	2.65	1.62	2.37	1.82	2.46	1.92
French bean	0.22	5.76	0.17	6.93	0.22	7.44	0.23	9.04	0.20	9.32	0.19	9.78
Groundnut (*Arachis hypogaea*)	19.9	0.94	18.2	1.1	22.0	1.3	23.7	1.54	25.68	1.65	27.94	1.68
Lupin (*Lupinus angustifolius*)	0.76	0.84	1.06	1.05	1.56	0.78	1.05	1.18	0.76	1.3	0.93	1.73
Lentil (*Lens culinaris*)	2.03	0.61	2.56	0.68	3.43	0.81	3.85	0.93	4.52	1.08	6.58	1.15
Pigeon pea (*Cajanus cajan*)	3.04	0.54	3.61	0.78	4.24	0.74	4.72	0.7	6.67	0.73	7.02	0.96
Soybean (*Glycine max*)	37.4	1.41	52.9	1.71	62.5	2.18	91.6	2.24	117.72	2.62	123.55	2.85
Vetch (*Vicia sativa*)	1.52	1.24	1.29	1.21	0.93	1.12	0.89	1.43	0.52	1.71	0.56	1.64

Area: Million hectares (M ha) Yield: Ton/hectares (t ha^{-1}), Stagnari et al. (2017)

Table 2.4 Acreage harvested area under major legume and production status of various regions during 1974–2017

	Asia		Africa		Europe		America		Oceania	
	Acreage harvested area (%)	Production status (%)	Acreage harvested area (%)	Production status (%)	Acreage harvested area (%)	Production status (%)	Acreage harvested area (%)	Production status (%)	Acreage harvested area (%)	Production status (%)
Bambara bean	–	–	+612	100	–	–	–	–	–	–
Chickpea	+37	85	+30	4.7	−35	1.3	Appeared (NA), +1 (SA)	4.2	–	4.3
Cowpea	+402	3	+168	94.8	+153	0.6	Appeared (NA), (SA)	1.6	–	–
Dry bean	+25	44.7	+207	19.8	−84	2.6	+16 (NA), −20 (SA)	32.6	+1778	0.2
Faba bean	−59	44.5	+7	28.6	−54	16.8	Disappeared (NA), −53 (SA)	4.3	+75,085	5.7
French bean	+66	88	–	3.6	−18	6.1	−39 (NA), +129 (SA)	2.1	+122	0.3
Groundnut	+6	65.1	+69	26.3	+16	–	−10 (NA), −22(SA)	8.6	−39	0.1
Lupin	−89	0.1	−82	3.5	−63	15.8	(NA),+577 (SA)	3.8	+315	76.8
Lentil	+72	54.4	−20	3.1	−45	1.6	+3376(NA), −75(SA)	37.4	Appeared	3.5
Pigeon pea	+108	85	+226	13	–	–	−83 (SA)	2	–	–
Soybean	+116	11.8	+642	0.7	+291	1.9	+71 (NA), +882 (SA)	85.6	−10	0
Vetch	−73	15.6	+109	17.8	−80	54.2	–	10.6	+4757	1.8

NA North America, *SA* South America

2.5 Bioactive Compounds from Legumes as Food

Legumes are excellent source of protein (20–45%) rich in essential amino acids, carbohydrates, dietary fats, oils, vitamins and minerals. The bioactive compound and amino acid contents of several legumes have been summarized in Table 2.5 that illustrates soybean protein to contain high level of essential amino acids *viz.*, cysteine, methionine and tryptophan than other leguminous proteins. The high protein content of legumes is because of their association with nitrogen fixing bacteria in the root nodules which convert the nitrogen gas into ammonium that is further incorporated in protein synthesis. These legumes are considered as "poor man's meat" and used in various commercial products such as tofu, miso, natto, soy paste,

Table 2.5 Bioactive compounds and amino acid content of legumes

	Chickpea	Cowpea	Common bean	Faba bean	Lentil	Pigeon pea	Soybean	Bambara groundnut
Polyphenols (%)*	0.5	0.2	1.0	0.8	0.8	0.2	0.4	0.75
Tannins (%)*	0	0.1	0.5	0.5	0.1	0	0.1	0.1
Phytic acid (%)*	0.5	0.9	1.1	1.0	0.6	0.1	1.0	0.8
Arginine (%)	1.8	1.6	0.7	1.5	2.2	2.2	7.2	4.0
Aspartic acid (%)	2.3	2.8	0.8	2.9	3.1	2.9	11.7	5.0
Histidine (%)	0.5	0.7	0.2	0.7	0.8	0.6	2.5	2.2
Serine (%)	1.0	1.2	0.3	1.3	1.3	1.1	5.1	3.2
Glutamic acid (%)	3.4	4.5	1.3	3.6	4.4	4.2	18.7	16.5
Proline (%)	0.8	1.1	0.3	1.0	1.2	1.0	5.5	3.2
Glycine (%)	0.8	1.0	0.3	0.9	1.1	1.1	4.2	3.3
Alanine (%)	0.8	1.1	0.3	1.0	1.2	1.1	4.3	3.5
Lysine (%)	1.3	1.6	0.5	1.6	2.0	1.8	6.4	3.0
Threonine (%)	0.7	0.9	0.3	1.0	1.0	0.9	3.9	2.5
Valine (%)	0.8	1.1	0.3	1.2	1.4	1.2	4.8	3.8
Isoleucine (%)	0.8	1.0	0.3	1.0	1.2	1.0	4.5	3.8
Leucine (%)	1.4	1.8	0.6	1.9	2.0	1.8	7.8	6.8
Tyrosine (%)	0.5	0.8	0.2	0.7	0.8	0.7	3.1	3.2
Phenylalanine (%)	1.0	1.4	0.3	1.3	1.4	1.1	4.9	4.3
Tryptophan (%)	0.2	0.7	0.1	0.3	0.3	0.3	1.3	0.7
Cystine (%)	0.3	0.5	0.1	0.3	0.4	0.4	1.3	0.5
Methionine (%)	0.3	2.0	0.1	0.4	0.2	0.3	1.3	2.0

(%)*: % dry matter, (%): g/100 g of protein (FAO 2016; Leonard 2012; Yao et al. 2015; Gulewicz et al. 2014)

soy sauce, textured vegetable protein (TVP), curry, dhal, fermented products (idli, dosa, papad), roasted/boiled snacks, sprouted beans, milk, curd, yoghurt, yuba, soup, peanut butter, peanut bar, flour, cakes and puddings, noodles, sweets etc.

The functional properties of legumes play a key role during the development of aforementioned food products. The functional properties are the physical and chemical characteristics of a protein which influence its behaviour in food system. The functional behaviour is affected by various factors such as shape, size, amino acid composition and sequence, hydrophobicity, structure, net charge, molecular rigidity against environmental parameters that include temperature, pH and salt concentration etc. These functional properties are essential to determine the possible uses of these as food ingredients. Functional properties of legumes like oil and water binding capacities, bulk density, protein solubility, gelling nature, emulsion stabilization and foaming properties provide information on how a particular ingredient would behave in a food matrix, play a prominent role during the cooking/processing, storage of food, and, in turn, associate with sensory properties. These properties are considered due to different conformational changes or interactions between food components, such as interactions between proteins, proteins, and polysaccharides, lipids, phenolic compounds, or phytic acid (Lara-Rivera et al. 2017). Furthermore, they constitute the functional base of diverse products and act as source of beneficial compounds that have a protective effect on the development of various diseases.

2.5.1 Food Products from Legumes

Legumes are being processed to develop nutritious, inexpensive, acceptable and easily prepared food products. The preparation of several products, their nutritional profile and health benefits due to their consumption are discussed below:

2.5.1.1 Tofu

Soybean is well known for its oil and protein content and hence comes under high human consumption. Tofu is cheese-like product made by coagulating hot soymilk with food grade chemicals like magnesium chloride, magnesium sulphate, calcium chloride, acetic acid, calcium sulphate and citric acid. The amino acid composition of tofu is glutamic acid (18.7%), aspartic acid (11.7%), leucine (7.8%), arginine (7.2%), lysine (6.4%), proline (5.5%), serine (5.1%), phenylalanine (4.9%), isoleucine (4.5%), valine (4.8%), alanine (4.3%), glycine (4.2%), threonine (3.9%), tyrosine (3.1%), histidine (2.5%), methionine (1.3%), cystine (1.3%) and tryptophan (1.3%) (Jubayer et al. 2013). Tofu is cheaper than other milk products like cheese and paneer, but highly nutritious with several health benefits like lower total cholesterol, triglycerides and low-density lipoprotein which lowers the risk of hypertension, cardiovascular diseases and atherosclerosis (Jubayer et al. 2013). Anderson et al. (1995) have conducted 38 controlled clinical studies, among which 30 studies were conducted with hyper-cholesterolemic subjects. The results suggested that mean intakes of 47 g/d, ranging from 17–124 g, of isolated or textured soy protein resulted in significant reduction in total cholesterol by 9.3%, low-density lipoprotein cholesterol by 12.9%, and triglycerides by 10.5%, with an insignificant change in high-density lipoprotein cholesterol levels, compared to animal protein.

2.5.1.2 Natto

It is a traditional Japanese food prepared by fermenting soybean with *Bacillus subtilis* var. natto, a Gram positive spore-forming bacterium (Park et al. 2012). During fermentation process, the bacteria release extracellular enzymes that act upon soybeans to produce mucilage like substance known as nattokinase. This substance has approximately four-times higher fibrinolytic activity than plasmin in clot lyses and thus it has been promoted as alternative medicine as blood thinner and clot-buster (Chen et al. 2018). Natto is identified to possess antithrombotic and fibrinolytic properties along with potential to prevent cardiovascular disease which emphasis its therapeutic benefits. Park et al. (2012) have evaluated the fibrinolytic and antithrombotic activities of natto on platelet aggregation in mice. They found that platelet aggregation was induced by collagen and adenosine 5′diphosphate (ADP). Normal levels of natto was administered to mice for 4 weeks and during that period the researchers observed prolonged partial thromboplastin time (PATT) and shortened euglobulin clot lysis time (ECLT). Besides, it was also confirmed that normal level consumption of natto decreased total cholesterol in serum that lead to improved blood circulation.

2.5.1.3 Peanut Butter

It is prepared from peanut that contains proteins, antioxidants, polyphenols, fibres, vitamins (B and E) and minerals. Peanuts are enriched with all the 20 amino acids predominated with arginine, and are an exceptional source of resveratrol, phytosterols and flavonoids that obstruct the absorption of cholesterol from diet. It is also a major source of Co-enzyme Q_{10} with antioxidant property that protects heart tissues from free radical damage (Arya et al. 2016). Peanut butter constitutes various micronutrients such as copper, manganese, magnesium, folate, iron, potassium and zinc. Peanut butter consumption reduces risk of heart attack and increases the blood circulation and lowers the blood pressure (Arya et al. 2016). Guasch-Ferré et al. (2017) have conducted a study to find out the effect of peanut butter consumption on incidence of cardiovascular diseases in human population that include, women (1,69,310 members) and men (41,526 members). The study proved that high consumption of nuts is inversely associated with coronary heart and cardiovascular diseases. The plausible reason for such a result might be presence of high dietary fibre, unsaturated fatty acids, vitamins, minerals and various bioactive compounds in the nuts which are the raw material for the butter.

2.5.1.4 Miso

It is prepared by fermenting soybeans, wheat or barley, malted rice and salt with *Aspergillus orzae* or *Aspergillus soyae*. The chief ingredient, soybeans, contains abundant proteins (glycinin and β-conglycinin), polyunsaturated fatty acids, lecithin, saponin, vitamin E and isoflavones such as daidzein, genistein, glycitein and daidzin (Yoshinaga et al. 2012). Fermentation of soybeans results in a high nutrient content of isoflavones and amino acids. Miso has been reported to efficiently prevent hypertension, cancer and hypercholesterolemia in mice (Okouchi et al. 2019). When the mice diet was supplemented with meso for 8 weeks, reduction in weight of white adipose tissue was observed and because of this reason most of the Japanese added meso to their diet to prevent obesity. Owing to its antioxidant activity, miso also plays a major role in preventing aging caused due to oxidative stress (Jayachandran and Xu 2019). Beside these products, edamame, tempeh, fuyu, soy milk, sufu, rich in isoflavone and other bioactive compounds are being produced and consumed globally owing to their potential benefits in human health.

2.6 Applications of Bioactive Peptides in Healthcare

Bioactive peptides derived from legumes (chickpea, soybean, lentil, lupin, common bean proteins etc.) are used to control several pathological conditions such as cancer, cardiovascular diseases, immunological conditions etc. thereby rendering therapeutic applications (Dirisala et al. 2017). These peptides exert beneficial effects on human health beyond its nutritional value (Sravani et al. 2015). Storage proteins, lectins and protease inhibitors are well known sources of bioactive peptides.

2.6.1 Bioactive Peptides Derived from Legume

Bioactive peptides are derived through cleavage or modification of parent proteins available in animal (meat, milk, sea foods etc.) (Sampath Kumar et al. 2012a; Nazeer and Sampath Kumar 2012) as well as plant sources that include cereals (rice, wheat, corn, barley etc.), legume (beans, soybean, pea etc.) pseudocereals (amaranthus and buckwheat), sunflowers, brassica species etc. (Daliri et al. 2017). These are liberated from the parent protein during their transit through small intestine due to the action of several enzymes released by the microbial flora retaining in the intestine. Other ways of peptide production includes bacterial fermentation, germination and food processing. Fermentation is a processing technique to protect as well to improve the health promoting characteristics and nutritional quality of the legumes. Several bacteria such as *Bacillus subtilis, Lactobacillus* etc. are being used in manufacturing soybean fermented products. Germination is a natural hydrolytic process involving intrinsic enzymes and enhancement in the nutrient content of the food legumes depends upon the soaking time, temperature and humidity (Huang et al. 2014). During this process the stored proteins in the cotyledon are mobilized for the production of peptides. Lima et al. (2017) reported that germination not only improved the anti-proliferative property of soya protein on C-33 cervical cancer cells and HeLa cells but also enabled the peptides to bring alterations in the cell signalling pathway thereby causing apoptosis.

The occurrence of various bioactive peptides in the food sources improve the protein quality and enhance the functionality of the food consumed and thus can be commercialized as therapeutic and nutraceutical products (Kumar et al. 2019). The bioactive peptides constitute amino acids that range between 2 and 20 residues and the biological activity of the bioactive peptides *viz.*, antimicrobial, anticancer, anti-oxidant, anti-hypertensive, anti-inflammatory, anti-proliferative, anti-coagulant, anti-obesity, anti-cholesterolemic depends upon the size and sequence of amino acid of the peptide fragment (Sampath Kumar et al. 2012b; Naidu et al. 2016). Table 2.6, elucidates the biological activities of bioactive peptides derived from legumes with a special emphasis on the sequence of the peptides responsible for the activities.

2.6.2 *Antimicrobial Activity of Bioactive Peptides*

Antimicrobial peptides are small peptides with wide spectral range against bacteria, protozoans, fungi etc. which target the membrane of the microbe causing cell lysis (Shaheena et al. 2019). Antimicrobial peptides with proline and tryptophan lyse the cells by targeting their intracellular pathways, forming hydrogen bonds with membrane and disrupting protein synthesis. Moreover, structural modification affects the function of the peptide hence, a clear understanding on structure and activity of the peptide is mandatory. For instance, substitution of L-Proline at last proline (P206) of P1 (LLYQEPVLGPVRGPF (L-Pro)IIV) with L-Proline resulted into loss of activity as well as the structural features of Bovine β-casein (Bonomi et al. 2011). Chickpea and soybean are highly consumed legumes with maximum antimicrobial activity. For instance, soypeptide with amino acid sequence of IKAFKEATKVDKVVVLWTA has antimicrobial activity against Gram negative as well as Gram positive bacteria (Najafian and Babji 2015). Similarly, lima bean peptide FVNQPYLLYSVHMK exhibits high antimicrobial and antioxidant activities, wherein hydrophobic amino acid *viz.*, Leu, Val and Phe; hydrophilic amino acids such as His, Pro and Lys, and aromatic amino acids, Phe and Tyr were responsible for the antioxidant activity of the peptide (Najafian and Babji 2015). Gahanea et al. (2018) reported that fluorenylmethyloxycarbonyl (Fmoc) phenylalanine, has anti-bacterial activity against Gram-positive bacteria including methicillin-resistant *Staphylococcus aureus* (MRSA). Another peptide, vulgarinin (7 kDa) derived from *Phaseolus vulgari* exhibited antibacterial action against *Bacillus megaterium*, *B. subtilis*, *Mycobacterium phlei* and *Proteus vulgaris*.

Table 2.6 Bioactive peptides from various legumes and their activities

Source	Bioactive peptide	Activity	Peptide size (kilo dalton)	Peptide sequence	
Common beans (*Phaseolus vulgaris*)	Peptide	Antibacterial and antifungal	2.2 & 6	Antifungal peptide KTCENLADTYKGPCFTTGSCDDHCK Antibacterial peptide LRLKKYKVPQL KVGIN, KVAGT, VRT, PGDL, LPMH, EKF, IRL KTCENLSDSFKGPCIPDGNCNKHCKEKEHLLSGRCRDDFRCWCTRNC	Carlos et al. (2015) Rogozhin et al. 2018
	Peptide PvD1	Antibacterial and antifungal	6		
	Defensin	Antibacterial	7.3		
	Vulgarinin	Antibacterial, antifungal, and antiviral	7.0		
Lentil (*Lens culinaris*)	Defensin	Antibacterial	5.4		
Chickpea (*Cicer arietinum*)	Cicerin	Antifungal and antiviral	8.2	Antiviral peptide ILMAFSIDSPDSLEN DVILMCFSIDSPDSLEN	Chang and Yang 2013
	Arietin	Antifungal and antiviral	5.6		

(continued)

Table 2.6 (continued)

Source	Bioactive peptide	Activity	Peptide size (kilo dalton)	Peptide sequence	
Soyabean (*Glycine max*)	Lunasin	Anticancer and cytotoxic	14	Anticancer peptide GLTSK, LSGNK, GEGSGA, MTEEY, and MPACGSS ASKWQHQQDS CRKQLQGVNL TPCEKHIMEK IQGRGDDDDD ANEIYFSFQRFNETNLILQR GLTSK, LSGNK, GEGSGA, MPACGSS, MTEEY ARQSHFANAQP RQSHFANAQP	Finkina et al. (2008) Marcela et al. (2017) Shoombuatong et al. (2018)
	Bowman-Birk Inhibitor	Cytostatic and cytotoxic	5.45		
	Lunasin-like peptide	Anti-proliferative	5		
Faba bean (*Vicia faba*)	VFTI-G1 (Bowman-Birk type trypsin inhibitor)	Anticancer	15		
Pea (*Pisum sativum*)	TI1B (Bowman-Birk isoinhibitor)	Anticancer	–		
Lentil (*Lens culinaris*)	Bowman-Birk isoinhibitors	Anticancer	6–9		
Common beans	Peptide	Anticancer	–		
Chickpea (*Cicer arietinum*)	Peptide	Anticancer	1.15		

Common bean is well known for the production of antimicrobial peptides. Antifungal peptides from common bean showed inhibitory activity against several plant pathogens *viz.*, *Rhizoctonia solani*, *Mycosphaerella arachidicola*, *Setosphaeria turcica* and *Verticillium dahlia* as well as fungi that can be pathogenic for humans such as *Helminthosporium maydis*, *Candida albicans* and *Fusarium oxysporum* (Chan et al. 2012). The Antimicrobial peptides also exhibit antiviral activity by preventing the binding of the virus to the membrane by electrostatically binding to the negative charges of glycosaminoglycans (Salas et al. 2015). Alternative way includes binding of the defensin to viral glycoproteins preventing the virus from binding with the surface of host cells. Mata et al. (2017) studied antiviral effect of Antimicrobial peptides by obstructing viral interaction with specific cellular receptors, during binding of HSV and the putative B5 cell surface membrane protein displaying a heptad repeat alpha-helix fragment.

2.6.3 Anticancer Activity of Bioactive Peptides

Bioactive peptides are capable of generating physiological effect against cancer cells and can induce cell death by different mechanisms like apoptosis, inhibiting angiogenesis and affecting the tubulin-microtubule equilibrium (Rao et al. 2018). Lunasin is a recently isolated 43 amino-acid bioactive peptide from soybean (GM2S-1 gene) with cancer-preventive activity. Studies on lunasin revealed that this peptide arrests the cell division and even causes mitotic arrest in mammalian cell lines, apparently by binding to kinetochore regions of the centromere and blocking microtubule attachment. Owing to the mechanism of action, lunasin could be a useful cancer therapeutic peptide provided it is specifically delivered to cancer cells (Vuyyuri et al. 2018). Besides, it also binds specifically to the deacetylated core histones H3/H4 which according to the current hypotheses on lunasin's mechanism of action is critical for the anticancer effects of lunasin (Inaba et al. 2014). It contains multiple functional domains that modulate gene expression through effects on integrin signaling and histone acetylation. Recent studies suggest that lunasin effects on integrin signaling in cancer stem cells reduce expression of stemness factors with a concomitant reduction in metastatic potential (Hsieh et al. 2016). Metastasis develops commonly when the cancer cells from the main tumor breaks and enters into the blood or lymphatic system.

A few studies have been reported till date on anti-metastatic activity of various pulses. The anti-metastatic potential of chickpea, lupin, lentil, peas, common bean, cowpea and faba bean has been studied by employing HT29 cells (Lima et al. 2016). The highest anti-metastatic activity was observed in the peas extract with albumin and globulin fraction at 100 μg/mL while lowest activity was shown in lupin extract. The anti-metastatic activity of pulses was evaluated through two ways of treatments – one way is by treating the mice melanoma cells with extract and the second way is by supplementing the mice diet with 100 mg/kg of extract (Banerji et al.

1998). The reduction in migration of metastatic lung melanoma cells have been observed from the ways identifying the anti-metastatic potential of pulses.

Besides peptides, lectins which are carbohydrate-binding proteins are found to exhibit anti-cancer properties. Binding to tumour membranes, lectins slow down the cell proliferation, stimulate the immune system and induce apoptosis. Similarly, Bowman-Birk inhibitors (BBI) produced from soybean and classified under protein protease inhibitors are reported to target specific anti-cancer pathways by inhibiting the generation of reactive oxygen species. Pulses generally have low levels of lipids, yet, the fatty acids such as oleic butyrate and linoleic acid have been associated with anti-cancer activities (Marqus et al. 2017).

2.6.4 Bioactive Peptide Activity Against Cardiovascular Disease

Globally, the second most dreadful disease after cancer is cardiovascular disease (CVD). Angiotensin converting enzyme (ACE) inhibitory peptides from lupin and other legumes plays a crucial role in regulating blood pressure in humans by catalyzing the hydrolysis of inactive angiotensin decapeptide to vasoconstrictor angiotensin II, an octapeptide that regulates blood pressure (He et al. 2013). In addition, ACE also cleaves bradykinin (vasodilator) into inactive fragments to increase the blood pressure in the body. Apart from the legume peptides, the phenolic extracts from mung bean were identified to possess anti-inflammatory and anti-proliferative activities besides efficiency to induce apoptosis in HepG2 and HeLa cells (Hafidh et al. 2012). The mung beans extract upregulated IFN-β and TNFα by inducing expression of tumour suppressor genes, apoptotic genes and IFNγ. A similar work was conducted on phenolic extracts of pea sprouts and lentil to identify anticancer activity in pea and lentil (Busambwa et al. 2016).

2.6.5 Antidiabetic Activity of Bioactive Peptide

Several studies revealed that high consumption of legumes could regulate lipid and lipoprotein metabolism, attenuate postprandial glycemia and insulinemia, diminish oxidative stress and increase insulin sensitivity (Cicero et al. 2017). Type 2 diabetes is a multi-clustering metabolic disorders accompanied with developing insulin resistance, abnormal glucose homeostasis, increased oxidative stress, impaired lipid and lipoprotein metabolism and sub-clinical inflammation which lead to long term pathogenic conditions such as nephropathy, retinopathy, neuropathy, disability and death in diabetic patients (Constantino et al. 2013; Tiwari 2015; Tiwari et al. 2019). Traditionally, a few medicinal plants were used in the treatment of diabetic patients. Recently, α-amylase inhibitory peptides, isolated from lentil and beans which reduce digestion and absorption of dietary carbohydrates and modulate postprandial glycemic response are being used. Some legumes bioactive peptides including conglutin γ and 7S globulin α chain have exclusive properties to regulate lipid metabolism and normalize lipid and lipoprotein levels (Mirmiran et al. 2014). The hypoglycemic effect of pinto, kidney beans, chickpeas, split peas, lentils, and soybeans (via inhibition of β-glucosidase and α-amylase activity) has been reported as being similar to those of antidiabetic drugs (Ganesan and Xu 2017). Legumes which are rich source of bioactive compounds exhibit anti-diabetic effects by interacting with specific receptors. Thus, regular consumption of legumes is considered to be the best strategy for maintaining proper health.

2.7 Conclusion

The legumes are excellent food source for bioactive compounds with profound applications in food, nutraceutical and health industries. Bioactive peptides could be either consumed directly or incorporated as one of the ingredients in dietary supplements, functional foods, and even in pharmaceuticals, with a rationale to deliver specific health benefits as anticancer, antimicrobial, anti-inflammatory and anti-diabetic agents. Peptides are novel anticancer agents that could specifically target cancer cells with less toxicity to normal cells, which will offer new avenues for cancer prevention and treatment. Moreover, consumption of legumes reduces digestion and absorption of dietary carbohydrates and modulates postprandial glycemic response. Hence, it is highly essential to develop research to find new peptide sequences and their mechanism of action for their applications in food and nutraceutical industries.

References

Anderson JW, Johnstone BM, Cook N (1995) Meta-analysis of the effects of soy protein intake on serum lipids. N Engl J Med 333:276–282. https://doi.org/10.1056/NEJM199508033330502

Anonymous (2013) Report of expert group on pulses. Department of Agriculture & Cooperation Government of India, Ministry of Agriculture. p 139

Annual Group Meet of AICRP on MULLaRP & ARID Legumes (2019) Press Information Bureau, Government of India. https://pib.gov.in/newsite/PrintRelease.aspx?relid=190042

Arya SS, Salve AR, Chauhan S (2016) Peanuts as functional food: a review. J Food Sci Technol 53:31–41

Banerji A, Fernandes A, Bane S, Ahire S (1998) The field bean protease inhibitor has the potential to suppress B16F10 melanoma cell lung metastasis in mice. Cancer Lett 129:15–20. https://doi.org/10.1016/S0304-3835(98)00090-1

Bonomi F, Brandt R, Favalli S, Ferranti P, Fierro O, Frøkiær H, Maria E (2011) Structural determinants of the immunomodulatory properties of the C-terminal region of bovine βcasein. Int Dairy J 21:770–776. https://doi.org/10.1016/j.idairyj.2011.04.012

Burbano C, Muzquiz M, Ayet G, Cuadrado C, Pedrosa MM (1999) Evaluation of antinutritional factors of selected varieties of Phaseolus vulgaris. J Sci Food Agric 79:1468–1472

Busambwa K, Sunkara R, Diby N, Offei-Okyne R, Boateng J, Verghese M (2016) Cytotoxic and apoptotic effects of sprouted and non-sprouted lentil, green and yellow split-peas. Int J Cancer Res 12:51–60. https://doi.org/10.3923/ijcr.2016.51.60

Champ MM (2002) Non-nutrient bioactive substances of pulses. Br J Nutr 88:307–319

Chan YS, Wong JH, Fang EF, Pan WL, Ng TB (2012) An antifungal peptide from *Phaseolus vulgaris* cv. brown kidney bean. Acta Biochim Biophys Sin Shanghai 44:307–315. https://doi.org/10.1093/abbs/gms003

Chang KY, Yang J-R (2013) Analysis and prediction of highly effective antiviral peptides based on random forests. PLoS One 8:e70166. https://doi.org/10.1371/journal.pone.0070166

Chaudhary A, Marinangeli CPF, Tremorin D, Mathys A (2018) Nutritional combined greenhouse gas life cycle analysis for incorporating Canadian yellow pea into cereal-based food products. Nutrients 10:490. https://doi.org/10.3390/nu10040490

Chen H, McGowan EM, Ren N, Lal S, Nassif N, Shad-Kaneez F, Qu X, Lin Y (2018) Nattokinase: a promising alternative in prevention and treatment of cardiovascular diseases. Biomark Insights 13:1177271918785130. https://doi.org/10.1177/1177271918785130

Christenhusz MJ, Byng JW (2016) The number of known plants species in the world and its annual increase. Phytotaxa 261:201–217. https://doi.org/10.11646/phytotaxa.261.3.1

Cicero AF, Fogacci F, Colletti A (2017) Potential role of bioactive peptides in prevention and treatment of chronic diseases: a narrative review. Br J Pharmacol 174:1378–1394. https://doi.org/10.1111/bph.13608

Clemente A, Mackenzie DA, Johnson IT, Domoney C (2004) Investigation of legume seed protease inhibitors as potential anti-carcinogenic proteins. In: Muzquiz M, Hill GD, Pedrosa MM, Burbano C (eds) proceedings of the fourth international workshop on antinutritional factors in legume seeds and oilseeds. EAAP publication no. 110, Wageningen, pp 137–142

Constantino MI, Molyneaux L, Limacher-Gisler F, Al-Saeed A, Luo C, Wu T, Twigg SM, Yue DK, Wong J (2013) Long-term complications and mortality in young-onset diabetes: type 2 diabetes is more hazardous and lethal than type 1 diabetes. Diabetes Care 36:3863–3869. https://doi.org/10.2337/dc12-2455

Daliri EB, Oh DH, Lee BH (2017) Bioactive Peptides. Foods (Basel, Switzerland) 6:32. https://doi.org/10.3390/foods6050032

Dillard CJ, German JB (2000) Phytochemicals: nutraceuticals and human health. J Sci Food Agric 80:1744–1756. https://doi.org/10.1002/1097-0010(2000915)

Dirisala VR, Nair RR, Srirama K, Reddy PN, Sambasiva Rao KRS, Kumar NSS, Parvatam G (2017) Recombinant pharmaceutical protein production in plants: unraveling the therapeutic potential of molecular pharming. Acta Physiol Plant 39:18

FAO (2016) International year of legumes: nutritious seeds for a sustainable future. Food and Agriculture Organisation of the United Nations and World Health Organisation. FAO, Rome. www.fao.org/pulses-2016

Finkina EI, Shramova EI, Tagaev AA, Ovchinnikova TV (2008) A novel defensin from the lentil Lens culinaris seeds. Biochem Biophys Res Commun 371:860–865. https://doi.org/10.1016/j.bbrc.2008.04.161

Gahanea AY, Ranjan P, Singh V, Sharma RK, Sinha N, Sharma M, Chaudhry R, Ashwani K (2018) Thakur Fmoc-phenylalanine displays antibacterial activity against Gram-positive bacteria in gel and solution phases. Soft Matter 28:2234–2244. https://doi.org/10.1039/c7sm02317k

Ganesan K, Xu B (2017) Polyphenol-rich lentils and their health promoting effects. Int J Mol Sci 18(11):2390. https://doi.org/10.3390/ijms18112390

Greiner R, Konietzny U (2006) Phytase for food application. Food Technol Biotechnol 44:125–140

Guasch-Ferré M, Liu X, Malik VS, Sun Q, Willett WC, Manson JE, Rexrode KM, Li Y, Hu FB, Bhupathiraju SN (2017) Nut consumption and risk of cardiovascular disease. J Am Coll Cardiol 70(20):2519–2532. https://doi.org/10.1016/j.jacc.2017.09.035

Gulewicz P, Martinez-Villaluenga C, Kasprowicz-Potocka M, Frias J (2014) Non-nutritive compounds in fabaceae family seeds and the improvement of their nutritional quality 120 functional food - improve health through adequate food by traditional processing-a review. Pol J Food Nutr Sci 64:75–89

Hafidh RR, Abdulamir AS, Jalilian FA BFA, Abas F, Sekawi Z (2012) Novel molecular, cytotoxical, and immunological study on promising and selective anticancer activity of Mung bean sprouts. BMC Complement Altern Med 12:208

He HL, Liu D, Ma CB (2013) Review on the angiotensin-I-converting enzyme (ACE) inhibitor peptides from marine proteins. Appl Biochem Biotechnol 169(3):738–749. https://doi.org/10.1007/s12010-012-0024-y

Hsieh CC, Wang CH, Huang YS (2016) Lunasin attenuates obesity-associated metastasis of 4t1 breast cancer cell through anti-inflammatory property. Int J Mol Sci 17:2109. https://doi.org/10.3390/ijms17122109

http://www.fao.org/faostat/en/#data/QC/visualize. Accessed 13 Aug 2019

Huang X, Cai W, Xu B (2014) Kinetic changes of nutrients and antioxidant capacities of germinated soybean (*Glycine max* L.) and mung bean (*Vigna radiata* L.) with germination time. Food Chem 143:268–276. https://doi.org/10.1016/j.foodchem.2013.07.080

Inaba J, McConnell EJ, Davis KR (2014) Lunasin sensitivity in non-small cell lung cancer cells is linked to suppression of integrin signaling and changes in histone acetylation. Int J Mol Sci 15:23705–23724. https://doi.org/10.3390/ijms151223705

International Year of Pulses 2016-IYP2016. Retrieved 14 December, 2015

Jayachandran M, Xu B (2019) An insight into the health benefits of fermented soy products. Food Chem 271:362–371. https://doi.org/10.1016/j.foodchem.2018.07.158

Jubayer MF, Uddin MB, Faruque MO (2013) Standardization parameters for production of tofu using WSD-Y-1 machine. J Bangladesh Agril Univ 11:307–312

Kamran F, Reddy N (2018) Bioactive peptides from legumes: functional and nutraceutical potential. Recent Adv Food Sci 1:134–149

Kouris-Blazos A, Belski R (2016) Health benefits of legumes and pulses with a focus on Australian sweet lupins. Asia Pac J Clin Nutr 25:1–17. https://doi.org/10.6133/apjcn.2016.25.1.23

Kumar A, Agarwal DK, Kumar YS, Reddy M, Chintagunta AD, Saritha KV, Pal G, Kumar SPJ (2019) Nutraceuticals derived from seed storage proteins: implications for health wellness. Biocatal Agric Biotechnol 17:710–719. https://doi.org/10.1016/j.bcab.2019.01.044

Lajolo FM, Genovese MI, Pryme IF, Dale M (2004) Beneficial (antiproliferative) effects of different substances. In: Muzquiz M, Hill GD, Pedrosa MM, Burbano C (eds) Proceedings of the fourth international workshop on antinutritional factors in legume seeds and oilseeds, EAAP publication no. 110, Wageningen, pp 123–135

Lara-Rivera AH, García-Alamilla P, Lagunes-Gálvez LM, Macias RR, López PMG, Natera JFZ (2017) Functional properties of *Lupinus angustifolius* seed protein isolates J Food Qual 8675814:8 pages. https://doi.org/10.1155/2017/8675814

Le Berre-Anton V, Bompard-Gilles C, Paya F, Rouge P (1997) Characterization and functional properties of the a-amylase inhibitor (a-AI) from kidney bean (*Phaseolus vulgaris*) seeds. Biochim Biophys Acta 1343:31–40

Leonard E (2012) Cultivating good health. In: Grains and legumes nutrition council. Cadillac Printing, Adelaide, pp 3–18

Lima AIG, Mota J, Monteiro SAVS, Ferreira RMSB (2016) Legume seeds and colorectal cancer revisited: protease inhibitors reduce MMP-9 activity and colon cancer cell migration. Food Chem 197:30–38. https://doi.org/10.1016/j.foodchem.2015.10.063

Lima A, Oliveira J, Saúde F, Mota J, Ferreira RB (2017) Proteins in soy might have a higher role in cancer prevention than previously expected: soybean protein fractions are more effective MMP-9 inhibitors than non-protein fractions, even in cooked seeds. Nutrients 9:201. https://doi.org/10.3390/nu9030201

Marcela GM, Eden CS, Rosalva ME (2017) Bioactive peptides from legumes as anticancer therapeutic agents. Int J Cancer Clin Res 4:081. https://doi.org/10.23937/2378-3419/1410081

Marqus S, Pirogova E, Piva TJ (2017) Evaluation of the use of therapeutic peptides for cancer treatment. J Biomed Sci 24:21

Mata ECG, Mourao CBF, Rangel M, Schwartz EF (2017) Antiviral activity of animal venom peptides and related compounds. J Venom Anim Toxins Incl Trop Dis 23:3. https://doi.org/10.1186/s40409-016-0089-0

Mejia EG, Dia VP (2010) The role of nutraceutical proteins and peptides in apoptosis, angiogenesis, and metastasis of cancer cells. Cancer Metastasis Rev 29:511–528. https://doi.org/10.1007/s10555-010-9241-4

Messina MJ (1999) Legumes and soybeans: overview of their nutritional profiles and health effects. Am J Clin Nutr 70:439–450. https://doi.org/10.1093/ajcn/70.3.439s

Mirmiran P, Bahadoran Z, Azizi F (2014) Functional foods-based diet as a novel dietary approach for management of type 2 diabetes and its complications: a review. World J Diabetes 5:267–281. https://doi.org/10.4239/wjd.v5.i3.267

Moller NP, Scholz-Ahrens KE, Roos N, Schrezenmeir J (2008) Bioactive peptides and proteins from foods: indication for health effects. Eur J Nutr 47:171–182. https://doi.org/10.1007/s00394-008-0710-2

Mudryj AN, Yu N, Aukema HM (2014) Nutritional and health benefits of pulses. Appl Physiol Nutr Metab 39:1197–1204. https://doi.org/10.1139/apnm-2013-0557

Muzquiz M, Varela A, Burbano C, Cuadrado C, Guillamon E, Pedrosa MM (2012) Bioactive compounds in legumes: pronutritive and antinutritive actions. Implications for nutrition and health. Phytochem Rev 11:227–244

Naidu KND, Ramu VD, Kumar NSS (2016) Anti-inflammatory & anti-helminthic activity of ethanolic extract of *Azadirachta indica* leaves. Int J Green Pharm 10:S1–S4. https://doi.org/10.22377/ijgp.v10i04.783

Najafian L, Babji AS (2015) Isolation, purification and identification of three novel antioxidant peptides from patin (Pangasius sutchi) myofibrillar protein hydrolysates. LWT-Food Sci Technol 60:452–461

Nazeer RA, Sampath Kumar NS (2012) Fatty acid composition of horse mackerel (*Magalaspis cordyla*) and croaker (*Otolithes ruber*). Asian Pac J Trop Dis:S933–S936

Nishida C, Uauy R, Kumanyika S, Shetty P (2004) The Joint WHO/FAO expert consultation on diet, nutrition and the prevention of chronic diseases: process, product and policy implications. Public Health Nutr 7:245–250. https://doi.org/10.1079/PHN2003592

Okouchi R, Sakanoi Y, Tsuduki T (2019) Miso (fermented soybean paste) suppresses visceral fat accumulation in mice, especially in combination with exercise. Nutrients 11(3):560. https://doi.org/10.3390/nu11030560

Park KJ, Kang JI, Kim TS, Yeo IH (2012) The antithrombotic and fibrinolytic effect of natto in hypercholesterolemia rats. Prev Nutr Food Sci 17(1):78–82. https://doi.org/10.3746/pnf.2012.17.1.078

Pusztai A, Bardocz S, Martín-Cabrejas M (2004) The mode of action of ANFs on the gastrointestinal tract and its microflora. In: Muzquiz M, Hill GD, Pedrosa MM, Burbano C (eds) Proceedings of the fourth international workshop on antinutritional factors in legume seeds and oilseeds, EAAP publication No. 110, Wageningen, pp 87–100

Rao MR, Willey RW, Sharma D, Green JM (1980). Pigeonpea genotype evaluation for intercropping. International workshop on pigeonpea, ICRISAT/ICAR, 15–19 December, ICRISAT, India

Rao S, Chinkwo KA, Santhakumar AB, Blanchard CL (2018) Inhibitory effects of pulse bioactive compounds on cancer development pathways. Diseases 6:72

Rimbach G, Pallauf J (1997) Cadmium accumulation, zinc status, and mineral bioavailability of growing rats fed diets high in zinc with increasing amounts of phytic acid. Biol Trace Elem Res 57:59–70

Rochfort S, Panozzo J (2007) Phytochemicals for health, the role of pulses. J Agric Food Chem 55:7981–7994. https://doi.org/10.1021/jf071704w

Rogozhin E, Ryazantsev D, Smirnov A, Zavriev S (2018) Primary structure analysis of antifungal peptides from cultivated and wild cereals. Plants (Basel, Switzerland) 7:74

Salas CE, Badillo-Corona JA, Ramírez-Sotelo G, Oliver-Salvador C (2015) Biologically active and antimicrobial peptides from plants. Biomed Res 2015:102129

Sampath Kumar NS, Nazeer RA, Jaiganesh R (2012a) Functional properties of protein hydrolysates from different body parts of horse mackerel (*Magalaspis cordyla*) and croaker (*Otolithes ruber*). Mediterr J Nutr Metab 5:105–110. https://doi.org/10.1007/s12349-011-0078-3

Sampath Kumar NS, Nazeer RA, Jaiganesh R (2012b) A *in vivo* antioxidant activity of peptide purified from viscera protein hydrolysate of horse mackerel *(Magalaspis cordyla)*. Int J Food Sci Technol 47:1558–1562. https://doi.org/10.1111/j.1365-2621.2012.03002.x

Schley PD, Field CJ (2002) The immune-enhancing effects of dietary fibres and prebiotics. Br J Nutr 87:221–230

Shaheena S, Chintagunta AD, Dirisala VR, Kumar NS (2019) Extraction of bioactive compounds from Psidium guajava and their application in dentistry. AMB Express 9(1):208

Shoombuatong W, Schaduangrat N, Nantasenamat C (2018) Unraveling the bioactivity of anticancer peptides as deduced from machine learning. EXCLI J 17:734–752

Singh AK, Bhatt BP (2013) Effects of foliar application of zinc on growth and seed yield of late-sown lentil. Indian J Agril Sci 83:622–626

Singh AK, Bhatt BP, Upadhya A, Singh BK, Kumar S, Sundaram PK, Chndra N, Bharati RC (2012) Improvement of faba bean (*Vicia faba* L.) yield and quality through biotechnological approach: a review. Afr J Biotechnol 11:15264–15271. https://doi.org/10.5897/AJB12.1926

Singh AK, Meena MK, Bharati RC, Gade RM (2013) Effect of sulphur and zinc management on yield, nutrient uptake, changes in soil fertility and economics in rice (*Oryza sativa*)-lentil (*Lens culinaris*) cropping system. Indian J Agril Sci 83:344–348

Singh D, Patel AK, Baghel SK, Singh MS, Singh A, Singh AK (2014) Impact of front line demonstration on the yield and economics of chickpea (*Cicer arietinum* L.) in Sidhi District of Madhya Pradesh. J AgriSearch 1:22–25. https://doi.org/10.20546/ijcmas.2019.806.277

Singh AK, Singh SS, Prakash V, Kumar S, Dwivedi SK (2015) Pulses production in India: present status, bottleneck and way forward. J AgriSearch 2:75–83

Singh RP, Chintagunta AD, Kumar SPJ (2019) Varietal replacement rate: prospects and challenges for global food security. Glob Food Sec online on 6th October 2019 100324. https://doi.org/10.1016/j.gfs.2019.100324

Sravani D, Aarathi K, Kumar NSS, Krupanidhi RVD, Venkateswarlu TC (2015) In vitro anti-inflammatory activity of *Mangifera indica* and *Manilkara zapota* leaf extract. Res J Pharm Tech 8:1477–1480

Srinivasan A, Giri AP, Harsulkar AM, Gatehouse JA, Gupta VS (2005) A Kunitz trypsin inhibitor from chickpea (*Cicer arietinum* L.) that exerts anti-metabolic effect on podborer (*Helicoverpa armigera*) larvae. Plant Mol Biol 57:359–374

Stagnari F, Maggio A, Galieni A, Pisante M (2017) Multiple benefits of legumes for agriculture sustainability: an overview. Chem Biol Technol Agric 4:2. https://doi.org/10.1186/s40538-016-0085-1

Stevens PF (2008) "Fabaceae". Angiosperm Phylogeny Website. Version 7 May, 2006. Retrieved 28 April, 2008

Tiwari P (2015) Recent trends in therapeutic approaches for diabetes management: a comprehensive update. J Diabetes Res 2015(340838):1–11. https://doi.org/10.1155/2015/340838

Tiwari P, Katyal A, Khan MF, Ashraf GM, Ahmad K (2019) Lead optimization resources in drug discovery for diabetes. Endocr Metab Immune Disord Drug Targets 19:754–774

UN Food and Agriculture Organization (2018) Pulse production in 2017; World regions/Production quantity (from pick lists). Retrieved 5 May, 2019

United Nations D. The International Year of Pulses. United Nations. Retrieved 14 December, 2015

Vuyyuri SB, Shidal C, Davis KR (2018) Development of the plant-derived peptide lunasin as an anticancer agent. Curr Opin Pharmacol 41:27–33

What is a pulse? Pulse Canada, Retrieved 26th June 2016

World Health Organization (2003) The world health report 2003: shaping the future. World Health Organization

Yao DN, Kouassi KN, Erba D, Scazzina F, Pellegrini N, Casiraghi MC (2015) Nutritive evaluation of the Bambara groundnut Ci12 landrace [Vigna subterranea (L.) Verdc. (Fabaceae)] produced in Côte d'Ivoire. Int J Mol Sci 16:21428–21441

Yoshinaga M, Toda N, Tamura Y, Terakado S, Ueno M, Otsuka K, Numabe A, Kawabata Y, Uehara Y (2012) Japanese traditional miso soup attenuates salt-induced hypertension and its organ damage in Dahl salt-sensitive rats. Nutrition 28:924–931

Zambrowicz A, Timmer M, Polanowski A, Lubec G, Trziszka T (2013) Manufacturing of peptides exhibiting biological activity. Amino Acids 44:315–320. https://doi.org/10.1007/s00726-012-1379-7

Chapter 3
Novel Dietary and Nutraceutical Supplements from Legumes

Savita Budhwar and Manali Chakraborty

Abstract Increasing global population along with malnutrition is a crucial matter of concern worldwide. Worldwide population of 7.6 billion is expected to touch the bar of 8.6 billion by 2030 and 11.2 billion in 2100 according to ongoing studies. Wastage of food is believed to be one of the major causes of hunger issue in upcoming future beside imbalanced nutritional status, though decline of stunting and wasting have been observed in past few years. Due to lack of untapped knowledge regarding resourceful agricultural products they are not exploited properly. Development of economically sustainable approach towards food security is much appreciated to manage malnutrition. Proper usage and consumption of the underutilized legumes in regular diet along with normal food can be a suitable and potential replacement of synthetic drugs taken frequently. The major points in this manuscript are the compared analysis of the nutrient as well as anti- nutrient composition of legumes along with their nutraceutical properties. Exploitation of such health beneficial factors present in legumes through value addition can be a revolutionary turn towards health management.

Keywords Legumes · Food security · Nutrition security · Nutraceuticals · Value added products · Bioactive compounds · Therapeutic purposes · Malnutrition · Agroclimatic condition · Underutilization

S. Budhwar (✉) · M. Chakraborty
Department of Nutrition Biology, Central University of Haryana, Mahendergarh, Jant-Pali, Haryana, India
e-mail: savitadahiya@cuh.ac.in

P. Guleria et al. (eds.), *Sustainable Agriculture Reviews 45*, Sustainable Agriculture Reviews 45, https://doi.org/10.1007/978-3-030-53017-4_3

Abbreviations

LDL	Low density lipoproteins
ALDH	Aldehyde dehydrogenase
NAD	Nicotinamide adenine dinucleotide
DNA	Deoxyribonucleic acid
HIV	Human immunodeficiency virus
WHO	World Health Organization
GABA	gamma-Aminobutyric acid
ACE	Angiotensin I- converting enzyme
IVPD	In-vitro protein digestibility

3.1 Introduction

Grain legumes alone or in combination with cereals have always been a rich source of nutritious diet to maintain a healthy lifestyle, globally. Their utilization is not only limited up to human diet but also contributes as livestock feed. These beneficial properties of grains legumes besides maintaining sustainability of agricultural systems also play a key role to increase income rate, poverty alleviation as well as employment generation. Utilization of these legumes is not limited up to the usage as a whole grain. The by- products, also known as agricultural wastes, generated from the legumes also proved to be a great source of health beneficial nutrients (Yorgancilar and Bilgicli 2014). These underutilized agricultural wastes of legumes are found to be similarly capable of maintaining nutrition security, though more scientific studies are needed to screen out the untapped properties. More over in term of economical significance Leguminosae (Fig. 3.1) are considered to be the most important family in the Dicotyledonae (Harborne 1994; Rebello et al. 2014).

Leguminosae is considered to be one of the largest families among all other flowering plants that consist of 18,000 species and 650 genera (Polhill and Raven 1981; Annor et al. 2014). Legumes are considered to be second to the cereals that contribute as food crops in world agricultural system (Fig. 3.2). This family shows great adaption to grow under harsh condition too (International Legume Database and Information Service).

Based upon recent surveys global population will reach up to the mark of 9.6 billion people by 2050 (United Nations, Department of Economic and Social Affairs, Population Division, New York) and this increment is going to create hunger issues risking food security. Increment in the production rate of legumes and legume by- products can play central role to sustain food as well as nutrition security (Voisin et al. 2014).

Fig. 3.1 Various types of major legumes- pulses crops (Grain legumes for sustainable agriculture. Krishisewa 2013)

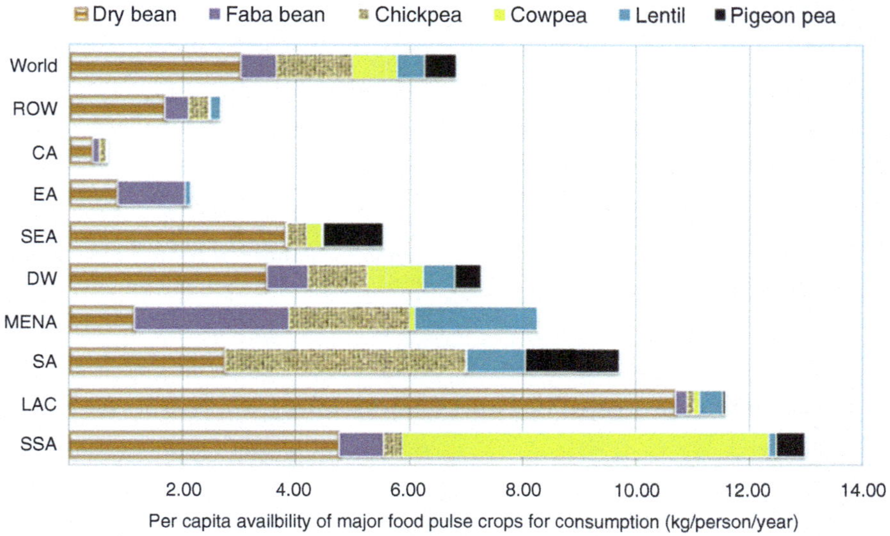

Fig. 3.2 Per capita availability of pulses for consumption (kg/person/year): In the above graph, consumption of major food pulse crops from 2006 to 2008 worldwide had been shown (FAO) (Maredia 2012)

Table 3.1 Harvested area and yield for legume grains on interval of 10 years (FAOSTAT data, 1974–2014)

Chickpea					
Year	1974	1984	1994	2004	2014
Harvested area (million ha)	10.6	9.85	9.96	10.5	14.8
Yield (t ha^{-1})	0.56	0.67	0.71	0.8	0.96
Cowpea					
Harvested area (million ha)	4.7	3.66	7.35	9.18	12.52
Yield (t ha^{-1})	0.35	0.35	0.35	0.35	0.35
Legumes					
Harvested area (million ha)	5.67	5.65	5.33	4.27	6.1
Yield (t ha^{-1})	0.54	0.58	0.65	0.82	0.84
Vegetables and leguminousnes					
Harvested area (million ha)	0.14	0.18	0.18	0.25	0.24[a]
Yield (t ha^{-1})	5.41	5.09	5.18	6.54	6.86[a]
Lentil					
Harvested area (million ha)	2.03	2.56	3.43	3.85	4.52
Yield (t ha^{-1})	0.61	0.68	0.81	0.93	1.08
Soybean					
Harvested area (million ha)	37.4	52.9	62.5	91.6	117.72
Yield (t ha^{-1})	1.41	1.71	2.18	2.24	2.62

[a]Data are referred to year 2013 (2014 data not available), FAOSTAT data

3.2 Production Area and Yield of Legumes

The production rate of pulses is 26.30 m ha with production of about 18.10 m ha during 2010–11. According to FAOSTAT data (1974–2014), harvested area of legume grains along with their yield has been increased or decreased as well. Also some legumes showed stagnant production rate through several years. Even though the rate has been increased for few legumes, it's not satisfactory when the year gap is been considered. Yield of some significant legume grains has been charted below (Table 3.1).

3.3 Health Promoting Factors of Legumes

Grain legumes are known to be a potent resource of protein with low cost effect available to below poverty lined population as compared to other grains and crops (Bouchenak and Lamri-Senhadji 2013). They are widely known as 'Poor man's meat'. The major common legumes easily available to the population of Asia, Latin America, Sub- Saharan and North Africa maintaining their food and nutrition security are chickpea, common bean, cowpea, faba bean, ground nut, lentil, pigeon pea (Messina 1999). Besides being nutrient rich, few legumes are also drought and heat tolerant but remain neglected, viz., tepary bean, lima bean, grass pea and bambara ground nut. Based upon recent surveys, a stagnant production of grain legumes is observed. More over it can be said that the production rate is declining per capita availability of grain legumes. It is very much crucial to take a step towards the alleviation of productivity, quality, and stability of production to make certain the availability of such nutritive legumes at affordable price throughout the season and environment (Battisti and Naylor 2009; Ndidi et al. 2014).

3.3.1 Nutrients and Anti-nutrients Present in Legume Grains

These crops are generally grown on marginal lands where rainfall is almost negligible (Kumar et al. 2009; Nedumaran et al. 2015). Food value of legumes is comparatively higher. Caloric value per unit weight is found to be equivalent to cereal grains. Protein rich legumes are also minerals and vitamin rich. Though lysine rich pulses lack sulphur rich amino acids and tryptophan, combination of cereals that lacks lysine fulfill that requirement. Cotyledon contains methionine and tryptophan that covers almost 93% of whole seed. Phosphorus and molybdenum level in soil influence amino acid composition of seed (Chen et al., 2010; Khalid and Elharadallou 2013). Adequate amount of crude fiber, protein and lipid components of pulses is also display hypercholesteronic effect (Mlyneková et al. 2014). High level of protein content in pulses also accomplishes the requirement of protein in vegetarian population instead of being dependent on non- vegetarian sources. Proper exploitation of pulses as neutraceuticals can be a revolutionary change in public health defining the switching over of treatment to prevention (Leonard 2012).

3.3.1.1 Micro and Macro Nutrients

Protein content of pulses varies in wide range, usually from 14.9–34.6%, located mainly in cotyledons and embryonic axis along with seed coat. Variability of protein is especially observed in seed collected from different branches (Yao et al. 2015). Embryo contains 2.5% protein. This protein content is mainly of two types, viz., enzymatic and structural protein also known as metabolic protein and the other one is storage protein. Normal cellular activities are mainly regulated by metabolic proteins whereas the storage protein is generally found in cell in discrete protein bodies (Maphosa and Jideani 2016). In Bengal gram (Chickpea) the protein content is from 14.9–29.6%. Among highly consumed legumes, the higher level protein content range is present in Mung bean that is 20.8–33.1. Pigeonpea possesses 18.8–28.5 protein range whereas Urdbean contains 21.2–31.3. Lentil has protein range somewhat similar to mung bean that is, 20.4–30.5 (Maphosa and Jideani 2016) (Tables 3.2, 3.3, 3.4, 3.5 and 3.6).

Table 3.2 Essential amino acid composition of different pulses (Maphosa and Jideani 2016)

Legumes	Lysine	Threonine	Valine	Leucine	Histidine
Chickpea	6.3	3.4	5.5	8.2	2.3
Pigeon pea	6.8	3.8	4.8	6.8	3.4
Mung bean	7.3	3.4	6.9	7.7	2.7
Urd bean	6.5	3.9	5.6	7.2	2.7
Lentil	5.1	3.0	5.1	5.5	2.1
Lathyrus	7.4	2.3	4.7	6.6	2.5
Pea	8.9	4.2	6.5	9.5	2.7

Legumes	Isoleucine	Methionine	Phenyl alanine	Tryptophan	Arginine
Chickpea	6.0	1.2	4.9	0.8	6.9
Pigeon pea	5.7	1.1	9.0	0.8	5.4
Mung bean	6.3	1.5	5.3	0.4	6.9
Urd bean	5.8	1.1	5.5	0.5	5.7
Lentil	5.8	0.6	4.0	0.6	7.0
Lathyrus	6.7	0.6	4.2	0.4	7.8
Pea	7.4	1.3	4.6	0.7	13.4

Though pulses showed wide range of variability in protein content, their low biological values seem to occur due to low methionine content. Presence of protease inhibitors and anti nutritional factors makes pulse protein less digestible comparative to other protein sources (Tamang et al. 2016)

Table 3.3 Biological value of grain legumes (Tamang et al. 2016)

Legumes	Biological value (%)
Chickpea	52–78
Pigeon pea	46–74
Mung bean	39–66
Urd bean	60–64
Lentil	32–58
Pea	48–49

Carbohydrate range of pulses belongs from 53.3% to 68%. Consumption of large quantity of pulses causes flatulence in human beings which leads to discomfort in the intestinal tract resulting in abdominal rumbling, cramps, pain and diarrhea (Karner 2016). This occurs due to absence of the hydrolytic enzyme alpha-1,6- galactosidase (digest members of raffinose family of sugars) in human mucosa. The undigested part of these accumulated oligosaccharides generates carbon dioxide, hydrozen and methane leading to flatus effect (Bliss et al. 2014). This effect can be manipulated through genetic modification leading to development of pulses with low raffinose family oligosaccharides. Soaking of seeds in water also seems to lower the oligosaccharide content significantly (Myriam et al. 2016)

Table 3.4 Carbohydrate content (%) of pulses (Myriam et al. 2016)

Pulses	Starch	Amylose	Total
Chickpea	32.2–50.0	31.8–45.8	60.1–61.2
Pigeonpea	40.4–48.2	38.6	57.3–58.7
Mungbean	37.0–53.6	13.8–35.0	53.1–61.2
Urdbean	32.2–47.9	43.9	56.5–63.7
Lentil	34.7–52.8	20.7–45.5	59.7
Pea	36.9–48.6	23.5–33.1	56.6
Cowpea	31.5–48.0	-	56.0–68.0

Table 3.5 Oligosaccharides % of pulses (Myriam et al. 2016)

Pulses	Sucrose	Raffinose	Stachyose	Verbascose	Total
Chickpea	0.7–2.9	0.7–2.4	2.1–2.6	0.4–4.5	3.5–9.0
Pigeonpea	2.7	1.0–1.1	2.7–3.0	4.0–4.1	3.5–10.2
Mungbean	0.3–0.2	0.3–2.6	1.2–2.8	1.7–2.8	3.9–7.2
Urdbean	0.7–1.5	0.0–1.3	0.9–3.0	3.4–3.5	3.0–7.1
Lentil	1.8–2.5	0.4–1.0	1.9–2.7	1.0–3.1	4.2–6.1
Pea	2.3–2.4	0.3–0.9	2.2–2.9	1.7–3.2	5.3–8.7
Cowpea	1.8–3.1	0.4–1.2	2.0–3.6	0.6–3.1	6.0–13.0

Pulses are also a superior source of crude fiber that ranges almost from 1.2–13.5%. Greater variability is observed mainly in chickpea, pigeon pea, mung bean and urdbean (Saxena et al. 2010). Major crude fibre present in chickpea and pea is cellulose whereas hemicellulose dominates in pigeon pea, urdbean and lentil

Table 3.6 Crude fiber % of pulses (Saxena et al. 2010)

Pulses	Lignin	Cellulose	Hemicellulose	Total
Chickpea	2.9–7.1	1.1–13.7	0.6–8.4	1.2–13.5
Pigeonpea	2.9	7.3	10.1	1.2–8.1
Mungbean	2.2–7.2	2.5–4.6	0.3–9.1	1.2–12.8
Urdbean	3.8	5.0	10.7	1.2–7.1
Lentil	2.6	4.1	6.0	3.8–4.6
Pea	0.5–0.9	0.9–4.9	1.0–5.1	4.6–7.0
Cowpea	0.6–1.8	-	1.7–4.0	

Almost 1.0–4.99% of total lipid content is found in pulses that may vary with variety, origin, location and climate. Higher amount of linoleic or linolenic acid are present in pulses

3.3.1.2 Anti-nutrient Factors and Their Effects upon Health

Besides being nutrient rich, legumes also contain anti nutritional factors or anti-nutritional compounds. These compounds might be protein in nature viz., hydrolase inhibitors and lectins along with non- protein nature content. Existence of these anti nutrients more often is considered to be for self protection of plants for the survival under natural conditions or more specifically it can be said that these anti nutrients are result of evolutionary adaption for survival (Brigide et al. 2014). Their toxic properties generally play defensive role against insects, fungi, predators and also in few cases stress conditions. Though presence of anti- nutrients shows adverse effects in some cases viz., quality of the food products leading to characteristic anthropocentric vision, in case of legumes and pulses present anti- nutrients also showed significant nutraceutical properties according to ongoing studies. Detailed study of these factors along with nutrients is very much necessary for the future prospectus (Marcello Duranti 2006).

Phenols present in the legume seeds play both the role of anti nutrient as well as bioactive compound viz., antioxidants. These health influencing factors show chelating property upon metal ions along with inhibition of peroxidation of lipids and scavenging of free radicals. The main contents seen in legumes are tannins, anthocyanins, phenolic acids, and flavonoids polyphenolic compounds viz., flavonol glycoside, anthocyanins and condensed tannins which is responsible for the colour of the seeds of legumes (Sanchez-Chino et al. 2015). Dark shade and pigmented seeds are seemed to have higher phenolic content that are found to be in red kidney beans (*Phaseolus vulgaris*) and black gram (*Vigna mungo*). Present level of phenolic compounds among legume grains also illustrates presence of antibacterial, antiviral, anti-inflammatory and anti- allergic properties. Phenols are also considered to be associated with lowering of the incidence of cancer, cardiovascular disease and diabetes. Based upon some studies fermentation enhance the antioxidant potential of legumes a lot. In some cases fermentation also helps to minimize the anti nutrient effect of phenols, present resulting in health beneficial consequences. Proper utilization of this maintenance between anti- nutrient and nutrient in legume seeds can provide prospective nutrient exploitation as potential nutraceutical (Shweta and Rana 2017; Shahwar et al. 2017; Niveditha and Sridhar 2012).

Isoflavones present in legume grains also shows antioxidant property being associated with lowering risk of osteoporosis, heart issues and cancer. Besides, it also seems to be used for therapeutic purposes like symptoms after menopause. Daidzein and genistein, the major isoflavone present in legume, are known to be the natural phytoestrogen that inhibit LDL oxidation leadingto reduction of the risk associated with atherosclerosis (Polak et al. 2015). Besides being an anti- cancer agent genistein also slow down the aggregation of platelet, production of leukotriene, DNA topoisomerase II, angiogenesis, decline of availability of sex hormones in body, initiation of apoptosis and differentiation in cancer cells. According to some researchers, daidzein is responsible for the induction of differentiation in B16 melanoma, and HL- 60 human leukemia cells. It also has a contribution in the inhibition of few enzymes, such as, ALDH- I which is an NAD dependent aldehyde dehydrogenase. Main role of ALDH- I is to catalyze the oxidation of acetyldehyde, prime product of alcohol metabolism that is also helpful for the therapeutic purpose of alcoholism (Shashank et al. 2015).

Seed hydrolase inhibitors play an important role in some legume grains, viz., pea, lentil, soybean etc. This inhibitor shows effect on many digestive enzymes, like chymotrypsin, trypsin, and amylase. Their effects persist if the above mentioned legumes are consumed uncooked. After cooking, these proteins get denatured and inactivated leading to positive nutritional effects in some cases. Inhibitors of Trypsin or chymotrypsin of both the Bowman- Birk and Kuntiz type and α- amylase inhibitors are the most characterized inhibitors present in legumes. Toxic nature of lectins present in legume seeds can be minimized by cooking process. Though their biological importance is still under controversies, this is also proved to a defense protein in legumes (Marcello Duranti 2006).

Stored phytic acid in legume seeds are in general present as form of phosphate and the storage location is the endosperm. This can inhibit the transcription of the viral genome that make phytic acid a potential anti- HIV agent. In spite of being an anti- nutrient it also contributes in many health promoting consequences, such as, anti carcinogenic act. This prevents the kidney stone formation along with the cure of cavities, plaque and tartar in the teeth by lowering demineralization that is reduction of the solubility of calcium, fluoride and phosphate. Its antioxidant property also contributes in the reduction of the risk regarding cardiac issues and diabetes mellitus (Xu and Chang 2007; Shahwar et al. 2017; Shashank et al. 2015). Brown coloured common bean seems to possess higher amount of phytic acid (1.1%) along with 3% of α- Galactosides and 0.5% tannins. Bengal gram consists 0.5% phytic acid content whereas pea tends to have 0.9% phytic acid and a large amount of α-Galactosides (5.9%). White coloured commonbean, faba bean, soybean possess same level of phytic acid (1.0%) (Gulewicz et al. 2014; Amarowicz and Pegg 2008).

Saponins are another important secondary plant metabolite that can be steroidal or triterpenoid. Saponins present in legumes are generally triterpenoids in nature. It shows anti carcinogenic nature especially in the case of colon cancer. Its role as immune stimulant is also the beneficial one beside being anti- inflammatory, antifungal, antiparasitic, hypercholesterolemic, hypoglycemic, immunomodulatory, etc.(Prakash et al. 2006; Shahwar et al. 2017).

Some oligosaccharides present in legumes also play a major role as a prebiotic by the promotion of the growth of bifido bacteria. Enzyme inhibitors present in legumes, viz., protease inhibitor, α- amylase inhibitor, α- glucosidase inhibitor, γ-aminobutyric acid, have potential contribution in various fatal conditions and disorders (Shweta and Rana 2017).

3.4 Prospective Nutraceutical Properties of Various Species of Legumes

Black soybean, also known as *Glycine* max L. consists of black seed coat generally is being used in oriental medicine practices for many years as tonic food and material. It is also believed to be a miracle in treatment of diseases like diabetes, hypertension and other conditions, such as, antiaging, beauty treatment, blood flow because of its active peptide compounds (Sefatie et al. 2013).

Pigeon pea (*Cajanus cajan* L. Millspaugh) is considered to be a crucial source of protein, starch, fat, crude fiber and minerals like, calcium, manganese. Scorched seed of this legume served with coffee can relieve headache and vertigo. Its fresh seed shows beneficial effect in incontinence in males. Besides these, immature seeds have medicinal properties to treat kidney ailments as well as hyperglycemic patients (Tiwari et al. 2013).

Another major legume, mung bean (*Vigna radiata*) believes to reduce risk of many diseases. *Vigna mungo* or black gram due to possessing crucial nutritional value is grabbing attention in field of food and nutrition. Consumption of black gram in diet especially during fever, affection for cough and liver issues along with the severe cases of diabetes is found to be beneficial (Shweta and Rana 2017).

Generally in the semi- arid tropics, Bengal gram (*Cicer arietinum*) is thought to be an important component of a cholesterol free nutrient rich diet. It is considered to be a perfect combination of beneficial micro and macro nutrients, viz., carbohydrates, protein, dietary fiber, vitamin and minerals. It contains few sterols, tocopherols and tocotrienols that show evidence of antiulcerative, antibacterial, antifungal, antitumor and anti- inflammatory properties besides opposing effect on cholesterol levels (Chibbar et al. 2010; Murty et al. 2010).

Plant sources are always been used as pharmaceuticals, nutraceuticals and for other purposes, viz., dyes, waxes, natural rubber, gums and so on. According to the estimated report of the World Health Organization's (WHO) Traditional Medicine Program, 80% of the global population use phytopharmaceuticals. Legumes are considered to be one of these plant sources, utilized as nutraceuticals (Table 3.7).

Table 3.7 Effectiveness of nutraceutical properties found in legume relatives of *Phaseolu* (Brad Morris 2003)

Legume (taxon)	Nutraceutical (plant part)	Use	Effectiveness
Hyacinth bean [*Lablab purpures* (L.) Sweet], Siratro [*Macroptilium atropurpureum* (DC.) Urban], *M. bracteatum* (Nees & C. Martius) Marechal & Baudet, Phasey bean [*M. lathyroides* (L.) Urban], Perennial horsegram [*Macrotyloma axillare* (E. Meyer) Verdc.], Horsegram [*M. uniflorum* (Lam.) Verdc.]	Myo- inositol (seed)	Panic disorder	Possible
Hyacinth bean, Phase bean	Genistein (Hypocotyl)	Menopause hot flashes	Possible
Winged bean (*Psophocarpus tetragonolobus* (L.) DC.	Beta-sitosterol	Benign prostatic hyperplasia, lowers low density lipoprotein cholesterol levels.	Likely
	Quercetin	Chronic, nonbacterial prostatitis	Possible

Healthy diets can combat malnourishment along with many other fatal death causing diseases. Pulse- cereal combined diets may provide health benefits and contribute in prevention of chronic diseases. Hypercholesterolemia, hypertension, anti-obesity effects of pulses are already well known to common people. Utilization of nutrient rich legume foods can be associated with therapeutic purposes instead of regular uses of synthetic drugs

3.5 Nutraceutical Properties of Legumes Towards Health Management

3.5.1 Legumes as Therapeutic Component

Beneficial role of legumes is uncountable. These traditional medicinal agents are spreaded all over the eastern and far eastern countries like India and China. Phytotherapeutic properties of legumes are properly used in the Mediterranean region especially. According to some researches diet including less non- vegetarian source and more cereal- legume combination is more favorable for a healthy life style.

Cardiovascular disease is one of the most frequently seen diseases in the present generation due to their unhealthy life style and diet. In this condition generally heart or blood vessels get affected. Fat gets deposited inside the arteries resulting in increased risk of blood clots and deficiency of oxygen rich blood to heart muscle due to blockage. Intake of dry legumes along with daily diet can control the lipid homeostasis leading to reduced risk of heart diseases. High fibre content and the low glycemic index properties of legumes are the major properties that can manage the risk associated. They are also claimed to be helpful in lowering elevated sugar level in diabetes by preventing insulin- resistance. This type of resistance is generally seen in case of diabetes type II. For normal individual, daily intake of legume helps to maintain their healthy life status and weight management by avoiding obesity due to low glycemic index.

Consumption of legumes also exhibit positive effects in prevention of cancer, especially colon cancer along with digestive tract diseases (Clemente and Olias 2017). Legume seed consumption is also beneficial for the individual suffering from liver issues. γ- aminobutyric acid, also known as GABA acts as a potential anti-hepatotoxic agent (Shweta and Rana 2017).

Not only the whole legume grain, but also the seed husk exhibits significant anti-oxidant as well as anti- hyperglycemic properties being an economical nutrient source as nutraceutical for hyperglycemic patients. It had been observed that methanolic extract of seed husk slow down hyperglycemic spikes and the load became almost similar to synthetic drug used for hyperglycemic purposes, viz., acarbose (Racete et al. 2009).

Legumes also play a key role to manage and somewhat prevent hypertension. They possess Angiotensin I- converting enzyme (ACE) inhibitor peptides (Wong and Ng 2005).

3.5.2 Bioactive Peptides Present in Legume Grains

7S globulin α' chain has been found in soybean that up- regulates the LDL receptors along with the reduction of plasma cholesterol and triglyceride level. This globulin chain also acts as anti- atheromatous agent. Faba bean, lupin and some other legume seeds also consist of storage proteins (undefined) that reduce plasma cholesterol and triglyceride level in body. BB serine- protease inhibitor is a type of trypsin/chymotrypsin inhibitor found in soybean, pea, and others. This inhibitor opposes the consequences of cancer, inflammatory, obesity, degenerative and autoimmune diseases. α- amylase inhibitor present in legumes also controls weight and obesity related disorders and diabetes as well. Hypoglycaemic and hypocholesterolaemic properties of legumes are considered to be responsible for the presence of the factor; conglutin γ. ACE inhibitor found in soybean also shows intrinsic activity such as, alkalase treatment or fermentation of soybean proteins as well as hypotensive property (Marcello Duranti 2006).

In spite of being an extra ordinary source of nutrient rich diet, legumes especially their seeds are also proven to be allergic in some cases. This is known to be causative agent for the allergic reactions generally through ingestion. Mainly anti- nutritional factors present in legumes are the responsible one for such allergies. Lupin is considered to be an allergen based upon a report in 2006 (European Commission). Studies revealed that the main reason for the allergy is α and β- conglutins and a less significant amount of γ and δ- conglutins. Peanut allergy is also a well known legume allergy and IgE mediated reaction to food is the reason. Though there is no such major chickpea allergen has been discovered, although several binding bands for IgE (10–70 kDa) have been screened out through immunoblotting technique (Ballabio et al. 2013). Still a lot of detailed investigation is needed for the legume sensitization to overcome possible allergic reactions and their health related issues (Jimenez-Lopez et al. 2018).

3.5.3 Use of Legume in Various Prospectuses

Legumes are utilized in various fields because of their health beneficial properties, especially due to high protein content. Some Asian dishes such as, tofu, nato miso, are being prepared using soybean (*Glycine max*). Roasted snacks, milk, yoghurt, sprouted beans, curd, yuba, sauce and many more are developed utilizing soybean. Many fermented products like idli, dosa, papad are made from Black gram (*Vigna mungo*). Legumes viz., peas, peanut/ground nut, adzuki beans, black- eyed peas, kidney beans, mung beans are utilized to formulate soup, fried cake, pudding, yoghurt or exploited as whole food. Chickpea has been consumed as foods to prepare falafel and hummus in Middle Eastern and Mediterranean region besides having as snacks, dhal, and curry. This is also used to formulate flour for bread making and fermented foods. Consumption of underutilized legumes is considered to be health beneficial as they contain high dietary fibre, protein, energy, anti- oxidant rich with much other significant potential.

Being highly nutritive, especially protein rich most of the legumes remain underutilized and neglected. Legumes like bambara ground nut, African locust bean,

African yam bean, pigeon pea, kidney bean, lima bean and marama bean are considered to be underutilized grain rather than deserving one. These economically affordable crops need to be given more concern in the developing or poor countries like sub Saharan Africa. With the increasing global population, it is very much urgent to show awareness about the cultivation of legumes to maintain food as well as nutrition security. Stagnant production of legumes is one of the reasons that mitigate the nutrient utilization to combat malnourishment. Besides solution to micro or macro nutrient deficiency, their untapped properties could be useful for the production of therapeutic, affordable, functional value added foods (Kalidass and Mahapatra 2014).

3.5.4 Myths Regarding Legume Processing and the Solutions

Due to lack of proper knowledge regarding legume consumption and utilization, there are too many myths, such as presence of anti- nutrients in legumes and pulses causes bloating and flatulence as their cooking purposes is not so feasible. Processing of legumes through some cooking or other traditional methods can minimize the content of anti- nutrients present in pulses and legumes. Soaking before cooking of these grains also helps to minimize the cooking time along with the lowering of anti- nutrients. Low consumption of legumes is mainly occurring due to less education delivered to consumers about methods (Olaleke et al. 2006) (Table 3.8).

Table 3.8 Processing techniques of legumes for proper utilization (Maphosa and Jideani 2017)

Limitation of legumes for consumption	Harmful effects	Processing techniques
Presence of trypsin and amylase inhibitors	Inhibitors slow down starch and protein digestibility	The level of inhibitors can be managed by boiling beans up to 80–90%. Fermentation also helps to reduce the contents.
Phytic acid content	Phytic acid present as phytate in legumes acts as chelating agents and decrease their bioavailability.	Fermentation, germination are the most convenient. Others include blanching, steaming and utilizing of gamma irradiation.
Lectins, saponins	Lessen availability of micro and macro nutrients in body.	Heat generated during cooking destroys these anti- nutrients along with fermentation.
Oligosaccharides	Bloating stomach	
Legumes contain low level of sulfur rich amino acids	Could not fulfill the need of sulfur	As cereals are considered to be rich source of sulfur containing amino acids. Legumes are taken in combination with cereals.

Being a high protein rich source, legumes and pulses are known to be perfect substitution for meat sources for vegetarians (Ebert et al. 2014)

3.5.5 Various Processing Techniques of Some Major Legumes

Due to generated heat during cooking nutrient as well as anti-nutrient loss occurs. Ordinary cooking for 15 min causes loss of 15% thiamine and 25% niacin present in lentils. Chickpea and faba bean also loose 44% and 53% thiamine, 78% and 79% riboflabin, 28% and 36% niacin for 150 min of cooking respectively. But in case of unsoaked beans, faba beans, lentils, beans, cowpea show loss of anti- nutrients such as, 41.5%, 46%, 47% and 46.4% oligosaccharide respectively for 60 min of cooking. In reverse with same cooking duration if the legumes are being soaked in distilled water for 12 hours, it shows almost 50–60% of oligosaccharide loss. in case of change of medium of soaking that is $NaHCO_3$ (0.2–0.5%), the loss is greater that is in the range of 60–70%. After soaking for 12 hours in distilled water if chickpea is cooked following ordinary procedure, it not only shows increment in the level of total soluble sugars (26–36%), reducing sugar (8–16%), and nonreducing sugar (26–38%), but also reflects reduced content of starch that is up to 13–22%. Starch digestibility rate gets increased significantly. According to studies, it has been considered that the increased level of sugars happens due to hydrolysis of starch and oligosaccharide to their simpler form on cooking. It may also rupture the starch granules by amylolysis that results in loss of starch content. When the cooking duration is extended up to 240 min, 49.5% increment is observed in total dietary fiber of chickpea as resistant starch and Maillard reaction products are generated during the process along with condensation of tannin protein products. Soaking of chickpea in citric acid, distilled water or 0.07% sodium bicarbonate solution for 9 hours causes loss of many micronutrients as well as some anti- nutrients such as, thiamin, riboflavin, niacin along with galactosides present in it. Simple water soaking causes mineral loss (calcium, copper, iron, magnesium, manganese, sodium, zinc) besides reduction of phytic acid content. Pressure cooking of legumes also decreases 50% in polyphenol content along with some amino acids like cysteine, lysine and protein level. But an improvement of IVPD from 72% to 84% has been observed. It also improve digestibility of legumes as well. Germination of legumes, especially chickpea for two and a half day increases total soluble sugar, reducing and non reducing sugar. But it lowers starch content while increases its digestibility rate. This increment in sugar level is known to be for the mobilization and hydrolysis of polysaccharides. Soaking can also result loss of vitamins present in legumes. Based upon ongoing studies, this loss of vitamin may depend upon several factors, such as, faster mass transfer from seed to water. Large amount of vitamin loss is observed in case of use of $NaHCO_3$ as soaking medium because of increased pH that makes vitamins more soluble (Santosh Satya et al., 2010).

3.6 Future Aspects

Due to cultivation of legumes in different agroclimatic conditions some changes are being noticed in the nutrient content in different species of the same genus. Soaking of these legumes prior to utilization as food product lowers the anti- nutrient content present in those. Though soaking for several hours might cause mineral as well as some other nutrient loss, smaller duration may slow down this loss. Addition of additives such as citric acid and sodium bicarbonate during soaking of legumes found to minimize the vitamin loss along with the retentions of other nutrients. Lemon juice can be a natural substitute of chemical additives viz., citric acid. Germination is also helpful to flush out the anti nutrient factors and to alleviate the nutrient contents present in legumes. Proper timing and duration of cooking purposes as well as the processing of these legumes is very much necessary to maintain the balance between food and nutrition security.

3.7 Conclusion

Proper knowledge about the nutrient rich greens is a crucial matter to maintain a healthy lifestyle rather than being dependent on the chemicals or synthetic products. In spite of availability of numerous varieties of traditional natural resources they remain underutilized due to some myths and misconceptions. Their untapped properties should be investigated properly to avoid their stagnant production along with the maintenance of the food and nutrition security. Legumes and pulses are one of these underrated agricultural resources. Combined diet of legumes- cereals provides perfect diet to avoid health issues. This protein rich grain also contributes as a significant protein source for vegetarians rather than consumption of meat or other non- vegetarian sources. Besides being a pool of micro as well as macro nutrients, legumes are also easily affordable to below poverty lined population. Presence of anti nutrient in legumes is matter of concern, but there are a lot of processing methods viz., germination, soaking, cooking etc. to minimize those contents. Not only are the legume grains, their by-products also found to be nutrient rich and health beneficial. Value addition of these agricultural wastes of legumes along with the seeds and grains as food products can boost the healthy factors and can be considered as nutraceuticals with pharmaceutical properties. Instead of using synthetic drugs for therapeutic purposes in nominal conditions frequently, it's better to consume legume rich diet along with cereals parallel to regular foods to stay away from further health complications.

Acknowledgements The authors duly acknowledge the Department of Nutrition biology, Central University of Haryana for the support and Vice Chancellor, Central University of Haryana for providing the motivation and infrastructure.

References

Amarowicz R, Pegg RB (2008) Legumes as a source of natural antioxidants. Eur J Lipid Sci Technol 110:865–878. https://doi.org/10.1002/ejlt.200800114

Annor GA, Zhen M, Boye JI (2014) Crops–legumes. In: Clark S, Jung S, Lamsal B (eds) Food processing: principles and applications. Wiley, Chichester, pp 305–337. https://doi.org/10.1002/9781118846315.ch14

Ballabio C, Pen E, Uberti F, Fiocchi A, Duranti M, Magni C, Restani P (2013) Characterization of the sensitization profle to lupin in peanut-allergic children and assessment of cross-reactivity risk. Pediatr Allergy Immunol 24:270–275. https://doi.org/10.1111/pai.12054

Battisti DS, Naylor RL (2009) Historical warnings of future food insecurity with unprecedented seasonal heat. Science 323:240–244

Bliss DZ, Savik K, Jung HG, Whitebird R, Lowry A, Sheng X (2014) Dietary fbre supplementation for fecal incontinence: a randomized clinical trial. Res Nurs Health 37(5):367–378. https://doi.org/10.1002/nur.21616

Bouchenak M, Lamri-Senhadji M (2013) Nutritional quality of legumes, and their role in cardiometabolic risk prevention: a review. J Med Food 16(3):185–198. https://doi.org/10.1089/jmf.2011.0238

Brigide P, Guidolin CS, Oliveira SM (2014) Nutritional characteristics of biofortifed common beans. Food Sci Technol (Campinas) 34(3):493–500. https://doi.org/10.1590/1678-457x.6245

Chen W, Sharma HC, Muehlbauer F (eds) (2010) Chickpea and lentil crop protection compendium. American Phytopathological Society, St Paul, 255 pp

Chibbar RN, Ambigaipalan P, Hoover R (2010) Molecular diversity in pulse seed starch and complex carbohydrates and its role in human nutrition and health. Cereal Chem 87:342–352. https://doi.org/10.1094/CCHEM-87-4-0342

Clemente A, Olias R (2017) Benefcial effects of legumes in gut health. Curr Opin Food Sci 14:32–36. https://doi.org/10.1016/j.cofs

Duranti M (2006) Grain legume proteins and nutraceutical properties. Fitoterapia 77:67–82

Ebert AW (2014 Jan) Potential of underutilized traditional vegetables and legume crops to contribute to food and nutritional security, income and more sustainable production systems. Sustainability 6(1):319–335

European Commission (2006) Directive 2006/142/EC of 22 December 2006, amending annex IIIa of directive 2000/13/EC of the European Parliament and of the council listing the ingredients which must under all circumstances appear on the labelling of foodstuffs. Off J Eur Union 368:110–111

Gulewicz P, Martinez-Villaluenga C, Kasprowicz-Potocka M, Frias J (2014) Non-nutritive compounds in fabaceae family seeds and the improvement of their nutritional quality by traditional processing–a review. Pol J Food Nutr Sci 64(2):75–89. https://doi.org/10.2478/v10222-012-0098-9

Harborne JB (1994) Phytochemistry of the leguminosae. In: Bisby FA, Southon IW (eds) Phytochemical dictionary of the leguminosae. Chapman and Hall, Boca Raton

Jimenez-Lopez JC, Foley RC, Breard E, Clarke VC, Lima-Cabelloa E, Floridoe JF, Singh KB, Alchea JD, PMC S (2018) Characterization of narrow-leaf lupin (*Lupinus angustifolius* L.) recombinant major allergen IgE-binding proteins and the natural β-conglutin counterparts in sweet lupin seed species. Food Chem 244:60–70. https://doi.org/10.1016/j.foodchem.2017.10.015

Kalidass C, Mahapatra AK (2014) Evaluation of the proximate and phytochemical compositions of an underexploited legume Mucuna pruriens var. utilis (Wall ex Wight) L.H. Bailey. Int Food Res J 21(1):303–308

Karner T (2016) Effect of palatable soluble fbre-containing carbohydrate food on postprandial blood glucose response in healthy individuals. Masters Thesis; Aarhus University, Denmark

Khalid II, Elharadallou SB (2013) Functional properties of cowpea (*Vigna Ungiculata* L.Walp), and lupin (Lupinus Termis) flour and protein isolates. J Nutr Food Sci 3:234. https://doi.org/10.4172/2155-9600.1000234

Krishisewa. 13th May, 2013. [https://www.krishisewa.com/articles/organic-agriculture/276-legumes-for-agriculture.html]

Kumar P, Joshi PK, Birthal PS (2009) Demand projections for food grains in India. Agric Econ Res Rev 22:237–243

Leonard E (2012) Cultivating good health. In: Grains and legumes nutrition council. Cadillac Printing, Adelaide, pp 3–18. ISSN 1039-6217

Maphosa Y, Jideani VA (2016) Physicochemical characteristics of Bambara Groundnut dietary fbres extracted using wet milling. S Afr J Sci 112(1/2):1–8. https://doi.org/10.17159/sajs.2016/20150126

Maphosa Y, Jideani VA (2017) The role of legumes in human nutrition https://doi.org/10.5772/intechopen.69127

Maredia M (2012 Feb 13) Global pulse production and consumption trends: the potential of pulses to achieve 'feed the future'food and nutritional security goals. In: Global pulse researchers meeting

Messina MJ (1999 Sep 1) Legumes and soybeans: overview of their nutritional profiles and health effects. Am J Clin Nutr 70(3):439s–450s. https://doi.org/10.1093/ajcn/70.3.439s

Mlyneková Z, Chrenková M, Formelová Z (2014) Cereals and legumes in nutrition of people with celiac disease. Int J Celiac Dis 2(3):105–109. https://doi.org/10.12691/ijcd-2-3-3

Morris B (2003) Bio-functional legumes with nutraceutical, pharmaceutical, and industrial uses. Economic Botany 57(2):254–261. Published by: Springer on behalf of New York Botanical Garden Press. Stable URL: http://www.jstor.org/stable/4256683. Accessed 20 Sept 2016 14:39 UTC

Murty CM, Pitaway JK, Ball MJ (2010) Chickpea supplementation in an Australian diet affects food choice, satiety and bowel function. Appetite 54:282–288. https://doi.org/10.1016/j.appet.2009.11.012

Myriam M, Grundy L, Edwards CH, Mackie AR, Gidley MJ, Buterwort PJ, Ellis PR (2016) Re-evaluation of the mechanisms of dietary fbre and implications for macronutrient bioaccessibility, digestion and postprandial metabolism. Br J Nutr 116(5):816–833. https://doi.org/10.1017/S0007114516002610

Ndidi US, Ndidi CU, Aimola IA, Bassa OY, Mankilik M, Adamu Z (2014) Effects of processing (boiling and roasting) on the nutritional and antinutritional properties of Bambara groundnuts (*Vigna subterranean* [L.] Verdc.) from Southern Kaduna, Nigeria. J Food Process 2014:1–9. https://doi.org/10.1155/20172129

Nedumaran S, Abinaya P, Jyosthnaa P, Shraavya B, Parthasarathy R, Bantilan C (2015) Grain Legumes Production, Consumption and Trade Trends in Developing Countries. Working Paper Series No 60. International Crops Research Institute for the Semi-Arid Tropics (ICRISAT). Telangana: International Crops Research Institute for the Semi-Arid Tropics, pp 1–57. https://doi.org/10.7910/DVN/V61SNB

Niveditha VR, Sridhar KR (2012) Antioxidant activity of raw, cooked and Rhizopus oligosporus fermented beans of Canavalia of coastal sand dunes of Southwest India. J Food Sci Technol 51(11):3253–3260. https://doi.org/10.1007/s13197-012-0830-9

Olaleke AM, Olorunfemi O, Emmanuel AT (2006) A comparative study on the chemical and amino acid composition of some Nigerian under-utilized legume flours. Pak J Nutr 5(1):34–38. ISSN 1680-5194

Polak R, Phillips EM, Campbell A (2015) Legumes: health benefts and culinary approaches to increase intake. Clin Diabetes 33(4):198–205. https://doi.org/10.2337/diaclin.33.4.198

Polhill RM, Raven PH (1981) Advances in legume systematics. Royal Botanic Gardens, Kew

Prakash D, Upadhyay G, Singh BN, Singh HB (2006) Antioxidant and free radical-scavenging activities of seeds and agri-wastes of some varieties of soybean (*Glycine max*). Food Chem 104:783–790. https://doi.org/10.1016/j.foodchem.2006.12.029

Racete SB, Spearie CA, Phillips KM, Lin X, Ma L, Ostlund RE (2009) Phytosterol-defcient and high-phytosterol diets developed for controlled feeding studies. J Am Diet Assoc 109(12):2043–2051. https://doi.org/10.1016/j.jada.2009.09.009

Rebello CJ, Greenway FL, Finley JW (2014) A review of the nutritional value of legumes and their effects on obesity and its related co-morbidities. Obes Rev 15(5):392–407. https://doi.org/10.1111/obr.12144

Sanchez-Chino X, Jomenez-Martinez C, Davila-Ortiz G, Alvarez-Gonzalez I, MadrigalBujaidar E (2015) Nutrient and non-nutrient components of legumes and its chemopreventive activity: a review. Nutr Cancer 67(3):401–410. https://doi.org/10.1080/01635581.2015.100472

Satya S, Kaushik G, Naik SN (2010) Processing of food legumes: a boon to human nutrition. Mediterr J Nutr Metab 3:183–195

Saxena KB, Kumar RV, SultanaR (2010) Quality nutrition through pigeonpea—a review. Health 2(11):1335–1344

Sefatie RS, Fatoumata T, Eric K, Shi YH, Guo-wei L (2013) In vitro antioxidant activities of protein hydrolysate from germinated black soybean (*Glycine max* L.). Adv J Food Sci Technol 5:453–459

Shahwar D, Mohsin T, Bhat MYK, Chaudhary S, Aslam R (2017) Health functional compounds of lentil (*Lens culinaris* Medik): a review. Int J Food Prop 20:1–15. https://doi.org/10.1080/10942912.2017.1287192

Shashank A, Tidke D, Ramakrishna S, Kiran G, Kosturkova GA (2015) Ravishankar. Nutraceutical potential of soybean: review. Asian J Clin Nutr 7:22–32. https://doi.org/10.3923/ajcn.2015.22.32

Shweta KM, Rana A (2017) Bioactive components of Vigna species: current prospective. Bull Environ Pharmacol Life Sci 6:1–13

Tamang JP, Shin DH, Jung SJ, Chae SW (2016) Functional properties of microorganisms in fermented foods. Front Microbiol 7:578. https://doi.org/10.3389/fmicb.2016.00578

Tiwari AK, Abhinay B, Babu KS, Kumar DA, Zehra A, Madhusudana K (2013) Pigeon pea seed husks as potent natural resource of anti-oxidant and anti-hyperglycaemic activity. Int J Green Pharm 9(4–5):36–49. https://doi.org/10.4103/0973-8258.120247

Voisin AS, Guéguen J, Huyghe C, Jeuffroy MH, Magrini MB, Meynard JM et al (2014) Legumes for feed, food, biomaterials and bioenergy in Europe: a review. Agron Sustain Dev 34:361–380

Wong JH, Ng TB (2005) Sesquin, a potent defensin-like antimicrobial peptide from ground beans with inhibitory activities toward tumor cells and HIV-1 reverse transcriptase. Peptide 26:1120–1126. https://doi.org/10.1016/j.peptides.2005.01.003

Xu BJ, Chang SKC (2007) A comparative study on phenolic profles and antioxidant activities of legumes as affected by extraction solvents. J Food Sci 72:159–166. https://doi.org/10.1111/j.1750-3841.2006.00260.x

Yao DN, Kouassi KN, Erba D, Scazzina F, Pellegrini N, Casiraghi MC (2015) Nutritive evaluation of the Bambara groundnut Ci12 landrace [*Vigna subterranea* (L.) Verdc. (Fabaceae)] Produced in Côte d'Ivoire. Int J Mol Sci 16:21428–21441. https://doi.org/10.3390/ijms160921428

Yorgancilar M, Bilgicli N (2014) Chemical and nutritional changes in biter and sweet lupin seeds (*Lupinus albus* L.) during bulgur production. J Food Sci Technol 51(7):1384–1389. https://doi.org/10.1007/s13197-012-0640-0

Chapter 4
Antioxidant Profile of Legume Seeds

Balwinder Singh, Jatinder Pal Singh, Amarbir Kaur, Amritpal Kaur, and Narpinder Singh

Abstract Legume seeds contain antioxidant compounds that prevent or slow down the oxidation process generally by donating electrons to free radicals. These compounds are primarily phenolic acids, flavonoids, saponins, tocopherols and vitamin C. Antioxidant potential of legume seeds is directly related with their chemical structure as well as the position of functional groups in various compounds. Most reports have suggested that legumes with colored seed coats possess strong antioxidant potential owing to the presence of high content of antioxidants.

Here we review information on the antioxidant components, impact of processing ways on antioxidants as well as their activities in legume seeds. Most researchers have primarily determined the in vitro antioxidant activity of legumes using 2,2-diphenyl-1-picrylhydrazyl radical scavenging activity, oxygen radical absorbance capacity and ferric reducing antioxidant power assays. Processing such as dehusking as well as dehulling, soaking, cooking, germination and fermentation changes the level and nature of antioxidants and their activities in grain legume seeds. The antioxidants present in legumes have been shown to reduce ageing as well as incidence of diseases including cancer, diabetes and cardiac problems. These antioxidants and their activities make them suitable dietary and functional food components for the prevention and management of human diseases.

Keywords Legume seeds · Antioxidants · Phenolics · Flavonoids · Tocopherols · Ascorbic acid · Saponins · Processing · Health benefits · Human diseases

---

B. Singh
Department of Biotechnology, Khalsa College, Amritsar, Punjab, India

J. P. Singh · A. Kaur · A. Kaur (✉) · N. Singh
Department of Food Science and Technology, Guru Nanak Dev University, Amritsar, Punjab, India

P. Guleria et al. (eds.), *Sustainable Agriculture Reviews 45*, Sustainable Agriculture Reviews 45, https://doi.org/10.1007/978-3-030-53017-4_4

4.1 Introduction

Legumes are the annual crops that yield 1–12 grains or seeds per pod. They are globally appreciated and grown worldwide under a wide range of agro-climates and regions (Singh et al. 2017a, b, c). Edible legumes include crops harvested for dry grain known as pulses that are consumed as whole, split and are used in cereal based different food products; crops used for extraction of oils such as peanuts and soybeans and crops harvested immature such as green beans and peas for food (Singh et al. 2017b, c). The appearance of different legumes is shown in Fig. 4.1. Legumes are considered as a significant source of nutrients like proteins, starch, dietary fiber and minerals (Magalhães et al. 2017). They are the preferred food sources in the Indian diet due to their economic value and benefitting nutritional uses (Singh et al. 2017a).

Legume seeds are used in different ways to prepare staple diets and are rich in antioxidant compounds having important physiological and metabolic functions (Singh et al. 2017b, c). They contain phenolics, flavonoids, tannins and saponins as bioactive constituents. These compounds are perceived to be beneficial for human health due to their antioxidant potential (Singh et al. 2017a, b, c). These compounds have bioactive potential and are very important from nutrition and food science point of view in commonly consumed legume seeds. Antioxidant potential of common food legume seeds has been reported by many researchers (Amarowicz and Shahidi 2017; Aguilera et al. 2011; Attree et al. 2015; Alshikh et al. 2015; Segev

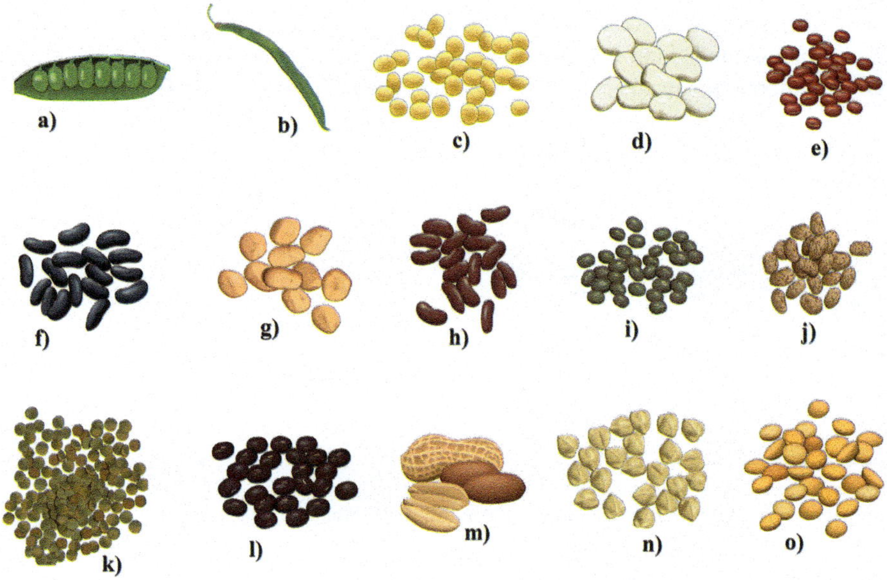

Fig. 4.1 Appearance of green pea (**a**), green bean (**b**), soybean (**c**), lima bean (**d**), adzuki bean (**e**), black bean (**f**), broad bean (**g**), kidney bean (**h**), pinto bean (**j**), lentil (**k**), black gram (**l**), peanut (**m**), chickpea (**n**) and split bean (**o**)

et al. 2010; Wang et al. 2016; Xu and Chang 2008; Zhao et al. 2014). The crux of these studies showed that legume seeds are useful as a great source of dietary antioxidant compounds. The varietal difference of legumes affects their antioxidant capacity and this varies broadly among different legume seed types (Singh et al. 2017b). Seed germination and technical processing influence natural endogenous antioxidants quantitatively and qualitatively in various leguminous seeds.

A considerable change in level of phenolics of leguminous seeds was noticed after processing (Singh et al. 2017b). The hull of legumes has majority of phenolics and their content and composition changes during various processing methods (López-Martínez et al. 2017). Germination and fermentation enhances the phenolic quantity, composition as well as antioxidant properties of some legumes and legume-based products commonly used for human consumption (Bartolomé et al. 1997). Germinated and sprouted legume seeds are recognized as an interesting source of natural antioxidants (Díaz-Batalla et al. 2006). Fermentation converts conjugate phenolic compounds into free forms in legumes and this bioconversion improves their health-link functionality (Torino et al. 2013). Cooking and boiling reduces the levels of antioxidant components present in seed coats of dry legumes and this might be due to their heat destruction and chemical rearrangements like binding or complex formation (Xu and Chang 2008; Volf et al. 2014). Edible legume seeds have been recommended in daily diet by many research scientists and health organizations for the reduction or management of degenerative ailments (Singh et al. 2017a, b, c).

Antioxidant components of legume seeds are quite important in making them important or essential ingredient in preparation of various functional foods (Singh et al. 2017b). In this chapter, information available in literature about antioxidant components of edible legume seeds and effect legume processing on these components are discussed.

4.2 Antioxidant Components of Legume Seeds

Legume seeds contain many antioxidant compounds that prevent or slow down the oxidation process generally by donating electrons to free radicals. The different types of antioxidants present in legume seeds and their general mechanism of reaction with free radicals is shown in Fig. 4.2. Comprehensive information on these compounds is listed below:

4.2.1 Phenolic Compounds

Phenolic compounds are aromatic compounds that contain one or more than one hydroxyl groups and are primarily made using shikimic acid and pentose phosphate pathways through phenylpropanoid metabolism (Balasundram et al. 2006). These

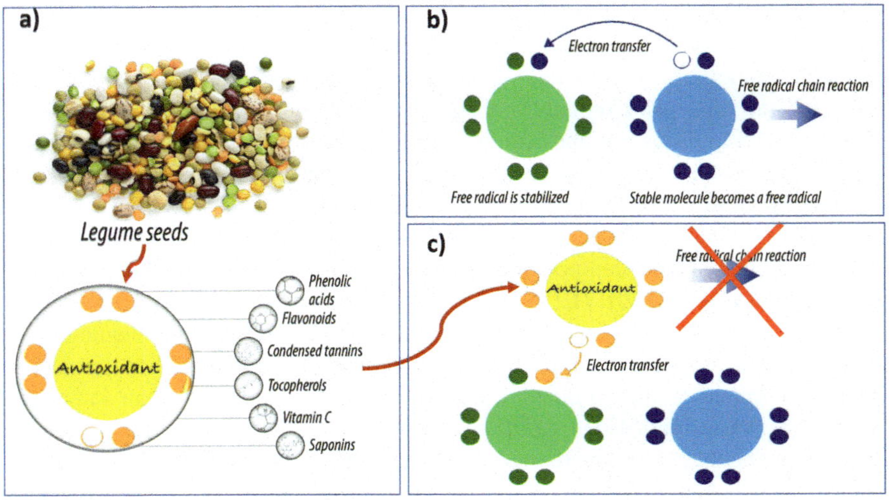

Fig. 4.2 Legume seeds contain many different types of antioxidants (**a**); mechanism of reaction of free radicals with stable molecules leading to free radical chain reaction (**b**); antioxidants react with free radicals to convert them into less innocuous forms mostly by donating electrons (**c**)

are structurally diverse with their chemical structures range between free phenolic acids and largely complex compounds. Phenolic acids, flavonoids and condensed tannins are the main phenolics in legumes as shown in Fig. 4.3 (Amarowicz and Pegg 2008; Gan et al. 2017; Lin et al. 2008; Singh et al. 2017b). Free-radical removing activities contribute to antioxidant potential of legumes. Phenolic compounds have a considerable role in physiological and metabolic functions (Singh et al. 2017a, b).

4.2.1.1 Phenolic Acids

Phenolic acids include derived products of hydroxybenzoic and hydroxycinnamic acids. Gallic, vanillic, syringic, gentisic, salicylic p-hydroxybenzoic, dihydroxybenzoic, 2,3,4-trihydroxybenzoic, protocatechuic are the main hydroxybenzoic acids reported with their concentrations and composition varying in different grain legume seeds (Attree et al. 2015; López-Amorós et al. 2006; Singh et al. 2017b; Wang et al. 2016; Xu et al. 2007). Caffeic, sinapic, ferulic, p-coumaric and chlorogenic acids are the hydroxycinnamic acids in seeds of legumes and they vary in content and composition among legume seeds types and varieties (Aguilera et al. 2010, 2011; López-Amorós et al. 2006; Singh et al. 2017b; Xu et al. 2007; Xu and Chang 2010). The phenolic components present in hull and defatted flour of ten legumes, yellow lentil, mung bean, small fava bean, smooth field pea, pigeon pea, baby lima bean, navy bean, cow pea, chickpea and white lupin were classified into

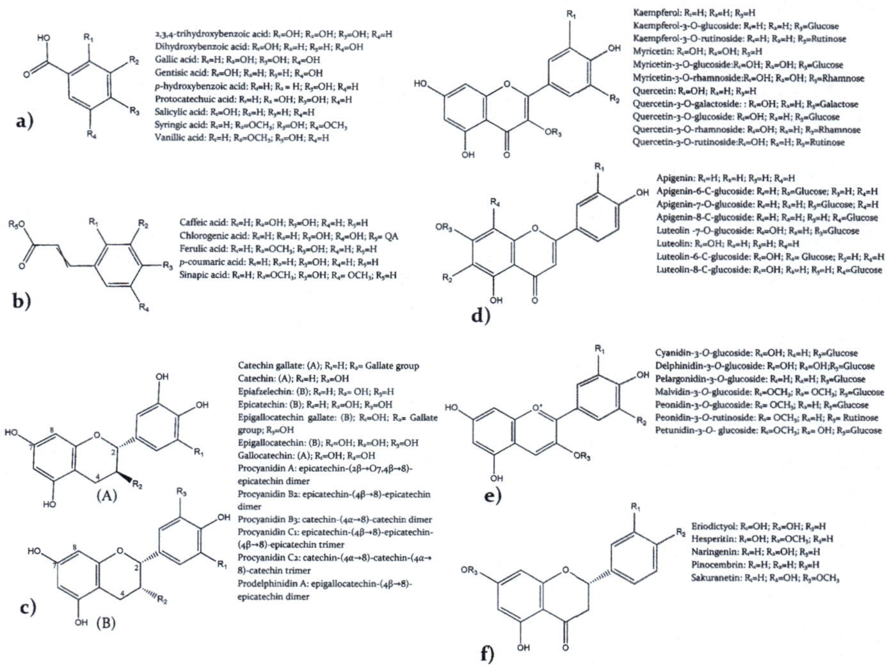

Fig. 4.3 Hydroxybenzoic acids (**a**); hydroxycinnamic acids (**b**); catechins and procyanidins (**c**); flavonols and flavones (**d**); anthocyanins (**e**); and flavanones (**f**) identified in legume seeds

soluble esters, free acids and residual constituents (Sosulski and Dabrowski 1984). The flours have soluble esters and their hydrolysis showed the existence of trans-p-coumaric, trans-ferulic, and syringic acid in almost each of the examined species. In mung bean, lentils, fava bean, field bean and pigeon pea, phenolic acids are present in lowest amounts as 2–3 mg per 100 g of material. Navy bean, cowpea, lupin and lima bean have the highest content of phenolic acids. Seed coat contains syringic, *p*-hydroxybenzoic, ferulic, protocatechuic, gallic, *trans*-ferulic and *trans*-*p*-coumaric acid in the soluble ester fraction. Gallic, protocatechuic, ferulic and p-coumaric are the common phenolic acids present in beans (Lin et al. 2008; Wang et al. 2016). Caffeic, gallic, ferulic, sinapic, p-coumaric and protocatechuic are the main phenolic acids identified in lentils (Alshikh et al. 2015). Polyphenols are mainly concentrated in the seed coat of legumes, although seed coat represents a very small portion, 10% approximately, of the total legume seed weight (Singh et al. 2017a, b). A study of bean coats conducted by Madhujith et al. (2004) depicted *p*-coumaric, vanillic, sinapic and caffeic acids as the significant phenolic acids. Legumes with dark colored seeds contain higher total phenolic content than the light-colored seeds (Segev et al. 2010; Chen et al. 2015a, b; Singh et al. 2017a, b).

4.2.1.2 Flavonoids

Flavonoids are the biggest category of phenolics found in leguminous seeds. The total flavonoids present in green pea, lentil, chickpea, yellow pea, black bean and red kidney bean, vary between 0.08 and 3.21 mg catechin equivalents per g and the maximum were detected in black and red kidney beans (Xu and Chang 2007). Flavonoids of legumes are contributed by catechins, procyanidins, anthocyanins, flavonols, flavones and flavanones (Attree et al. 2015; Amarowicz and Pegg 2008; Dueñas et al. 2004; Singh et al. 2017b). Catechins, procyanidins and anthocyanins are the flavonoids present in legumes having colored seed coats (Amarowicz and Pegg 2008; Chen et al. 2015a, b; Han et al. 2015; Segev et al. 2010; Singh et al. 2017a, b). Catechin, catechin gallate, gallocatechin, epicatechin, epigallocatechin, epigallocatechin gallate, epiafzelechin, prodelphinidin A, procyanidin A, procyanidin B2, procyanidin B3, procyanidin C1 and procyanidin C1 are the catechins as well as procyanidins identified in leguminous seeds (Aguilera et al. 2010, 2011; Alshikh et al. 2015; Amarowicz et al. 2008; Chen et al. 2015a; Singh et al. 2017b; Xu and Chang 2010). Adzuki beans contain procyanidin dimers and trimers and their content varied from 15.9 to 213 mg per g (Amarowicz et al. 2008). Red, black and brown bean hull extract contained procyanidin B2, C1, and C2 (Madhujith et al. 2004).

Beans, lentils and peas with colored seed coats contained three to sixfolds higher anthocyanin content than whole seeds (Xu et al. 2007; Singh et al. 2017b). The anthocyanins identified in grain legume seeds are glucosides of cyanidin, delphinidin, malvidin, petunidin, peonidin and pelargonidin and they are responsible for the pigmentation of seed coats (Attree et al. 2015; Kan et al. 2017; Singh et al. 2017b; Han et al. 2015). The flavonol glycosides identified in seeds of legumes includes quercetin-3-O-glucoside, quercetin-3-O-galactoside, quercetin-3-O-rutinoside, quercetin-3-O-rhamnoside, myricetin-3-O-glucoside, myricetin-3-O-rhamnoside, kaempferol-3-O-glucoside and kaempferol-3-O-rutinoside (Dueñas et al. 2016; Gan et al. 2016; Magalhães et al. 2017; Singh et al. 2017b). Pinocembrin, eriodictyol, naringenin, sakuranetin and hesperitin are the flavanones identified in legume seeds (Aguilera et al. 2011; García-Lafuente et al. 2014; Singh et al. 2017b). The main flavone glycosides present are apigenin-6-C-glucoside, apigenin-8-C-glucoside, apigenin-7-O-glucoside, luteolin-7-O-glucoside, luteolin-6-C-glucoside and luteolin-8-C-glucoside (Dueñas et al. 2016; Gan et al. 2016; Magalhães et al. 2017; Singh et al. 2017b). The total flavonoid content in common beans, soybean, lentils, yellow pea and green pea ranges from 0.92 to 4.24, 1.06 to 4.04, 3.04 to 4.54, 0.09 to 0.17 and 0.05 to 0.15 mg catechin equivalents per g, respectively (Xu et al. 2007). Dark such as black and red colored chickpeas contained ten to eleven folds more total flavonoid content than light (beige as well as regular cream) colored chickpeas (Segev et al. 2010).

4.2.1.3 Tannins

Tannins are oligomeric or polymeric flavanols primarily present in legumes with colored hulls including black beans, red beans and lentils (Alshikh et al. 2015; Amarowicz and Pegg 2008; Lin et al. 2008). These can be categorized as hydrolysable and condensed (non-hydrolysable) tannins. Catechin, epicatechin, gallocatechin and epigallocatechin are smaller constituents of condensed tannins identified in legume seeds (Jin et al. 2012; Mirali et al. 2014). Tannins are the polyphenolic compounds primarily identified in the testa of legumes and they vary in subunit composition among different legumes varieties. Condensed tannin content in beans, soybean, lentils, and peas ranged from 0.47 to 5.73, 1.06 to 4.04, 3.73 to 10.20 and 0.22 to 0.61 mg catechin equivalents per g, respectively (Xu et al. 2007). Legume seeds are suitable for newer uses as health promoting food ingredients owing to their bioactive phenolic constituents and beneficial effect on human health (Singh et al. 2017a, b). Phenolic compounds reported in leguminous seeds have strong antioxidant potential along with their ability to interact with proteins (Amarowicz and Shahidi 2017; Amarowicz and Pegg 2008; Singh et al. 2017a, b).

4.2.2 Tocopherols and Vitamin C

Tocopherols are the natural antioxidant compounds that protect oxidizable lipids by scavenging peroxyl radical in cell membranes and lipoprotein particles. They are the major biological antioxidants that scavenge free radical species and terminate lipid per-oxidation. Four tocopherols, α, β, γ & δ, isomers were identified and no tocotrienols were reported in leguminous seeds as shown in Fig. 4.4a (Amarowicz and Pegg 2008). The levels of α, β, γ & δ-tocopherols vary quite markedly among legume seeds (Grela and Günter 1995; Kalogeropoulos et al. 2010). α-Tocopherol is less abundant and both β and γ-tocopherols predominate in legume seeds. The largest level of γ-tocopherol was present in soybean, lentils and peas (Amarowicz and Pegg 2008; Grela and Günter 1995). δ-Tocopherol predominates in black-eyed and pinto beans (Kalogeropoulos et al. 2010). The level of γ-tocopherol in kidney bean, horse bean, common bean, lentil and field pea was 54.5, 50.3, 34.1, 66.5 and 87.2 mg per kg, respectively (Grela and Günter 1995). The level of α, γ and δ-tocopherols in soybean were reported to be highest and the values were 65.5, 237 and 62.4 mg per kg, respectively (Grela and Günter 1995). The levels of α, β, γ & δ-tocopherols in raw seeds of different lentil cultivars were reported in the range of 3.84–8.69, 1.94–3.81, 91.11–104.68 and 2.01–2.74 mg per g dry weight, respectively (Fernandez-Orozco et al. 2003). The γ and δ-tocopherol contents in raw cowpeas seeds were reported as 0.43 and 1.83 mg per 100 g whereas as in flour samples their contents were 0.31 and 1.35 mg per 100 g, respectively (Doblado et al. 2005). In pigeon pea seeds, α, β, γ & δ-tocopherol levels were reported as 1.06, 0.06, 9.31 and 0.27 mg per 100 g dry weight, respectively (Torres et al. 2006). Kalogeropoulos et al. (2010) reported tocopherol content of 0.82, 0.55 and 0.47 mg per 100 g fresh

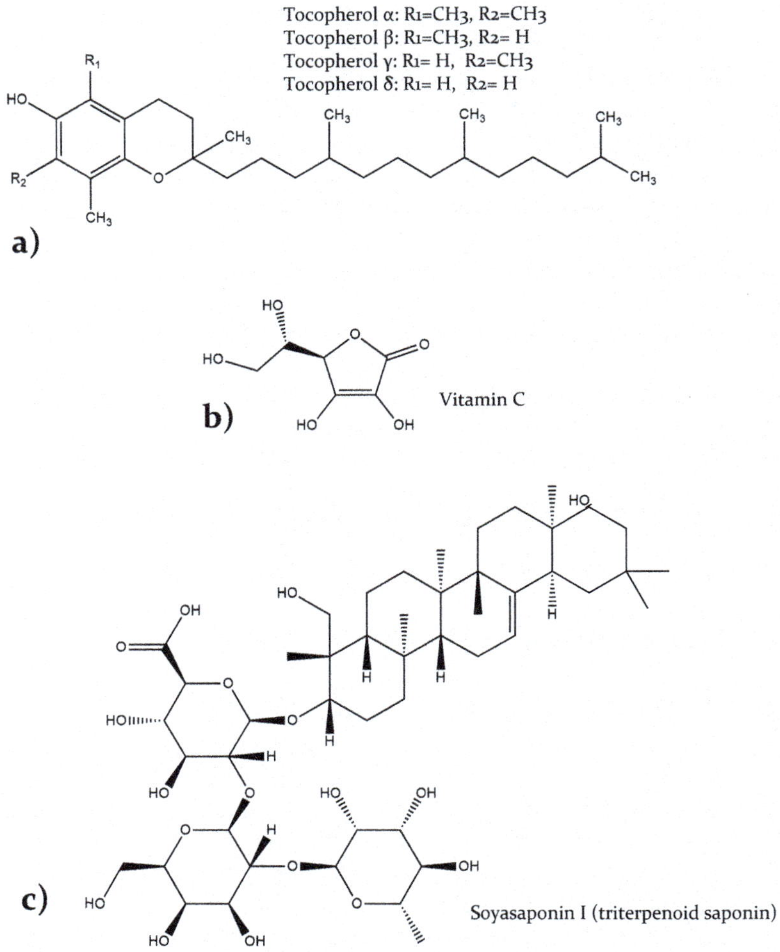

Fig. 4.4 General chemical structures of tocopherols (**a**); vitamin C (**b**); and saponins (**c**)

weight in small lentils, large lentils and white lupins, respectively. Raw lupins presented α, *γ* and *δ*-tocopherol content of 0.19, 20.1 and 0.25 mg per 100 g dry weight, respectively (Frias et al. 2005). γ-Tocopherol predominates in most of legumes and have role in detoxifying nitrogen dioxide and other nitrogen species (Amarowicz and Pegg 2008). Antioxidant potential of tocopherols is because of having direct radical scavenging effects (Kagan et al. 1990). They have a stabilizing function in cell membranes and are known as membrane antioxidants.

Vitamin C, also referred to as ascorbic acid, is a water dissolving compound which provides protection against oxidative damage (Fig. 4.4b). Vitamin C is commonly present as ascorbic acid and dehydroascorbic acid in plant foods (Favell 1998). Dehydrated legume seeds are poor source of vitamin C, having contents less than 1.0 mg per 100 g, and immature, soaked or sprouted legume seeds contains appreciable level of 8–40 mg per 100 g fresh weight of vitamin C (Moriyama and Oba 2008). Vitamin C content reported in mung beans, azuki beans, black gram and peanuts was 6.50, 2.61, 0.45 and 0.13 mg per 100 g dry weight, respectively (Moriyama and Oba 2008). In kidney bean, broad bean, green peas and soybean varieties, vitamin C content ranged from 0.10 to 0.84, 0.21 to 2.01, 0.32 to 7.40 & 0.72 to 3.50 mg per 100 g dry weight, respectively (Moriyama and Oba 2008). The amount of ascorbic acid varied among pea varieties from 0.40 to 1.48 μmol per g and constituted a major part of water-soluble antioxidant capacity of peas (Nilsson et al. 2004). In green pea, vitamin C was mainly present as dehydroascorbic acid.

4.2.3 Saponins

Saponins are non-volatile glycosylated compounds with surface active properties due to non-polar triterpene or steroid aglycone or sapogenin moiety linked with polar sugar molecules in their structures. Saponins have linkages between sapogenin and saccharide chain that is responsible for surface active properties and different types of biological activities (Fig. 4.4c). Saponins identified in legume seeds are mostly triterpene glycosides and these are recognised as monodesmosides, that contain one saccharide chain attached to sapogenin, and bidesmosides, that two saccharide chains separately attached to sapogenin (Lásztity et al. 1998; Singh et al. 2017c). Saponins have been reported in the seeds of edible legumes including beans, black grams, lentils, chickpeas, lupins, pigeon peas and peas (Singh et al. 2017c). Saponin levels in chick peas, kidney beans, navy beans and haricot beans were reported as 5.6, 1.6, 2.1 and 1.9 g per 100 g dry weight, respectively (Fenwick and Oakenfull 1983). Studies have reported anti-oxidant, anti-inflammatory, anti-cancer, hypocholesterolemic and immunomodulatory properties of saponins (Singh et al. 2017c). Saponin exhibited marginal radical scavenging activities compared to other antioxidant components such as phenolics and flavonoids present in legume seeds (Lee et al. 2011).

4.3 Impact of Processing on the Levels of Antioxidant Compounds

Legume seeds are usually not eaten raw but instead they are processed and cooked to get a desirable taste, flavor, texture, color and nutritional profile. Food legume seeds are subjected to different processing operations such as dehulling, soaking, milling, sometimes germination, fermentation and finally cooking. These operations result in edible products having more nutritional value and small amounts of antinutritional factors, growth inhibitors and haemagglutinins. Legume seed processing by different methods can impact the levels of antioxidant components *e.g.*

phenolic compounds, ascorbic acid, tocopherols and saponins. This might be due to leaching, heat degradation or formation of insoluble complex with other components (Singh et al. 2017b). Processing reduces the antinutritional components and affect the composition of antioxidants in common food legume seeds. Different legume processing methods that effects content and composition of antioxidants are:

4.3.1 Dehulling

The seed coats or hulls of some edible legume seeds are difficult to digest and are bitter in taste (Wang et al. 2009). Many grain legumes are consumed after dehusking or dehulling meaning removal of seed coats. A common practice in India for mung bean, urd bean, lentil, chickpea and pigeon pea seeds is that they are consumed in the form of dhal (Singh and Singh 1992). Dehulling reduces anti-nutritional constituents, cooking time, and improves taste, palatability, digestibility and bioavailability of nutrients in legumes (Wang et al. 2009; Deshpande et al. 1982). Dehulling results in considerable loss of legume bioactive constituents. Dehulling significantly reduces the level of phenolics mainly present in the hull of the food legumes (Wang et al. 2009; Deshpande et al. 1982). Legume seeds with high proportions of hulls have shown significant loss of phenolics after dehulling process. The level of tannins in ten cultivars of dry beans ranged from 33.7 to 282.8 mg catechin equivalents per 100 g, whereas in dehulled beans level ranged from 10.0 to 28.7 mg catechin equivalents per 100 g, respectively. Dehulling decreased tannin content by 68–95% in dry beans (Deshpande et al. 1982). The legumes seed hull contains phenolic acids such as protocatechuic, p-trans-ferulic, hydroxybenzoic, trans-p-coumaric and syringic and gallic acids in the soluble ester fraction (Sosulski and Dabrowski 1984). The dehulling and dehusking process had considerably lowered the phenolic acid content of flour from fava bean, pigeon pea, lentil, and mung bean but had slight effect on level of phenolics in chickpea, field pea and navy bean flour (Sosulski and Dabrowski 1984). Tannins are present primarily in the hulls of lentils and dehulling significantly decreases tannin content. The tannin content reported in raw seeds of lentils is 4.7 g per kg dry weight and tannin content was reduced to 0.13 g per kg dry weight in dehulled lentil seeds (Wang et al. 2009).

4.3.2 Soaking and Germination

Dry legume seeds are commonly soaked in hydrate before cooking. Soaking is a general strategy to make texture soft and reduce cooking time of dry grain legume seeds (Xu and Chang 2008). Soaking significantly decreased the total phenolic content in lentils, yellow pea, green pea and chickpea. After soaking, about 9–38% of total phenolic content in lentils and 2–12% of total phenolic content in peas and chickpeas were lost (Xu and Chang 2008). In lentils, loss of total phenolic content increased with increase in the hydration rate. Soaking removed procyanidins from lentils seeds (López-Amorós et al. 2006). Loss of phenolic components depends on the distribution and quantity of phenolics in cotyledon and hull of dry legume seeds. Soaking decreases the quantity of phenolic acids in beans, lentils and peas. Level of hydroxybenzoic acids except protocatechuic acid decreased in the soaked seeds of beans, peas and lentils compared to dry seeds (López-Amorós et al. 2006). Soaking is followed by germination in many common dry legume seeds. Soaking and germination process modifies the antioxidant compounds of legumes. Total phenolic content of dark colored legumes decreased significantly because of the removal of soluble molecules during soaking and germination (Lin and Lai 2006).

Germination is a successful process for the improvement of legume characteristics and sprouted legumes are preferred across the globe. Germination period, light and conditions play an important role in the metabolic changes of legumes (López-Amorós et al. 2006). Germination enhances the nutritional profile of pulses and legumes by increasing the bioactive components and reducing the anti-nutritional factors. Germination significantly changes the bioactive constituents and increases functionality of legumes (López-Amorós et al. 2006). Germination changes the composition of phenolic compounds in lentils. Activation of endogenous enzymes during germination brings structural changes in procyanidin-type compounds of lentils (Bartolomé et al. 1997). Germination process modifies phenolics of lentils, beans and peas quantitatively and qualitatively depending upon the legume type, temperature, time period and germination conditions. Lentils showed a considerable lowering in the antioxidant activity after germination, whereas beans and peas showed an increase in antioxidant activity (López-Amorós et al. 2006). Germination increases soluble phenolic components in case of germinated seeds compared to raw or cooked seeds. Tannin content decreased, while level of phenolics and flavonoids significantly increased after germination of lentils. The p-hydroxybenzoic and p-vanillic aldehydes were detected after germination in case of lentils, pea and beans. These compounds originated due to degradation of lignin present in seeds by enzymatic oxidation. The level of trans-ferulic and trans-p-coumaric acid was increased by germination of lentils in the presence of light.

Germination decreased p-hydroxybenzoic and increased p-coumaric and vanillic acid content in beans and had no effect on the ferulic acid content (Díaz-Batalla et al. 2006). Germination of beans increased the contents of isoflavone and quercetin derivatives as compared to raw or boiled seeds. The existence of flavonol glycosides such as kaempferol-3-rutinoside, kaempferol-3-glucoside, quercetin-3-rutinoside and quercetin-3-ramnoside in high concentration were reported in germinated beans after a week of germination in the presence of light, whereas they were not elucidated in dry beans (López-Amorós et al. 2006). Germinated beans are good sources of isoflavones such as genistein, daidzein, biochanin A and biochanin B with total isoflavone content of 29.92 μg per g compared to 1.27 μg per g reported in only one genistein derivative of dry beans (López et al. 2013).

Germination of dry beans in the presence of light accelerated the process of formation of flavonol (quercetin and kaempferol) glycosides. In another study, quercetin content was reported higher and kaempferol content was reported lower in germinated bean seeds than the non-germinated ones (Díaz-Batalla et al. 2006). The isoflavonoids such as daizein, 8.2–129.1 μg per g, coumesterol, 2.4–35.6 μg per g, and genistein, 2.6–9.7 μg per g, were observed in germinated bean seeds and they were not observed in dry beans (Díaz-Batalla et al. 2006). Soaking and germination can modify the tocopherol content in legumes along with polyphenolic compounds. Germination in dark, at 25 °C, for 4 days resulted in an increase in the level of α-tocopherol and decrease in the levels of β-, γ- & δ-tocopherols in lentils (Fernandez-Orozco et al. 2003). Germination increases α-tocopherol, decreases γ-tocopherol and does not change the level of δ-tocopherol in lupins (Frias et al. 2005).

Fernandez-Orozco et al. (2003) studied ascorbic acid content in raw, cooked and germinated lentil seeds and detected ascorbic acid, 0.435–0.715 mg per g dry weight, only in germinated lentil seeds. Ascorbic acid is biosynthesized during germination of lentil seeds. The synthesis of ascorbic acid enhanced with an enhancement in the germination period up to 120 h at optimum room temperature in chickpea (Sattar et al. 1991). The increase in total ascorbate, that included ascorbic acid and dehydroascorbic acid, pool was noticed in embryo axes and cotyledons after 1 day of germination, and ascorbic acid content increased by 1.5 and 2.9 times in cotyledons and embryo axes of germinated chickpea, respectively, as compared to dry chickpea seeds (Wojtyla et al. 2006). Vitamin C content increased to 4.5 fold in mung beans, 29.21 mg per 100 g, and threefold in black gram, 1.41 mg per 100 g, by soaking overnight, 8–16 h, in cold water at 20 °C in the dark (Moriyama and Oba 2008). In spite of this, in case of green peas significant loss of vitamin C was reported during the same soaking process. Legume seeds are often soaked and sprouted prior to cooking and soaking significantly diminishes the saponin content (Jood et al. 1986; Singh et al. 2017c).

4.3.3 Boiling, Steaming and Cooking

Legumes are commonly cooked using boiling or steaming process preceding consumption. Cooking changes the appearance, texture, firmness, flavor and chemical compositions of dry legume seeds. Boiling, steaming and pressure cooking mostly reduces phenolic compounds in legume by breakdown and leaching in to cooking water. Loss of phenolic compounds under thermal and pressure conditions is due to decomposition, chemical transformation and formation of phenolic-protein complexes (Xu and Chang 2008). Hydrophilic phenolic compounds get removed from the cell walls during soaking and germination and not while cooking of lentils, whereas lipophilic phenolic compounds are diffused from seed coats to cooking water and then to cotyledons during prolong cooking of lentils (Fernandez-Orozco et al. 2003).

Different cooking methods including boiling, microwave heating and steaming reduced a significant amount of total phenolic content in peas. However, the reverse happened in the green beans, where the total phenolic content increased after cooking, and this may be owing to the elevated extraction ability of phenolics after processing (Turkmen et al. 2005). Pressure cooking drastically reduces phenolic content by 90% in seed coats of common beans (Rocha-Guzmán et al. 2007). Siah et al. (2012) reported that boiling was a better cooking method than autoclaving for beans as former had less effect on active antioxidant components. Reduction in the level of kaempferol and quercetin ranging from 5% to 71% and 12% to 65%, respectively in common bean on autoclaving for 20 min was evident (Díaz-Batalla et al. 2006). Xu and Chang (2008) reported more loss of total phenolic content by pressure cooking, 68%, than regular cooking, 50–56%, in selected food legumes. Boiling as well as cooking decreased the level of phenolics by 73% in mung beans (Barroga et al. 1985). Cooking followed by discarding of cooking water resulted in loss of 30–40% of phenolic compounds in common beans (Bressani and Elias 1980). Impact of boiling and steaming on total phenolic content differed among common food legume seeds. In green pea, yellow pea and chickpea, losses of 45.9–50.8%, 43.5–46.4% and 29.2–37.5% in total phenolic content, respectively were noted by regular boiling under different process conditions (Xu and Chang 2008). This study further reported that pressure steaming process has less effect on the total phenolic content of chickpea, green pea and yellow pea and more effect on lentil as compared to regular steaming. Duenas et al. (2016) studied the influence of cooking on phenolics of dark beans. The anthocyanin content of dark beans was reduced by 68% after boiling for 60 min. Further boiling reduced the quercetin content in beans, while cooking increased the kaempferol derivatives content by 6%. Other phytochemicals were also negatively affected by boiling in case of dark beans.

Cooking significantly affects the level of tocopherols present in legumes (Kalogeropoulos et al. 2010). The level of tocopherols was reported to range 0.26 and 1.78 mg per 100 g fresh weight, with lower content in beans and larger in chickpeas (Kalogeropoulos et al. 2010). Cooking promoted losses of α- and γ-tocopherol content and total vitamin E activity in grain legume seeds. In Garbanzo beans, black beans, pinto beans, fava beans, lentils, split peas and bayo beans, a loss of 9%, 12%, 17%, 38%, 44%, 48% and 59%, respectively in total tocopherol contents was observed by cooking (Wyatt et al. 1998). Germination and cooking changes the tocopherols level in leguminous seeds (Amarowicz and Pegg 2008; Fernandez-Orozco et al. 2003). γ-Tocopherol is a predominant isomer reported in raw lentils and its level is reported to decrease by 50% in germinated and cooked lentils (Fernandez-Orozco et al. 2003). Boiled mung beans contained 1.4 fold higher amount of total vitamin C, 9.42 mg per 100 g dry weight, than dehydrated mung beans, 6.50 mg per 100 g dry weight. However in another study, boiled grains of broad beans, adzuki beans, green peas and soybean contained 70–100% lesser vitamin C contents than dehydrated ones (Moriyama and Oba 2008). In grass pea seeds, cooking decreased the level of vitamin C by 10–11% (Korus et al. 2002). Vitamin C content decreased by 14–43% in broad bean (Kmiecik et al. 1990), 30% in grass pea (Korus et al. 2002) and 65–85% in green bean by blanching (Kaack 1994). Total vitamin C content decreased by 45% in common pea after cooking of frozen samples (Vanderslice et al. 1990). Studies have reported that different processing and cooking methods decreased saponin content in grain legume seeds by liberation, degradation *e.g.* enzymatic and chemical and transformation (Singh et al. 2017c).

4.3.4 Fermentation

Fermentation improves shelf-life, healthful value and organoleptic properties of legumes. Fermentation is a great processing tool to enhance the content and composition of phenolics in legumes and to enhance the antioxidant activity (Hur et al. 2014; Limón et al. 2015). Complex polyphenolic compounds are broken into simpler and more biologically active phenolic constituents by fermentation (Dueñas et al. 2005). Solid state fermentation for 96 h enhanced the total content of non-anthocyanin phenolic compounds from 45.61 μg per g in unfermented extracts to 251.6 μg per g in water-dissolvable extracts of kidney beans (Limón et al. 2015). The levels of hydroxycinnamic compounds particularly ferulic acid in kidney beans was increased by solid state fermentation. Total phenolic content in non-fermented water-dissolvable extracts of lentils was 30 mg gallic acid equivalents per g extract. Solid state fermentation with *Bacillus subtilis* increased the total phenolic content in water-soluble extracts of lentils to 34–35 mg gallic acid equivalents per g after 96 h of fermentation (Torino et al. 2013). Fermented lentils exhibited higher contents of individual phenolic constituents (Bartolomé et al. 1997). The level of (+)-catechin and *p*-hydroxybenzoic acid were 7.31 and 1.48 mg per g in raw lentils and 17.53 and 7.39 mg per g, respectively in fermented lentils. Fermentation in lentils has improved their nutritive value and is more attractive than the raw lentils to the consumers. Significant quantities of 2,5-dihydroxybenzoic acid, 14.96 mg per g, p-hydroxyphenylpropionic acid, 3.90 mg per g, and tryptophol, 2.70 mg per g, were found in fermented lentils and these were not detected in case of raw lentils (Bartolomé et al. 1997). The concentration of free phenolics was reported to be improved by natural and induced fermentation of cowpea flour (Dueñas et al. 2005).

Fermented soybean foods predominantly contain isoflavones as aglycones. Hubert et al. (2008) reported the change of glycosylated isoflavones into aglycones by fermentation in soybean. In fermented soybeans, a decrease in the amount of isoflavone glucosides and increase in level of isoflavones were observed due to their liberation from glucosides by β-glucosidase (Murakami et al. 1984). Fermented chickpea seeds contained a higher total phenolic content in comparison to raw chickpea seeds (Fernandez-Orozco et al. 2009). β-tocopherol was initially absent, but was reported in flour after fermentation. Fermentation reduced the level of γ-tocopherol by one-third, 6.64 mg per 100 g dry weight, and level of α-, β-, and δ- tocopherols remained unaffected in pigeon pea (Torres et al. 2006). Preparation of flour sharply decreased γ and δ-tocopherol contents in cowpea. Further fermentation of cowpea flour with natural microbes and *Lactobacillus plantarum* sharply decreased γ- and δ-tocopherol contents (Doblado et al. 2005). Fermented lupin seeds contained a higher level of α-tocopherol, 32% increased, and a low level of γ- and δ-tocopherol, 6% & 44% decrease. In lupin flour after natural and inoculated fermentation, a sharp decrease, 47–96%, in all tocopherol isomers were observed and vitamin C completely disappeared (Frias et al. 2005)

4.4 Antioxidant Potential

Information regarding antioxidative and antiradical properties of phenolic compounds, tocopherols, vitamin C and saponins present in food legume seeds helps in understanding their beneficial functions. Legumes are excellent dietary sources of natural antioxidants to prevent oxidative stress mediated diseases (Singh et al. 2017a, b, c; Wang et al. 2016). The various health properties of antioxidants present in legume seeds is shown in Fig. 4.5. The synergistic effects of polyphenols and other antioxidants may contribute to overall total antioxidant activity in legume seeds. Oxidative stress is responsible for high occurrence of diseases including cancer, diabetes, obesity, neural disorders and heart problems. Many studies have correlated the antioxidant activity of foods with reduced incidence of chronic diseases in humans. Antioxidants prevent or delay oxidation of proteins, DNA and lipids by free radicals synthesized in cells during oxidation (Singh et al. 2017a). Antioxidant activity is a capacity of antioxidant compounds to eliminate free radicals. Antioxidants act as metal chelators, singlet oxygen quenchers, free radical terminators and hydrogen donors to neutralize free radicals (Hur et al. 2014). Health uses of food legumes have been due to the presence of polyphenols as antioxidant compounds. Antioxidant potential of polyphenols is dependent on their structures, such

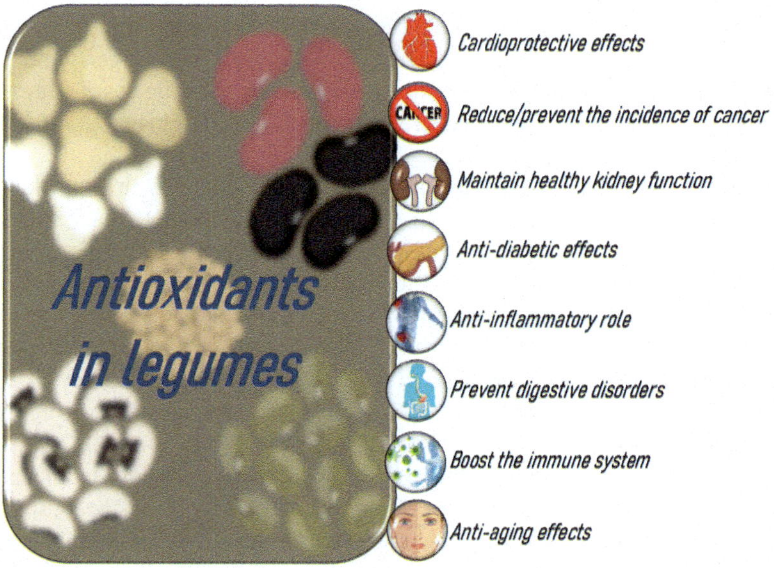

Fig. 4.5 Human health benefits of antioxidants present in legume seeds

as number as well as position of hydroxyl groups and degree of glycosylation (Aguilera et al. 2011; Singh et al. 2017b). Galloyl ester and hydroxyl group present in the structures of polyphenols are important in metal ion chelation.

Legumes are rich in antioxidant compounds with high antioxidant potential as investigated by ferric reducing antioxidant power, 2,2-diphenyl-1-picrylhydrazyl radical scavenging activity, trolox equivalent antioxidant capacity and oxygen radical absorbance capacity assays (Xu et al. 2007; Xu and Chang 2009; Wang et al. 2016). Legumes with dark colored seed coats exhibited higher 2,2-diphenyl-1-picrylhydrazyl activity, ferric reducing antioxidant power and trolox equivalent antioxidant capacity values compared to legumes with white or light colored seed coats. Flavonoids and tannins are the pigments that impart color to legume seeds and have plenty of antioxidant activities (Segev et al. 2010; Singh et al. 2017a, b). Flavonoids are present in common beans with black and red color seed coats and have good antioxidant activity (Tsuda et al. 1994). Beans with pigmented seed coats contain flavonoids and proanthocyanidins or condensed tannins and they have shown higher antioxidative properties than most of the commonly consumed fruits and vegetables (Gan et al. 2016). The seed coats of peanuts, lentils and peas are rich in flavonoids and they have shown higher antioxidant potential than the cotyledons (Attree et al. 2015; Dueñas et al. 2006). Lentils contain a considerable amount of catechins, procyanidins, flavonols and dihydroflavonols in their seed coats and have shown high antioxidant activity, oxygen radical absorbance capacity value of 66.97 μmol trolox equivalents per g dry weight (Aguilera et al. 2010). 2,2-diphenyl-1-picrylhydrazyl activity values for seed coat of six peanut varieties varied from 25.8 to 28.8 μmol trolox equivalents per g and for raw kernel and cotyledon values ranged from 10.8 to 28.9 and 1.50 to 2.72 μmol trolox equivalents per g, respectively (Attree et al. 2015). Condensed tannins are primarily concentrated in the hull of legume seeds and have higher antioxidant activity than simple and free phenolic compounds (Singh et al. 2017a, b). Flavonoids presented stronger antioxidative potential than non-flavonoid compounds due to their chemical structures. They form complexes with metal ions, reduce metal ion initiated lipid oxidation and scavenge hydroxyl and peroxyl radicals.

Catechins, procyanidins, anthocyanins and anthocyanidins are the flavonoids present in seed coat of legumes with strong antioxidative potential (Singh et al. 2017a, b). Colored Pinto beans contain catechins and procyanidins and showed more antioxidative potential than uncolored Cannellini beans (Aguilera et al. 2011). Lentils, chickpeas and kidney bean are rich source of polyphenols and have shown interesting amounts of antioxidant activity and radical scavenging capacity in different studies (Amarowicz and Pegg 2008; Zhao et al. 2014; Singh et al. 2017b). Lentils contain large content of phenolic compounds and have shown higher antioxidant potential as compared to chickpeas and peas (Singh et al. 2017a, b). Wang et al. (2016) positively correlated the phenolic contents of 14 selected beans from China with total antioxidant activities. They documented that black bean, spring bay bean, pearl bean and flower waist bean are high in phenolic compounds and have shown strong antioxidant activities in 2,2-diphenyl-1-picrylhydrazyl activity, ferric reducing antioxidant power, and trolox equivalent antioxidant capacity assays.

Alshikh et al. (2015) significantly correlated the total phenolic content with reducing power, trolox equivalent antioxidant capacity, 2,2-diphenyl-1-picrylhydrazyl activity and hydroxyl radical scavenging activity in lentils.

Tocopherols act as antioxidants by quenching singlet oxygen and transferring hydrogen atom to 6-hydroxyl group on the chroman ring. Antioxidant potential of ascorbic acid includes donation of hydrogen atom to lipid radicals, scavenging of free radicals and expulsion of molecular oxygen (Hur et al. 2014). Fernandez-Orozco et al. (2003) found that tocopherols and ascorbate, only in germinated seeds, contribute to the trolox equivalent antioxidant capacity and peroxyl radical-trapping capacity of lentils. Tocopherols, ascorbic acid and glutathione delivered about 33.8% of total antioxidant activity in germinated lentils and the values were 28 and 6 times higher in comparison with cooked and raw lentil seeds, respectively. Tocopherols, ascorbic acid and glutathione contributed 22.4% of total peroxyl radical-trapping capacity in germinated and cooked lentil seeds. The study reported participation of low molecular weight antioxidants such as tocopherols and vitamin C to the total antioxidant potential of lentils. The water-soluble antioxidant capacity of peas was mainly due to the content of ascorbic acid and antioxidant capacity varied among pea varieties (Nilsson et al. 2004). Luo et al. (2016) documented that the hulls of mung bean and adzuki bean are rich in saponins and these compounds contribute towards antioxidant properties of beans. Another study also reported antioxidant potential of saponins present in cotyledons, seed coats, and sprouts of black bean (Guajardo-Flores et al. 2012). The antioxidant potential of saponins from runner bean was measured against superoxide radicals and results of the study suggest that saponins can efficiently prevent harmful effects or damage caused by free radicals (Yoshiki et al. 1994).

Antioxidant potential of legumes seeds has a role in the prevention and management of many health related problems. Legume seeds as food are very useful source of dietary fiber and natural antioxidants for daily incorporation in the diet. The antioxidant components especially polyphenols and saponins of legume seeds have anti-diabetic, anti-carcinogenic and anti-inflammatory properties. Legume consumption contributes in attenuating oxidative stress, improving serum lipoproteins and promoting vascular health (Singh et al. 2017a, b, c).

4.4.1 Effect of Processing on Antioxidant Potential

The changes in antioxidant activity observed during processing of legumes have been attributed to formation or breakdown of antioxidant components, loss of water-soluble antioxidant compounds, oxidative reaction during processing and solid mass losses during processing (Singh et al. 2017a, b, c; Gan et al. 2017; Aguilera et al. 2010, 2011; Xu and Chang 2008). The changes in 2,2-diphenyl-1-picrylhydrazyl activity values depend on the legume type and processing conditions (Xu and Chang 2008). Soaking, germination, cooking and fermentation changes the antioxidant potential of common food legumes. Different processing operations that can affect the antioxidant potential of legume seeds are:

4.4.1.1 Soaking and Germination

Legumes containing high concentration of free phenolics and soluble antioxidant components in seed coats are prone to lose such molecules by soaking (Singh et al. 2017b). Significant changes in antioxidative potential of such legumes were observed after soaking and germination. Leaching of water soluble compounds into soaking water decreases the antioxidant potential of legumes (Aguilera et al. 2010, 2011). Soaking decreases the antioxidant activity in lentils and beans due to reduction in level of flavonols and elimination of catechins and procyanidins (Aguilera et al. 2010, 2011). Germination modifies antioxidant components of legume seeds. Effect of germination on antioxidant activity varies among different food legumes as germination process works differently for each legume type. Germination of vetch and lentil seeds for prolonged period of 168 h have decreased peroxyl radical-trapping capacity values of buffered and methanolic extracts (Zielinski 2002). López-Amorós et al. (2006) documented a significant enhancement in antioxidant potential in peas and beans after germination due to changes in their phenolic composition. Beans showed a higher antiradical efficacy after 4–6 days of germination as compared to the fresh raw seeds. Peas showed a marked increase in antiradical capacity after 4 days of germination. However in the same study, germination modified the antioxidant potential of lentils in a negative fashion. Most of the studies reported that germination enhanced the antioxidant activity as compared to raw seeds and was attributed due to the increase of some antioxidant compounds such as vitamins and phenolics during germination (Gan et al. 2017). Germination has also enhanced the antioxidant activity and vitamin C and E contents in lupins (Frias et al. 2005).

4.4.1.2 Boiling, Steaming and Cooking

Soaking, boiling and steaming considerably reduces the phenolic compounds and decreases antioxidant potential (2,2-diphenyl-1-picrylhydrazyl activity values) in legume seeds (Xu and Chang 2008). Color change of soaking water was observed in yellow pea, green pea, chickpea and lentils. 2,2-diphenyl-1-picrylhydrazyl activity values were reduced in yellow pea, 6–14%, green pea, 9–18%, chickpea, 10–35%, and lentil, 8–10%, after soaking and removal of soaking water (Xu and Chang 2008). Some soluble antioxidant constituents were released into the soaking, boiling and steaming water. 2,2-diphenyl-1-picrylhydrazyl activity values were significantly reduced in yellow pea, 50–60%, green pea, 60–70%, chickpea, 85–95% and lentil after boiling, 9–30%. Boiling process has a destructive effect on the antioxidant components of legumes. Pressure boiling decreased more free radical scavenging activity in lentils than regular boiling. After steaming, free radical scavenging activities of yellow pea, green pea, chickpea and lentils reduced by 49–67%, 51–67%, 33–83% and 14–26%, respectively (Xu and Chang 2008). Changes in the free radical scavenging capacities observed were due to effect of heat on antioxidant components and loss of soluble antioxidants. 2,2-diphenyl-1-picrylhydrazyl activity values were reduced more in lentil and less in chickpea by pressure steaming compared to regular steaming (Xu and Chang 2008).

Cooked beans contained low content of phenolic compounds but they showed 2,2-diphenyl-1-picrylhydrazyl activity values comparable to seed coat and greater than crude cotyledons (Rocha-Guzmán et al. 2007). The total antioxidant capacity of fresh peas remained unchanged with every method of cooking whereas, the total antioxidant capacity of green bean observed by 2,2-diphenyl-1-picrylhydrazyl scavenging values were elevated during the cooking process (Berger et al. 2007). The process of soaking, steaming and boiling of green pea, chickpea, yellow pea and lentils resulted in noteworthy reduction of 2,2-diphenyl-1-picrylhydrazyl scavenging activity when related with fresh untreated samples (Xu and Chang 2008). Boiling and soaking treatment have decreased oxygen radical absorbance capacity values. However, boiling under pressure and steaming elevated the oxygen radical absorbance capacity value of legumes. The reduction in phytochemicals such as flavonoids and phenolics in dark beans after boiling reduced the antioxidant activity. Despite some flavonoids were lost during cooking, the level of fat soluble antioxidants increased by cooking, in case of lentils. Heat disruption of cell wall and cell membrane releases carotenoids and tocopherols from matrix. In this way, cooking helps to preserve the health uses of lentil seeds (Zhang et al. 2014). Steaming processes caused lesser changes in the antiradical capacity than boiling processes (Xu and Chang 2008). Steaming is recommended for preservation of antioxidant components along with reduction of cooking time.

4.4.1.3 Fermentation

Fermentation is a useful way to enhance nutritive value and antioxidant potential of food products. Fermentation enhances the antioxidant activity by enhancing the release of bioactive compounds in legumes (Torino et al. 2013; Hur et al. 2014). Fermentation produces the structural breakdown of plant cell walls and synthesis or liberate different antioxidants (Hur et al. 2014). Fermented lentils have shown higher antioxidant activity than raw lentils due to their larger amounts of catechins and phenolic acids (Bartolomé et al. 1997). Solid state fermented extracts and unfermented extracts of soybean exhibited highest antioxidant activities of 508–541 and 170 μg trolox equivalents per g, respectively (Limón et al. 2015). Fermentation improved the antioxidant activity of legumes by increasing the level of phenolic compounds (Fernandez-Orozco et al. 2009). However, a study conducted by Torres et al. (2006) documented 4% deduction in the total antioxidant potential of fermented pigeon pea compared to raw seeds.

Changes in the antioxidant activity of fermented legumes depends on fermentation process, microorganism involved, method of extract preparation and legume type used (Torino et al. 2013). Soybean fermented by *Lactobacillus plantarum* have shown higher antioxidant capacity than natural fermentation (Fernandez-Orozco et al. 2009). Phenolic compounds are structurally changed after bacterial breakdown. Lactic acid bacteria depolymerize the high molecular weight phenolic compounds into simple phenolics in controlled fermentation The phenolic glucosides are converted into their aglycone form by fermentation. Fermented soybean foods predominantly contains aglycones as the isoflavone structures and have shown higher antioxidative potential than unfermented soybean (Hubert et al. 2008; Hur et al. 2014). The conversion of glycosylated isoflavones into their aglycone structures increases the total antioxidant activity of fermented foods. Fermentation, spontaneous and inoculated, slightly improves the antioxidant activity in cowpea. Inoculated and spontaneous fermentation increases antiradical activity of cowpea seeds as observed by 2,2-diphenyl-1-picrylhydrazyl scavenging method (Dueñas et al. 2005). Microbial hydrolysis increases the level of phenolics and improve antioxidant potential during fermentation (Hur et al. 2014). Interaction among antioxidants produces synergistic or antagonistic effects in fermentation process and changes antioxidant activity of water-soluble fermented lentil extracts (Torino et al. 2013).

4.5 Conclusion and Future Prospects

Legumes are important sources of bioactive constituents as natural antioxidants. In this chapter, information on the antioxidants present in food legume seeds and their activities are discussed. Legumes are the popular foods in many countries of the world because of their beneficial impact on the humans. Antioxidant components present in legume seeds includes many classes of chemical compounds such as phenolic acids, flavonoids, condensed tannins, tocopherols, saponins and ascorbic acid. Antioxidant activities of aforementioned compounds along with nutritional value of proteins and dietary fibre present in legume seeds make them suitable agents for the production of functional foods. Moreover, antioxidants are important in management of serious health issues.

Although recent studies have elaborated information about bioactive components of many food legumes, still many loopholes exist in available literature on antioxidants of legumes. Further studies should be encouraged to get in depth information about content and composition of antioxidants present in different legume types along with the effect of legume processing and cooking ways on the antioxidants and their activities. Effect of soaking, germination, cooking and fermentation on the content and composition of antioxidants and their activities should be more investigated.

Acknowledgments Authors are thankful to University Grants Commission, New Delhi for providing financial assistance.

References

Aguilera Y, Dueñas M, Estrella I, Hernández T, Benitez V, Esteban RM, Martín-Cabrejas MA (2010) Evaluation of phenolic profile and antioxidant properties of Pardina lentil as affected by industrial dehydration. J Agric Food Chem 58(18):10101–10108

Aguilera Y, Estrella I, Benitez V, Esteban RM, Martín-Cabrejas MA (2011) Bioactive phenolic compounds and functional properties of dehydrated bean flours. Food Res Int 44:774–780

Alshikh N, de Camargo AC, Shahidi F (2015) Phenolics of selected lentil cultivars: antioxidant activities and inhibition of low-density lipoprotein and DNA damage. J Funct Foods 18:1022–1038

Amarowicz R, Pegg RB (2008) Legumes as a source of natural antioxidants. Eur J Lipid Sci Technol 110:865–878

Amarowicz R, Shahidi F (2017) Antioxidant activity of broad bean seed extract and its phenolic composition. J Funct Foods 38:656–662

Amarowicz R, Estrella I, HernANdez T, Troszyńska A (2008) Antioxidant activity of extract of adzuki bean and its fractions. J Food Lipids 15(1):119–136

Attree R, Du B, Xu B (2015) Distribution of phenolic compounds in seed coat and cotyledon, and their contribution to antioxidant capacities of red and black seed coat peanuts (*Arachis hypogaea* L.). Ind Crop Prod 67:448–456

Balasundram N, Sundram K, Samman S (2006) Phenolic compounds in plants and agri-industrial by-products: antioxidant activity, occurrence, and potential uses. Food Chem 99(1):191–203

Barroga CF, Laurena AC, Mendoza EMT (1985) Polyphenols in mung bean (*Vigna radiata* (L.) Wilczek): determination and removal. J Agric Food Chem 33(5):1006–1009

Bartolomé B, Estrella I, Hernandez T (1997) Changes in phenolic compounds in lentils (Lens culinaris) during germination and fermentation. Z Lebensm Unters Forsch A 205(4):290–294

Berger M, Küchler T, Maaßen A, Busch-Stockfisch M, Steinhart H (2007) Correlations of ingredients with sensory attributes in green beans and peas under different storage conditions. Food Chem 103(3):875–884

Bressani A, Elias LG (1980) The nutritional role of polyphenols in beans. In: Hulse JH (ed) Polyphenols in cereal and legumes. International Development Research Center, Ottawa, pp 61–68

Chen PX, Tang Y, Marcone MF, Pauls PK, Zhang B, Liu R, Tsao R (2015a) Characterization of free, conjugated and bound phenolics and lipophilic antioxidants in regular-and non-darkening cranberry beans (*Phaseolus vulgaris* L.). Food Chem 185:298–308

Chen PX, Dupuis JH, Marcone MF, Pauls PK, Liu R, Liu Q, Tang Y, Zhang B, Tsao R (2015b) Physicochemical properties and in vitro digestibility of cooked regular and nondarkening cranberry beans (*Phaseolus vulgaris* L.) and their effects on bioaccessibility, phenolic composition, and antioxidant activity. J Agric Food Chem 63(48):10448–10458

Deshpande SS, Sathe SK, Salunkhe DK, Cornforth DP (1982) Effects of dehulling on phytic acid, polyphenols and enzyme inhibitors of dry beans (*Phaseolus vulgaris*). J Food Sci 47(6):1846–1850

Díaz-Batalla L, Widholm JM, Fahey GC, Castaño-Tostado E, Paredes-López O (2006) Chemical components with health implications in wild and cultivated Mexican common bean seeds (*Phaseolus vulgaris* L.). J Agric Food Chem 54(6):2045–2052

Doblado R, Zielinski H, Piskula M, Kozlowska H, Muñoz R, Frías J, Vidal-Valverde C (2005) Effect of processing on the antioxidant vitamins and antioxidant capacity of *Vigna sinensis* var. carilla. J Agric Food Chem 53(4):1215–1222

Dueñas M, Estrella I, Hernández T (2004) Occurrence of phenolic compounds in the seed coat and the cotyledon of peas (*Pisum sativum* L.). Eur Food Res Technol 219(2):116–123

Dueñas M, Fernández D, Hernández T, Estrella I, Muñoz R (2005) Bioactive phenolic compounds of cowpeas (*Vigna sinensis* L). Modifications by fermentation with natural microflora and with Lactobacillus plantarum ATCC 14917. J Sci Food Agric 85(2):297–304

Dueñas M, Hernández T, Estrella I (2006) Assessment of *in vitro* antioxidant capacity of the seed coat and the cotyledon of legumes in relation their phenolic contents. Food Chem 98:95–103

Dueñas M, Sarmento T, Aguilera Y, Benitez V, Mollá E, Esteban RM, Martín-Cabrejas MA (2016) Impact of cooking and germination on phenolic composition and dietary fibre fractions in dark beans (*Phaseolus vulgaris* L.) and lentils (Lens culinaris L.). LWT Food Sci Technol 66:72–78

Favell DJ (1998) A comparison of the vitamin C content of fresh and frozen vegetables. Food Chem 62(1):59–64

Fenwick DE, Oakenfull D (1983) Saponin content of food plants and some prepared foods. J Sci Food Agric 34:186–191

Fernandez-Orozco R, Zieliński H, Piskuła MK (2003) Contribution of low-molecular-weight antioxidants to the antioxidant capacity of raw and processed lentil seeds. Mol Nutr Food Res 47(5):291–299

Fernandez-Orozco R, Frias J, Zielinski H, Muñoz R, Piskula MK, Kozlowska H, Vidal-Valverde C (2009) Evaluation of bioprocesses to improve the antioxidant properties of chickpeas. LWT Food Sci Technol 42(4):885–892

Frias J, Miranda ML, Doblado R, Vidal-Valverde C (2005) Effect of germination and fermentation on the antioxidant vitamin content and antioxidant capacity of *Lupinus albus* L. var. Multolupa. Food Chem 92(2):211–220

Gan RY, Deng ZQ, Yan AX, Shah NP, Lui WY, Chan CL, Corke H (2016) Pigmented edible bean coats as natural sources of polyphenols with antioxidant and antibacterial effects. LWT Food Sci Technol 73:168–177

Gan RY, Lui WY, Wu K, Chan CL, Dai SH, Sui ZQ, Corke H (2017) Bioactive compounds and bioactivities of germinated edible seeds and sprouts: an updated review. Trends Food Sci Technol 59:1–14

García-Lafuente A, Moro C, Manchón N, Gonzalo-Ruiz A, Villares A, Guillamón E, Rostagno M, Mateo-Vivaracho L (2014) In vitro anti-inflammatory activity of phenolic rich extracts from white and red common beans. Food Chem 161:216–223

Grela ER, Günter KD (1995) Fatty acid composition and tocopherol content of some legume seeds. Anim Feed Sci Technol 52:325–331

Guajardo-Flores D, García-Patiño M, Serna-Guerrero D, Gutiérrez-Uribe JA, Serna-Saldívar SO (2012) Characterization and quantification of saponins and flavonoids in sprouts, seed coats and cotyledons of germinated black beans. Food Chem 134:1312–1319

Han KH, Kitano-Okada T, Seo JM, Kim SJ, Sasaki K, Shimada KI, Fukushima M (2015) Characterisation of anthocyanins and proanthocyanidins of adzuki bean extracts and their antioxidant activity. J Funct Foods 14:692–701

Hubert J, Berger M, Nepveu F, Paul F, Dayde J (2008) Effects of fermentation on the phytochemical composition and antioxidant properties of soy germ. Food Chem 109(4):709–721

Hur SJ, Lee SY, Kim YC, Choi I, Kim GB (2014) Effect of fermentation on the antioxidant activity in plant-based foods. Food Chem 160:346–356

Jin AL, Ozga JA, Lopes-Lutz D, Schieber A, Reinecke DM (2012) Characterization of proanthocyanidins in pea (*Pisum sativum* L.), lentil (*Lens culinaris* L.), and faba bean (*Vicia faba* L.) seeds. Food Res Int 46(2):528–535

Jood S, Chauhan BM, Kapoor AC (1986) Saponin content of chickpea and black gram: varietal differences and effects of processing and cooking methods. J Sci Food Agric 37:1121–1124

Kaack K (1994) Blanching of green bean (*Phaseolus vulgaris*). Plant Foods Hum Nutr (Formerly Qualitas Plantarum) 46(4):353–360

Kagan V, Serbinova E, Packer L (1990) Antioxidant effects of ubiquinones in microsomes and mitochondria are mediated by tocopherol recycling. Biochem Biophys Res Commun 169(3):851–857

Kalogeropoulos N, Chiou A, Ioannou M, Karathanos VT, Hassapidou M, Andrikopoulos NK (2010) Nutritional evaluation and bioactive microconstituents (phytosterols, tocopherols, polyphenols, triterpenic acids) in cooked dry legumes usually consumed in the Mediterranean countries. Food Chem 121(3):682–690

Kan L, Nie S, Hu J, Wang S, Cui SW, Li Y, Xu S, Wu Y, Wang J, Bai Z, Xie M (2017) Nutrients, phytochemicals and antioxidant activities of 26 kidney bean cultivars. Food Chem Toxicol 108:467–477

Kmiecik W, Lisiewska Z, Jaworska G (1990) Vitamin C level in fresh, frozen and canned broad bean in relation to their varieties and degree of maturity. Roczn PZH 41:17–24

Korus A, Lisiewska Z, Kmiecik W (2002) Effect of freezing and canning on the content of selected vitamins and pigments in seeds of two grass pea (*Lathyrus sativus* L.) cultivars at the not fully mature stage. Mol Nutr Food Res 46(4):233–237

Lásztity R, Hidvégi M, Bata A (1998) Saponins in food. Food Res Int 14(4):371–390

Lee JH, Jeon JK, Kim SG, Kim SH, Chun T, Imm JY (2011) Comparative analyses of total phenols, flavonoids, saponins and antioxidant activity in yellow soy beans and mung beans. Int J Food Sci Technol 46(12):2513–2519

Limón RI, Peñas E, Torino MI, Martínez-Villaluenga C, Dueñas M, Frias J (2015) Fermentation enhances the content of bioactive compounds in kidney bean extracts. Food Chem 172:343–352

Lin PY, Lai HM (2006) Bioactive compounds in legumes and their germinated products. J Agric Food Chem 54(11):3807–3814

Lin LZ, Harnly JM, Pastor-Corrales MS, Luthria DL (2008) The polyphenolic profiles of common bean (*Phaseolus vulgaris* L.). Food Chem 107(1):399–410

López A, El-Naggar T, Dueñas M, Ortega T, Estrella I, Hernández T, Gómez-Serranillos MP, Palomino OM, Carretero ME (2013) Effect of cooking and germination on phenolic composition and biological properties of dark beans (*Phaseolus vulgaris* L.). Food Chem 138(1):547–555

López-Amorós ML, Hernández T, Estrella I (2006) Effect of germination on legume phenolic compounds and their antioxidant activity. J Food Compos Anal 19(4):277–283

López-Martínez LX, Leyva-López N, Gutiérrez-Grijalva EP, Heredia JB (2017) Effect of cooking and germination on bioactive compounds in pulses and their health benefits. J Funct Foods 38:624–634

Luo J, Cai W, Tong W, Xu B (2016) Phytochemical distribution in hull and cotyledon of adzuki bean (*Vigna angularis* L.) and mung bean (*Vigna radiata* L.), and their contribution to antioxidant, anti-inflammatory and anti-diabetic activities. Food Chem 201:350–360

Madhujith T, Amarowicz R, Shahidi F (2004) Phenolic antioxidants in beans and their effects on inhibition of radical-induced DNA damage. J Am Oil Chem Soc 81(7):691–696

Magalhães SC, Taveira M, Cabrita AR, Fonseca AJ, Valentão P, Andrade PB (2017) European marketable grain legume seeds: further insight into phenolic compounds profiles. Food Chem 215:177–184

Mirali M, Ambrose SJ, Woods SA, Vandenberg A, Purves RW (2014) Development of a fast extraction method and optimization of liquid chromatography– mass spectrometry for the analysis of phenolic compounds in lentil seed coats. J Chromatogr B 969:149–161

Moriyama M, Oba K (2008) Comparative study on the vitamin C contents of the food legume seeds. J Nutr Sci Vitaminol 54(1):1–6

Murakami H, Asakawa T, Terao J, Matsushita S (1984) Antioxidative stability of tempeh and liberation of isoflavones by fermentation. Agric Biol Chem 48(12):2971–2975

Nilsson J, Stegmark R, Åkesson B (2004) Total antioxidant capacity in different pea (Pisum sativum) varieties after blanching and freezing. Food Chem 86(4):501–507

Rocha-Guzmán NE, González-Laredo RF, Ibarra-Pérez FJ, Nava-Berumen CA, Gallegos-Infante JA (2007) Effect of pressure cooking on the antioxidant activity of extracts from three common bean (*Phaseolus vulgaris* L.) cultivars. Food Chem 100(1):31–35

Sattar A, Atta S, Akhtar MA, Wahid M, Ahmad B (1991) Biosynthesis of ascorbic acid and riboflavin in radiated germinating chickpea. Int J Vitam Nutr Res 61(2):149–154

Segev A, Badani H, Kapulnik Y, Shomer I, Oren-Shamir M, Galili S (2010) Determination of polyphenols, flavonoids, and antioxidant capacity in colored chickpea (*Cicer arietinum* L.). J Food Sci 75(2):S115–S119

Siah SD, Konczak I, Agboola S, Wood JA, Blanchard CL (2012) *In vitro* investigations of the potential health benefits of Australian-grown faba beans (*Vicia faba* L.): chemopreventative capacity and inhibitory effects on the angiotensin-converting enzyme, α-glucosidase and lipase. Br J Nutr 108(S1):S123–S134

Singh U, Singh B (1992) Tropical grain legumes as important human foods. Econ Bot 46(3):310–321

Singh B, Singh JP, Shevkani K, Singh N, Kaur A (2017a) Bioactive constituents in pulses and their health benefits. J Food Sci Technol 54(4):858–870

Singh B, Singh JP, Kaur A, Singh N (2017b) Phenolic composition and antioxidant potential of grain legume seeds: a review. Food Res Int 101:1–16

Singh B, Singh JP, Singh N, Kaur A (2017c) Saponins in pulses and their health promoting activities: a review. Food Chem 233:540–549

Sosulski FW, Dabrowski KJ (1984) Composition of free and hydrolyzable phenolic acids in the flours and hulls of ten legume species. J Agric Food Chem 32(1):131–133

Torino MI, Limón RI, Martínez-Villaluenga C, Mäkinen S, Pihlanto A, Vidal- Valverde C et al (2013) Antioxidant and antihypertensive properties of liquid and solid state fermented lentils. Food Chem 136:1030–1037

Torres A, Frias J, Granito M, Vidal-Valverde C (2006) Fermented pigeon pea (Cajanus cajan) ingredients in pasta products. J Agric Food Chem 54(18):6685–6691

Tsuda T, Ohshima K, Kawakishi S, Osawa T (1994) Antioxidative pigments isolated from the seed of Phaseolus vulgaris L. J Agric Food Chem 42:248–251

Turkmen N, Sari F, Velioglu S (2005) The effect of cooking methods on total phenolics and antioxidant activity of selected green vegetables. Food Chem 93:713–718

Vanderslice JT, Higgs DJ, Hayes JM, Block G (1990) Ascorbic acid and dehydroascorbic acid content of foods-as-eaten. J Food Compos Anal 3(2):105–118

Volf I, Ignat I, Neamtu M, Popa VI (2014) Thermal stability, antioxidant activity, and photo-oxidation of natural polyphenols. Chem Pap 68(1):121–129

Wang N, Hatcher DW, Toews R, Gawalko EJ (2009) Influence of cooking and dehulling on nutritional composition of several varieties of lentils (*Lens culinaris*). LWT Food Sci Technol 42(4):842–848

Wang YK, Zhang X, Chen GL, Yu J, Yang LQ, Gao YQ (2016) Antioxidant property and their free, soluble conjugate and insoluble-bound phenolic contents in selected beans. J Funct Foods 24:359–372

Wojtyla Ł, Garnczarska M, Zalewski T, Bednarski W, Ratajczak L, Jurga S (2006) A comparative study of water distribution, free radical production and activation of antioxidative metabolism in germinating pea seeds. J Plant Physiol 163(12):1207–1220

Wyatt CJ, Carballido SP, Mendez RO (1998) α-and γ-tocopherol content of selected foods in the Mexican diet: effect of cooking losses. J Agric Food Chem 46(11):4657–4661

Xu BJ, Chang SKC (2007) A comparative study on phenolic profiles and antioxidant activities of legumes as affected by extraction solvents. J Food Sci 72(2):S159–S166

Xu B, Chang SK (2008) Effect of soaking, boiling, and steaming on total phenolic content and antioxidant activities of cool season food legumes. Food Chem 110(1):1–13

Xu B, Chang SK (2009) Total phenolic, phenolic acid, anthocyanin, flavan-3-ol, and flavonol profiles and antioxidant properties of pinto and black beans (*Phaseolus vulgaris* L.) as affected by thermal processing. J Agric Food Chem 57:4754–4764

Xu B, Chang SK (2010) Phenolic substance characterization and chemical and cell based antioxidant activities of 11 lentils grown in the northern United States. J Agric Food Chem 58(3):1509–1517

Xu B, Yuan S, Chang S (2007) Comparative analyses of phenolic composition, antioxidant capacity, and color of cool season legumes and other selected food legumes. J Food Sci 72:S167–S177

Yoshiki Y, Kim JH, Okubo K (1994) Saponins conjugated with 2,3-dihydro- 2,5-dihydroxy-6-methyl-4H-pyran-4-one from Phaseolus coccineus. Phytochemistry 36:1009–1012

Zhang B, Deng Z, Tang Y, Chen PX, Liu R, Ramdath DD, Liu Q, Hernandez M, Tsao R (2014) Effect of domestic cooking on carotenoids, tocopherols, fatty acids, phenolics, and antioxidant activities of lentils (*Lens culinaris*). J Agric Food Chem 62(52):12585–12594

Zhao Y, Du SK, Wang H, Cai M (2014) *In vitro* antioxidant activity of extracts from common legumes. Food Chem 152:462–466

Zielinski H (2002) Peroxyl radical-trapping capacity of germinated legume seeds. Mol Nutr Food Res 46(2):100–104

Chapter 5
Application of Legume Seed Galactomannan Polysaccharides

Harikrishna Naik Lavudi and Sateesh Suthari

Abstract The present article is focused on the legume seed galactomannans and their multipurpose applications. Galactomannans are natural and abundant polysaccharides available in legume seed endosperm with multidimentional aspects which are cost-effective and eco-friendly. The chemical structure is formed by a linear chain of a β-(1–4)-D-mannan backbone with single D-galactose branches attached at α-(1–6) position and the ratios of galactose and mannose are different. The galactomannans are versatile materials for various applications such as binding, emulsifying, gelling, water retention capacity, suspending, thickening and formation of films. These can easily mixed with xanthans by physical association, exhibits synergistic effect and forms complex mixtures due to entangle nature. This property could be effective in present scenario. Here, the paper reviews the galactomannans, structural aspects and applications in different industries. These polysaccharides are widely used in food, thermoplastic, textile industry, rubber and as cosmetics, encapsulating agents in pharmaceutical industries. Their structure can be modified, sulphated and carboxy methylated and used in nanotechnology for drug delivery and drug carriers.

Keywords Legumenosae · Drug delivery · Endosperm · Galactomannan · Gelling · Hydrocolloids · Macromolecule · Polysaccharide · Synergistic · Xanthans

H. N. Lavudi (✉)
Department of Plant Sciences, University of Hyderabad, Hyderabad, Telangana, India

S. Suthari
Department of Plant Sciences, University of Hyderabad, Hyderabad, Telangana, India

Department of Botany, Vaagdevi Degree & PG College, Warangal, Telangana, India

P. Guleria et al. (eds.), *Sustainable Agriculture Reviews 45*, Sustainable Agriculture Reviews 45, https://doi.org/10.1007/978-3-030-53017-4_5

Abbreviations

EFSA	European Food Safety Authority
G/M ratio	galactose/mannose ratio
ICU	intensive care unit
LBG	locust bean gum
PHGG	partially hydrolyzed guar gum
PR	plasmon resonance

5.1 Introduction

Seed Galactomannans are, heterogeneous polysaccharides, largely found in legume seed endosperm (Kapoor 1992; Edwin and Andrade 2002; Srivastava and Kapoor 2005; Suvakanta et al. 2014). The galactomannans are well known for multipurpose utility and neutral macromolecules (Reid and Bewley 1979). In aqueous media, these galactomannans are viscous in nature, have emulsifying, gelling and stabilizing properties, and preferred as hydrocolloids. They are inexpensive, eco-friendly, non-polluting, and possess various applications (Srivastava and Kapoor 2005; Stephen and Churns 1995). Legume (*sensu lato* Legumnosae = Fabaceae) seed endosperm is the main source of galactomannans.

Seed galactomannans have highly variable chemical structure in the nature. The degree of mannose and galactose ratio is highly variable and it varies from species to species (Dea and Morrison 1975; Dey 1978; Winter et al. 1984). These can influence the solubility and stiffness of the chain (Wu et al. 2012). The galactomannans play a vital role in seed imbibitions to germinate the seed due to the accumulation of moisture at the early stage of seed swelling (Reid and Bewley 1979). The galactomannans are catabolized and transferred to embryo as a carbon source and protects the seed.

5.1.1 Major Source of Galactomannans

Legume endosperm is an important naturally available source of galactomannans (Table 5.1). These are cell wall storage polysaccharides, majorly present in the hull, inner seed coat and kernels which are helpful in seed germination.

Table 5.1 Legume species studied for different aspects of galactomannan chemistry

Botanical name	References
Abarema langsdorfii Benth.	Buckeridge et al. (1995)
Acacia bahiensis Benth.	Buckeridge et al. (1998)
Acacia seyal Delile	Buckeridge et al. (1998)
Anadenanthera colubrina (Vell.) Brenan	Buckeridge et al. (1995)
Anadenanthera falcata (Benth.) Speg.	Buckeridge et al. (1995)
Adenanthera pavonina L.	Ceraqueira et al. (2009) (Fig. 5.1)
Caesalpinia ferrea C.Mart.	Clayton et al. (2010)
Caesalpinia pulcherrima (L.) Sw.	Ceraqueira et al. (2009)
Caesalpinia spinosa (Molina) Kuntze	Ceraqueira et al. (2009)
Calliandra bavipes Benth.	Buckeridge et al. (1995)
Calliandra bracteosa Benth.	Buckeridge et al. (1995)
Cassia grandis L.f.	Harsha and Kapoor (2003)
Ceratonia siliqua L.	Garcia-Ochoa and Casas (1992)
Chamaecrista absus (L.) H.S. Irwin and Barneby	Kapoor and Mukherjee (1969)
Cyamopsis tetragonoloba (L.) Taub.	Anderson (1949)
Dichrostachys cinerea (L.) Wight and Arn.	Harikrishna et al. (2018) (Fig. 5.2)
Gleditsia triacanthos L.	Bourbon et al. (2010)
Leucaena leucocephala (lam.) de wit	Unrau (1961)
Leucaena pulverulenta (Schltdl.) Benth.	Buckeridge et al. (1995)
Leucochloron incuriale (Vell.) Barneby and J.W.Grimes	Buckeridge et al. (1995)
Mimosa aspericarpa (Hoehne) Burkart	Ganter and Reicher (1999)
Mimosa bimucronata (DC.) Kuntze	Ganter and Reicher (1999)
Mimosa flocculosa Burkart	Ganter and Reicher (1999)
Mimosa platyphylla Benth.	Buckeridge et al. (1995)
Mimosa pudica L.	Harikrishna et al. (2017) (Fig. 5.3)
Mimosa scabrella Benth.	Vendruscolo et al. (2009)
Mimosa taimbensis Burkart	Ganter and Reicher (1999)
Parkinsonia aculeata L.	Garros-Rosa et al. (2006) (Fig. 5.4)
Peltophorum pterocarpum (DC.) K.Heyne	Nwokocha and Peter (2014)
Prosopis juliflora (Sw.) DC.	Figueiredo (1983)
Senna alexandrina mill.	Manjoosha and Kapoor (2001)
S. spectabilis (DC.) Irwin and Barneby	Kapoor et al. (1998)
Styphnolobium japonicum (L.) Schott	Bourbon et al. (2010)
Trigonella foenum-graecum L.	Brummer et al. (2003)

5.2 Chemical Structure of Seed Galactomannans

Seed galactomannans are naturally available material and have highly variable chemical structures. Galactomannans can be varied to a wide degree by altering galactose/mannose ratio and the distribution of galactose along the mannose chain

Fig. 5.1 *Adenanthera pavonina*: (**a**) Habitat; (**b**) Inflorescence; (**c**) Pods; (**d**) Seeds

(Winter et al. 1984). Galactomannas are commonly formed by a linear chain of β-1,4-D-mannopyranosides branched with single units of D-galactopyranosides through α-1,6 linkages (Fig. 5.5; Dea and Morrison 1975; Dey 1978) and have high molecular weight. Srivastava and Kapoor (2005) extensively studied the structural aspects of galactomannans and compared with other compounds based on the G/M ratios. The galactomannans chemical structure is species specific i.e. G/M ratio differs from species to species.

5.3 Properties and Functions of Galactomannans

The galactomannans are considered as versatile polysaccharides (Reid and Bewley 1979), basically neutral polymers. The solubility of galactomannans in water can be increased by the substitution of single galactose side chain and exhibits high viscosity even at lower concentrations (Vendruscolo et al. 2009). The degradation of galactomannans is possible under high acidic and alkaline conditions only

Fig. 5.2 *Dichrostachys cinerea*: (**a**) Habitat; (**b**) Inflorescence; (**c**) Pods; (**d**) Seeds

(Srivastava and Kapoor 2005). These gums accumulate moisture at the initial stage of seed swelling and play an important role in seed germination by transferring starch energy to embryo and later on, then galactomannans could be catabolized (Reid 1971) and protect the seed. It was proved that galactomannans initially take up large quantity of water during imbibition for the germination of seed (Reid and Bewley 1979).

The utility of galactomannan gums are being increased day-by-day due to the expeditious growth in the consumption of novelty food and instant meal. The consumers are also aware of the importance of galactomannans and trying to enhance the fiber content and bring down the fat in their daily diet (Williams and Phillips 2003a). These have not been evolved much and less exploited at Industrial level. According to the report of the European Food Safety Authority (EFSA), the utility of *Cassia* gum complying with new specifications as an excellent material for food additive. These galactomannans enhance the dietary fiber and reduce the fat and is documented well in the literature (Trowell et al. 1976; Gupta et al. 2001).

Fig. 5.3 *Mimosa pudica*: (**a**) Habitat; (**b**) Inflorescence; (**c**) Pods; (**d**) Seeds

5.3.1 Rheology and Viscous Nature

Several galactomannans like Guar gum, LBG (locust bean gum), Fenugreek gum and Tara gums have significant commercial applications are majorly used in commercial purpose because of their thickening and gelling properties. These polysaccharides are extensively studied for different aspects like rheology and viscous nature (Morries 1984; Morries et al. 1981). These compounds can easily interact with other polysaccharides like agarose, carrageenan, and xanthan and form highly viscous solutions (Yalpani 1988).

5.4 Applications of Seed Galactomannans

Seed galactomannans are highly diversified polysaccharides and widely distributed in nature. Comparatively these are cost-effective, eco-friendly and non-toxic. The chemical properties of these polysaccharides made them flexible materials for different applications (Table 5.2). The properties of seed galactomannans include

Fig. 5.4 *Parkinsonia aculeata*: (**a**) Habitat; (**b**) Inflorescence; (**c**) Pods; (**d**) Seeds

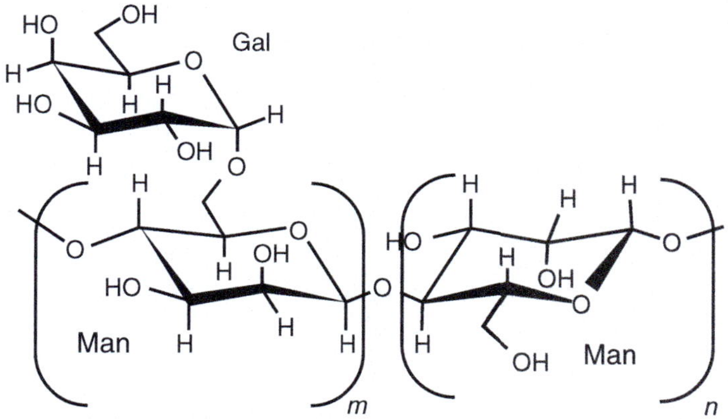

Fig. 5.5 Common structure of galactomannan

Table 5.2 Application of seed galactomannans from Leguminosae

Application	References
Water retention capacity	
Preparation of nano particles	Soumya et al. (2010)
Food and pharmaceuticals	Stephen and Churns (1995); Bressolin et al. (1997); Srivastava and Kapoor (2005)
Drug delivery agents, sustained-release tablets, coating agents	Efentakis and Kouttis (2001)
Toiletries preparation, thickener in toothpastes, conditioner in shampoos, dyes, denture fixture powders	Williams and Phillips (2003a, b); Sharma et al. (2008); Vendruscolo et al. (2009)
Thickening	Sostar and Schneider 1998
Emulsifying	Tammishetti and Thimma (2001)
Binding	Frias and Sgarbier (1998); Schneider and Soster-Turk (2003)
Suspending, gelling	Williams and Phillips (2003a, b); Friend (2005)
Formation of films	Varshosaz et al. (2006)
Pharma industry	Thakur et al. (2009); Cerqueira et al. (2009, 2011)
Cosmetics, food, drilling, explosives, paper, petroleum, pharmaceuticals and textiles	Clark (1986)
Printing industry	Sostar and Schneider (1998)
Sheet formation, folding, dense surface for printing	Schneider and Soster-Turk (2003)
Food coating agents, adhesives, ice-creams, food processing units	Stephen and Churns (1995)
Cured meat preparation, frozen, tinned meat, beverages, diary products	Cerqueira et al. (2011); Kapoor (1992); Srivastava and Kapoor (2005)
Surface coating agents, binder, tablets disintegrator	Baveja et al. (1991); Pauly et al. (1999); Varshosaz et al. (2006)
Drug	Tauseef and Sasi Kumar (2011); Pal et al. (2007)
Laxative, inflammatory diseases	Friend (2005)
Non-cytotoxic doses	Theisen (2001)
Cancer chemopreventive agent	Gamal-Elden et al. (2006)
Anti-coagulant, anti-viral activity	Herold et al. (1995)
Probiotics	Macfarlane and Cummings (1999); Grizard et al. (2001); Murphy (2001); Edwards (2003); Rushdi et al. (2004); Slavin and Greenberg (2003)
Prebiotics	Tuohy et al. (2001)
Chemotaxonomy	Buckedidge et al. (1995); Buckedidge and Dietrich (1990); Bailey (1971); Reid (1985); Harborne et al. (1971); Tindale and Roux (1974); El Tinay et al. (1979); Brain (1990); Brain and Maslin (1996); Ezeagu and Gowda (2006); Srivastava and Kapoor (2005); Harikrishna et al. (2017)

water retention capacity, thickening, emulsifying, binding, suspending, gelling, formation of films and in pharmaceutical industries (Sostar and Schneider 1998; Tammishetti and Thimma 2001; Schneider and Soster-Turk 2003; Williams and Phillips 2003a, b; Friend 2005; Varshosaz et al. 2006; Thakur et al. 2009; Cerqueira et al. 2009, 2011). Due to their versatile nature, the galactomannans are utilized in different ways like cosmetics, food, drilling, explosives, paper, petroleum, pharmaceuticals and textiles (Clark 1986) (Fig. 5.6). The bonding properties of the galactomannans can be enhanced by interactions with monomers and polymers due to the synergistic interactions of numerous hydroxyl (-OH) groups. These can also be used in various forms for human consumption due to their gel nature. Presently, the international trends demand the seed gums as an alternative source. A recent study explored that the galactomannans are used for nano-particles preparation (Soumya et al. 2010).

Due to high solution viscosity and mixed gel formation in aqueous media with other proteins and polysaccharides, the galactomannans are used in various industries like food and pharmaceuticals (Stephen and Churns 1995; Bressolin et al. 1997; Srivastava and Kapoor 2005). The separated endosperm from roasted seed is further milled into desirable mesh size. Finally, the crude gum procured from seed using dry milling process requires further purification with polar solvents like ethanol by precipitation (Srivastava and Kapoor 2005).

In recent days, many natural polymers have been used successfully in several applications such as drug delivery agents, sustained-release tablets and coating agents. These materials include the galactomannan from guar gum, *Mimosa scabrella, Gleditsia triacanthos*, *Sesbania* gum, etc. (Efentakis and Kouttis 2001). Gums and mucilages have broad range of applications in food and biomedical industries. The utility pattern depends on their specific physico-chemical properties that they provide, inexpensive with compare to the synthetic polymers. Gums and

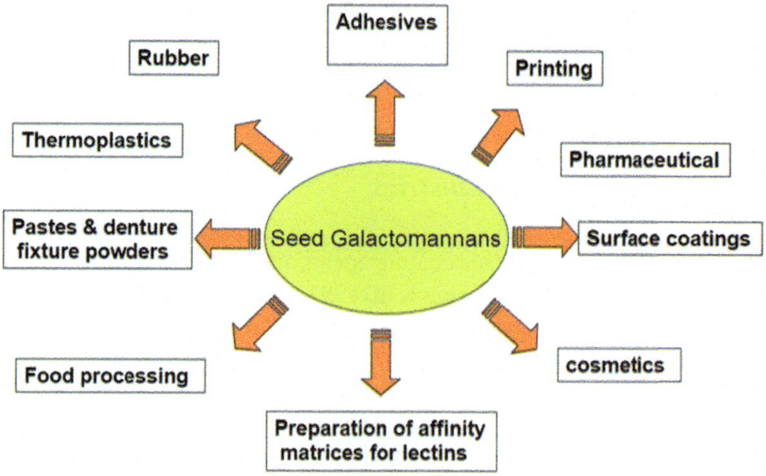

Fig. 5.6 Schematic representation of galactomannas application (Harikrishna et al. 2017)

mucilages are having high water-solubility, swellability, large availability, non-toxicity, cheap and are very important substances for pharmaceutical formulations. These are abundantly available in many leguminous plants.

5.4.1 Industrial Applications

The galactomannans are highly useful in rubber industries, thermoplastic, food industries and cosmetics, toiletries preparation, as thickener in toothpastes, as conditioner in shampoos, dyes in textile industries, denture fixture powders (Williams and Phillips 2003a, b; Sharma et al. 2008; Vendruscolo et al. 2009) and used in printing industries for the sizing and finishing (Sostar and Schneider 1998). Galactomannans enhance the sheet formation, folding and dense surface for printing (Schneider and Soster-Turk 2003).

5.4.2 Food Industries

Galactomannas are highly useful in food industries for different purposes such as food coating agents, adhesives, ice-creams and food processing units (Stephen and Churns 1995). These can be used as thickener in creams and preparation of milk desserts in dairy industries, stabilizer in sorbets, ice-cream and cheese industries. These have high moisture retention, enhance flavor and used as fat substitutes. It is also used in diabetic products preparation, like coffee whiteners and baby milk formulations. These can improve the bread quality in bakery product industries. Galactomannans are also useful in powdered products, desserts and hot milk puddings preparation, jellied products and syrups, and also used in the preparation of cured meat, frozen, tinned meat, beverages as thickener and dairy products (Kapoor 1992; Srivastava and Kapoor 2005; Cerqueira et al. 2011).

5.4.3 Pharmaceutical Industries

Galactomannans are good surface coating agents and binder or used as disintegrator for the preparation of tablets (Baveja et al. 1991; Pauly et al. 1999; Varshosaz et al. 2006) and drugs (Tauseef and Sasi Kumar 2011; Pal et al. 2007). Mainly used as ingredient in some bulk-forming laxatives and useful in the treatment of inflammatory diseases (Friend 2005). They can be used for nano-particles preparation in drug delivery (Tammishetti and Thimma 2001) and easily interact with other compounds which will be highly helpful for the preparation of tablets and additives in pharmaceutical industries. Since the last two decades, galactomannan based nanoparticles have been extensively improved and explored for pharmaceutical applications (Soumya et al. 2010).

5.4.4 Cancer Chemoprevention

Cancer chemoprevention is the use of medication to cure or prevent the cancer. It aims to halt or reverse the development and progression of precancerous cells through the use of non-cytotoxic doses of nutrients and pharmacological agents (Theisen 2001). Now-a-days, the identification of novel effective cancer chemopreventive agents has become a significant strategy in the prevention of cancer (Gamal-Eldeen et al. 2006). Sulphated polysaccharides, which are attributed to the presence of negatively charged sulphate groups have been reported to have *in vitro* anti-coagulant and anti-viral activities (Herold et al. 1995). Previous studies reported that the Guar gum has the ability to take them out of the body by binding toxic compounds (Frias and Sgarbier 1998). C-glycosylated derivatives have been prepared from guar gum and then an additional sulphation of that derivative to prepare sulphated derivative and further studied the anti-inflammatory and cancer chemopreventive properties of these derivatives, so that they can be used as alternatives in the health and food industries to provide cancer prevention in high risk population (Gamal-Eldeen et al. 2006).

5.4.5 Probiotics

Probiotics are primarily carbohydrates and non-digestible food materials. These can stimulate the activity and growth of probiotic bacteria in the gut region and can readily be added into enteral feeds with potential benefit (Macfarlane and Cummings 1999; Grizard et al. 2001; Murphy 2001). Probiotics are usually some strains of *Bifidobacteria*, *Lactobacilli* and *Streptococci* and have beneficial health effects like to reduce eczema in atopic infants and duration of rotavirus diarrhea and having protective actions against intestinal diseases (Edwards 2003). Fermentable dietary fibers modify the gut environment with the production of sort-chain fatty acids and change the gut microflora. Previous studies reported that guar gum enriched enteral nutrition which can decrease diarrhea frequency in ICU patients as a potential prebiotic, helped in lowering plasma glucose and cholesterol levels, and increasing plasma calcium level (Rushdi et al. 2004). This indicates that guar gum can be used as a potential prebiotic. PHGG (Partially Hydrolyzed Guar Gum) increases *Bifidobacterium* concentration in the gut (Slavin and Greenberg 2003) and acts as a prebiotic (Tuohy et al. 2001).

5.5 Chemotaxanomy

Galactomannans play a vital role in chemotaxanomy of *Leguminosae*. The G/M ratios in galactomannans regulate the solubility properties. The legume species have shown different G/M ratios. The compound which has higher galactose percentage content performs good cold water solubility and viscosity is also very high (Buckedidge et al. 1995).

Buckedidge and Dietrich (1990) analyzed 23 plant taxa of Leguminosae seeds. Although the possibility of using galactomannan as a taxonomic evelutionary marker was pointed out, the number of plant species is still very low to have an accurate view of the distribution of galactomannan within the family. The yield and M/G ratio are two important features which can be used as effective tools in the biochemical taxonomy of Leguminosae. According to Bailey et al. (1971) high galactomannan contents and high mannose to galactose ratios are characteristics of the Caesalpinioideae (Ceasalpiniaceae *sensu stricto*), whereas low contents and low mannose to galactose ratios are features of the Fabaceae (Papilionoideae) (Bailey et al. 1971; Reid 1971). Therefore, the galactomannon distribution throughout Leguminosae might reflect the chemotaxonomy/systematics evidences for the elucidation of plant taxonomic problems and evolutionary pattern of the taxon (Buckeridge et al. 1995). There is considerable research on biochemical and biosystematic aspects on the Legume taxa of Africa, Australia and Brazil (Harborne et al. 1971; Tindale and Roux 1974; El Tinay et al. 1979; Brain 1990; Brain and Maslin 1996; Ezeagu and Gowda 2006), but for Indian taxa there was few reports available on the galactomannan biochemistry (Srivastava and Kapoor 2005; Harikrishna et al. 2017) and no such reports on biochemical taxonomy and its biomedical applications were found.

5.6 Commercial Seed Galactomannans

Source of the galactomannas are widely used in commercial purposes (Vipul et al. 2013) such as guar gum, fenugreek gum, tara gum, locust bean gum (Wu et al. 2009), karaya gum (Hana et al. 2017). Most popular commercial seed galactomannans and their M/G ratio are presented in Table 5.3.

Literature survey revealed that due to lack of information on galactomannans and other biochemical parameters such as proteins from the Mimosaceae is not possible to interpret of focus on classification and evolution on Mimosaceae (*sensu stricto*) and Leguminosae (*sensu lato*). The studies on the seed galactomannans and protein analysis for biochemical and biosystematics aspects would open new vistas in Indian legume taxa.

Table 5.3 Most popular commercial seed galactomannans and their M/G ratio

Trade name	Source	Family	M/G ratio
Guar	*Cyamopsis tetragonoloba*	Fabaceae	2:1
Fenugreek	*Trigonella foenum-graecum*		1:1
Locust bean	*Ceratonia siliqua*		4:1
Tara	*Caesalpinia spinosa*		3:1

(*M/G ratio* mannose/galactose ratio)

5.7 Importance of Galactomannans in Nanotechnology

Nanobiotechnology becomes one of the rapidly growing fields in food, drug discovery industries, protection, entrapment and delivery of highly sensitive bioactive compounds. The size of nanoparticles ranges from 1 to 100 nm. Nanoparticles are potential carriers for different kinds of drugs and cosmetics (Akagi et al. 2006). Nanoparticles are multifunctional and can be used in cosmetics, medical, pharmaceutical, biosensors and food supplements (Gupta and Verma 2014). Based on the source, different types of nanoparticles are prepared, silver, gold, copper, palladium nanoparticles, etc. Copper nanoparticles used as catalytic activity, anti-fouling, gas sensor, biocidal, wound dressing, wound healing and in solar cells (Jung et al. 2006). Silver nanoparticles are good catalysts for accelerating some biochemical reactions. Gold nanoparticles also having large surface bio-conjugation and have many optical properties with localized Plasmon Resonance (PR) used in biomedicine (Tedesco et al. 2010; Yasmin et al. 2014). Nanoprecipitation and sonication methods are major techniques extensively used for the preparation of nanoparticles (Gupta and Verma 2014).

Rapid advancements in nanotechnology in recent times have opened up new avenues for industrial and consumer sectors for which they have been developed as the hotbed of a new industrial revolution. For example, the food sector, globally worth over 4 trillion US$ per annum (Murray 2007). Food based applications of nanobiotechnologies offer a wide range of benefits. Nanoformulation drugs can also improve the uptake, absorption, biocompatibility and bioavailability of nutrients and supplements in the body. Antibacterial nano-coating agents on food preparation surfaces can help maintain hygiene conditions. Helmut Kaiser Consultancy (2004) reported that the nano-food sector is led by the USA, Japan, China and India could be the biggest future markets for nano-food products. Galactomannans play significant role in upcoming years. Seed galactomannan is a natural biopolymer which is widely used for the preparation of polysaccharide based nanoparticles (Taheri and Razavi 2015). Guar gum, gum karaya and their derivatives have been used in many applications such as in food industry, drug-delivery and health care products due to their abundant availability, inexpensive and eco-friendly. Guar gum is the best example for the preparation of natural polysaccharide (galactomannan) nanoparticles, also works as a good cross linking agent in drug delivery and shows anti-cancer properties (Al-Saidan et al. 2004; Soumya et al. 2010; Rakesh et al. 2014). Carboxymethyl guar gum, nanoparticles and sulphated nanoparticles are also used in pharmaceutical and drug delivery industries (Gupta and Verma 2014).

5.8 Conclusions

The present review paper explains the overall information of different aspects of the galactomannans present in the family Legumenosae, their characteristics, chemical composition, structure of the galactomannas and applications in various fields. The present study indicates that the galactose:mannose ratio of galactomannans influences the characteristics of the family and also demonstrate the commercial availability of the galactomannans in the family Legumenosae. Due to various properties and applications of galactomannans, further investigation is required for the extraction and purification and its broad range of applications in various fields. It will be a good platform for chemotaxonomy, food technology, pharmaceutical chemistry, nanotechnology, food processing and fields of biomedical applications.

Acknowledgements The financial assistance by University of Hyderabad and UGC-SAP is highly acknowledged. The authors thankful to Late Prof. K. Seshagirirao, Department of Plant Sciences, University of Hyderabad, the Management and the Principal, Vaagdevi Degree & PG College, Hanamkondafor immense support and encouragement.

References

Akagi T, Higashi M, Kaneko T, Kida T, Akashi M (2006) Hydrolytic and enzymatic degradation of nanoparticles based on Amphiphilic poly (gammaglutamicacid)-graft-L-phenylalanine copolymers. Biomacromolecules 7:297–303

Al-Saidan SM, Krishnaiah YSR, Satyanarayana V, Bhaskar P, Karthikeyan RS (2004) Pharmacokinetic evaluation of guar gum-based three-layer matrix tablets for oral controlled delivery of highly soluble metoprolol tartarate as a model drug. Eur J Pharm Biopharm 58:697–703

Anderson E (1949) Endosperm mucilages of legume. Ind Eng Chem Res 41:2887–2890

Bailey RW, Harborne JB, Boulter D, Turner BL (1971) Polysaccharides in the leguminosae. In: Herborne JB, Boulter D, Turner BL (eds) Chemotaxonomy of the leguminosae. Academic, New York/London, pp 503–541

Baveja SK, Rao KVR, Arora J (1991) Chemical investigations of some galactomannan gums as matrix tablets for sustained drug delivery. Indian J Chem 30B:133–137

Bourbon AI, Pinheiro AC, Ribeiro C, Miranda C, Maia JM, Teixeira JA, Vicente AA (2010) Characterization of galactomannans extracted from seeds of Gleditsia triacanthos and Sophora japonica through shear and extensional rheology: comparison with guar gum and locust bean gum. Food Hydrocoll 24:184–192

Brain P (1990) Immunology and phylogeny II: further studies on *Acacia*. S Afr J Sci 86:195–199

Brain P, Maslin BR (1996) A serological investigation of the classification of *Acacia* sibgenus *Phyllodinea* (Leguminosae: Mimosoideae). Biochem Syst Ecol 24:379–392

Bressolin TM, Sander PC, Reicher F, Sierakowski MR, Rinaudo M, Ganter JLMS (1997) Viscometric studies on xanthan and galactomannan. Carbohydr Polym 33:131–138

Brummer Y, Cui W, Wang Q (2003) Extraction, purification and physicochemical characterization of fenugreek gum. Food Hydrocoll 17:229–236

Buckeridge MS, Dietrich SMC (1990) Galactomannan from Brazilian legume seeds. Rev Bras Bot 13:109–112

Buckeridge MS, Panegassi VR, Rocha DC, Dietrich SMC (1995) Seed galactomannan in the classification and evalution of the leguminosae. Phytochemistry 38(4):871–875

Buckeridge MS, Mayworm M, Nascimento AS, Salantino A (1998) Seeds of species from the 'caatinga': proteins, oils and fatty acid contents. Braz J Bot 21:3–4

Ceraqueira MA, Pinheiro AC, Souza BWS, Lima AMP, Ribeiro C, Miranda C, Teixeira JA, Moreira RA, Teixeira MA, Moreira RA, Coimbra MA, Goncalves MP, Vicente AA (2009) Extraction, purification and characterization of Galactomannans from non-traditional sources. Carbohydr Polym 75:408–414

Cerqueira MA, Lima AMP, Souza BWS, Teixeira JA, Moreira RA, Vicente AA (2009) Functional polysaccharides as edible coatings for cheese. J Agric Food Chem 57:1456–1462

Cerqueira MA, Bourbon A, Pinheiro A, Martins J, Souza B, Teixeira J, Vicente AA (2011) Galactomannans us in in the development of edible films/coatings for food applications. Trends Food Sci Technol 22(12):662–671

Clark RK (1986) In: Glass JE (ed) Water soluble polymers. American Chemical Society, Washington, DC, p 171

Clayton F d S, Lucyszyn N, Ferraz FA, Sierakowski MR (2010) Caesalpinia ferrea var. ferrea seeds as a new source of partially substituted galactomannan. Carbohydr Polym 82:641–647

Dea ICM, Morrison A (1975) Chemistry and interactions of seed galactomannans. Adv Carbohydr Chem Biochem 35:241–312

Dey PM (1978) Biochemistry of plant galactomannans. Adv Carbohydr Chem Biochem 35:341–376

Edwards CA (2003) GUMS: dietary importance. In: Encyclopedia of food sciences and nutrition. Academic, San Diego, pp 3007–3012

Edwin GA, Andrade TC (2002) Testing procedures for galactomannan purification. Polym Test 21:551–556

Efentakis M, Kouttis A (2001) Release of furosemide from multiple unit and single unit preparations containing different viscosity grade of sodium alginate. Pharm Dev Technol 6(1):91

El Tinay AH, Karamall KA, El Amin HM, Shigidi MTA, Ishag KEA (1979) Serotaxonomic studies on Sudan acacias. J Exp Bot 30:607–615

Eziagu EI, Gowda LR (2006) Protein extractability, fractionation and amino acid composition of some leguminous seeds found in Nigeria. J Food Biochem 30(1):1–11

Figueiredo AA (1983) Extração, identificação e caracter s ticas do polissacar d o da semente de algaroba (*Prosopis juliflora* DC). Cienc Tecnol Aliment 3:82–89

Frias AC, Sgarbier VC (1998) Guar gum effects on food intake, blood serum lipids and glucose levels of Wistar rats. Plant Foods Hum Nutr 53:15–28

Friend DR (2005) New oral delivery system for treatment of inflammatory bowel disease. Adv Drug Deliv Rev 375:247–265

Gamal-Eldeen AM, Amer H, Helmy WA (2006) Cancer chemopreventive and anti-inflammatoy activities of chemically modified guar gum. Chem Biol Interact 161:229–240

Ganter JLMS, Reicher F (1999) Water-soluble galactomannans from seeds of *Mimosaceae* spp. Bioresour Technol 68:55–62

Garcia-Ochoa F, Casas JA (1992) Viscosity of locust bean (*Ceratonia siliqua*) gum solutions. J Sci Food Agric 59(1):97–100

Garros-Rosa I, Reicher F, Carmen LOP, Sierakowski MR, Moreira RA (2006) Characterization of the galactomannans from *Parkinsonia aculeata* seeds and their application on affinity chromatography. Polimeros 16(2):99–103

Grizard D, Dalle M, Barthomeuf C (2001) Changes in insulin and corticosterone levels may partly mediate the hypolipidemic effect of guar gum and low-molecular weight pectin in rats. Nutr Res 21:1185–1190

Gupta AP, Verma DK (2014) Preparation and characterization of carboxymethyl guar gum nanoparticles. Int J Biol Macromol 68:247–250

Gupta JJ, Yadav BPS, Gupta HK, Sahoo SK, Agrahar D (2001) Nutritive value of detoxified jack bean (*Canavalia ensiformis*) seed for young chicks. Indian J Anim Sci 71(12):1169–1171

Hana P, Ivana C, David P, Otakar H, Josef J, Lucy V (2017) Study of effects and conditions on the solubility of natural polysaccharide gum karaya. Food Hydrocoll 68:148–156

Harborne JB, Boulter D, Turner BL (1971) Chemotaxonomy of the leguminosae. Academic, New York

Harikrishna NL, Seshagirirao K, Goycoolea FM (2017) Extraction, purification and characterization of water soluble galactomannans from Mimosa pudica seeds. Euro Biotechnol J 1(4):303–309

Harikrishna NL, Seshagirirao K, Goycoolea FM (2018) Extraction and physicochemical characterization of galactomannans from Dichrostachys cinerea seeds. Food Hydrocoll 82:451–456
Harsha J, Kapoor VP (2003) *Cassia grandis* Linn. f. seed galactomannan: structural and crystallographical studies. Carbohydr Res 338:1907–/1912
Helmut Kaiser Consultancy (2004) Study: nanotechnology in food and food processing industry worldwide 2003e2006e2010e2015. Available at. www.hkc22.com/Nanofood.html
Herold BC, Gerber SI, Polonsky T, Belval BJ, Shaklee PN, Holme K (1995) Identification of structural features of heparin required for inhibition of herpes simplex virus type 1 binding. Virology 206:1108–1116
Jung J, Oh H, Ji J, Kim S (2006) Metal nanoparticle generation using a small ceramic heater with a local heating area. J Aerosol Sci 37:1662–1670
Kapoor VP (1992) Rheological properties of seed galactomannan from *Cassia nodosa* Buchhem. Carbohydr Polym 25:79–84
Kapoor VP, Mukerjee S (1969) Nature of sugars in *Cassia absus* seed polysaccharide. Curr Sci 38:38
Kapoor VP, Francois RT, Joseleau J, Milas M, Chanzy H, Rinaudo M (1998) *Cassia spectabiliss* DC seed galactomannan: structural, crystallographical and rheological studies. Carbohydr Res 306:231–241
Macfarlane GT, Cummings JH (1999) Probiotics and prebiotics: can regulating the activities of intestinal bacteria benefit health? Br Med J 318:999–103
Manjoosha C, Kapoor VP (2001) Structure of a galactomannan from the seeds of *Cassia angustifolia* Vahl. Carbohydr Res 332:439–444
Morris ER (1984) Rheology of hydrocolloids. In: Phillips GO, Wedlock DJ, Williams PA (eds) Gums and stabilisers for the food industry 2. Pergamon Press, Oxford, pp 57–78
Morris ER, Cutler AN, Ross-Murphy SB, Rees DA, Price J (1981) Concentrations and shear rate dependence of viscosity in random coil polysaccharide solutions. Carbohydr Polym 1:5–21
Murphy O (2001) Non-polyol low-digestible carbohydrates: food applications and functional benefits. Br J Nutr 85(Suppl 1):S47
Murray S (2007) Food: the world's biggest industry. www.forbes.com/2007/11/11/growth-agriculture-business-forbeslife-food07-cx_sm_1113bigfood.html
Nwokocha LM, Peter AW (2014) Isolation and characterization of a novel polysaccharide from seeds of *Peltophorum pterocarpum*. Food Hydrocoll 41:319–324
Pal S, Mal D, Singh RP (2007) Synthesis and characterization of cationic guar gum: a high performance flocculating agent. J Appl Polym Sci 105(6):3240–3245
Pauly M, Freis O, Pauly G (1999) Galactomannan and xyloglucan: bioactive polysaccharides. Cosmet Toiletries 114:65–78
Rakesh S, Siddhartha M, Biswanath S (2014) Effect of ionic crosslink on the release of metronidazole from partially carboxymethylated guar gum tablet. Carbohydr Polym 106:414–421
Reid JS (1971) Reserve carbohydrate metabolism in germinating seeds of *Trigonella foenum-graecum*L. (Leguminosae). Planta 106:131–142
Reid JSG, Bewley JD (1979) A duel role for the endosperm and its galactomannan reserves in the germinative physiology of fenugreek (*Trigonella foenum-graecum* L.), an endospermic leguminous seed. Planta 147:145–150
Reid JSG, Dey PM, Dixon RA (1985) In: Dey PM, Dixon RA (eds) Galactomannans in biochemistry of storage carbohydrates in green plants. Academic, London, pp 265–288
Rushdi TA, Pichard C, Khater YH (2004) Control of diarrhea by fiber-enriched diet in ICU patients on enteral nutrition: a prospective randomized controlled trial. Clin Nutr 23:1344–1352
Schneider R, Sostar-Turk S (2003) Good quality printing with reactive dyes using guar gum and biodegradable additives. Dyes Pigments 57(1):7–14
Sharma BR, Dhuldhoya NC, Merchant SU, Merchant UC (2008) A glimpse of galactomannans. Sci Tech Enterpr 3:1–10
Slavin JL, Greenberg NA (2003) Partially hydrolyzed guar gum: clinical nutrition uses. Nutrition 19:549–552

Sostar S, Schneider R (1998) Guar gum as an environment-friendly alternative thickener in printing with reactive dyes. Dyes Pigments 39(4):211–221
Soumya RS, Swapankumar G, Abraham ET (2010) Preparation and characterization guar gum nanoparticles. Int J Biol Macromol 46:267–269
Srivastava M, Kapoor VP (2005) Seed galactomannans: an overview. Chem Biodivers 2:295–317
Stephen AM, Churns SC (1995) Food polysaccharides and their application. Marcel Dekker, New York, pp 1–18
Suvakanta D, Narsimha MP, Pulak D, Joshabir C, Biswajit D (2014) Optimization and characterization of purified polysaccharide from Musa sapientum L. as a pharmaceutical excipient. Food Chem 149:76–83
Taheri A, Razavi SMA (2015) Fabrication of cress seed gum nanoparticles, an anionic polysaccharide, using desolvation technique: an optimization study. Bio Nano Sci 5:1040–1116
Tammishetti S, Thimma R (2001) Barium chloride crosslinked corboxymethyl guar gum beads for gastrointestinal drug delivery. J Appl Polym Sci 82(12):3084–3090
Tauseef S, Sasi Kumar S (2011) Pharmaceutical and pharmacological profile of guar gum an overview. Int J Pharm Pharm Sci 3(5):38–40
Tedesco S, Doyle H, Blasco J, Redmond G, Sheehan D (2010) Oxidative stress and toxicity of gold nanoparticles in *Mytilus edulis*. Aquat Toxicol 100:178–186
Thakur S, Chauhan GS, Ahn JH (2009) Synthesis of acryloyl guar gum and its hydrogel materials for use in the slow release of L-DOPA and L-tyrosine. Carbohydr Polym 76:513
Theisen C (2001) Chemoprevention: what's in a name? J Natl Cancer Inst 93:743–774
Tindale MD, Roux DG (1974) An extended phytochemical survey of Australian species of *Acacia*: chemotaxonomic and phylogenetic aspects. Phytochemistry 13:829–839
Trowell HC, Southgate DAT, Wolever TMS, Leeds AR, Gassull MA, Jenkins DJA (1976) Dietary fiber redefined. Lancet 1:967
Tuohy KM, Kolida S, Lustenberger AM (2001) The prebiotic effects of biscuits containing partially hydrolyzed guar gum and fructo-oligosaccharides- a human volunteer study. Br J Nutr 86:341
Unrau AM (1961) The constitution of a galactomannan from the seed of *Leucaena glauca*. J Organomet Chem 26:3097–3101
Varshosaz J, Tavakoli N, Eram SA (2006) Use of natural gums and cellulose derivatives in production of sustained release metaprolol tablets. Drug Deliv 13:247–265
Vendruscolo CW, Ferrero C, Pineda EAG, Silveira JLM, Freitas RA, Jimenez-Castellanos MR, Bresolin MB (2009) Physicochemical and mechanical characterization of galactomannan from *Mimosa scabrella*; effect of dying method. Carbohydr Polym 76:86–93
Vipul DP, Jani GK, Moradiyaa NG, Randeriaa NP, Nagar BJ, Naikwadi NN, Variya BC (2013) Galactomannan: a versatile biodegradable seed polysaccharide. Int J Biol Macromol 60:83–92
Williams PA, Phillips GO (eds) (2003a) Gums and stabilisers for the food industry 11. The Royal Society of Chemistry, Cambridge, p x 370. ISBN 0-85404-836-7
Williams PA, Phillips GO (2003b) Introduction to food hydrocolloids. In: Phillips GO, Williams PA (eds) Handbook of hydrocolloids. CRC Press, New York, pp 1–19
Winter WT, Chien YY, Bouckris H (1984) Structural aspects of food galactomannans. In: Phillips GO, Wedlock DJ, Williams PA (eds) Gum and stabilisers for the food industry, vol 2. Pergamon, Oxford, pp 535–539
Wu Y, Cui W, Eskin NAM, Goff HD (2009) An investigation of four commercial galactomannans on their emulsion and rheological properties. Food Res Int 42:1141–1146
Wu Y, Li W, Cui SW, Goff HD (2012) A molecular modeling approach to understand conformation–functionality relationships of galactomannans with different mannose/galactose ratios. Food Hydrocoll 26:359–364
Yalpani M (1988) Polysaccharides synthesis, modifications and structure/property relationship. Elsevier, Amsterdam
Yasmin A, Ramesh K, Rajeshkumar S (2014) Optimization and stabilization of gold nanoparticles by using herbal plant extract with microwave heating. Nano Converg 1:12

Chapter 6
Legumes as Preventive Nutraceuticals for Chronic Diseases

Abdelkarim Guaadaoui, Meryem Elyadini, and Abdellah Hamal

Abstract Due to the increase of inadequate consumption habits, healthy diets are becoming an aspect of good practices. In this context, interest in food composition has expanded beyond nutrients to include bioactive compounds that could prevent many diseases. A balanced diet with varied foods constitutes a good source of various bioactive compounds and components. Legumes bioactive compounds might display a wide extend of useful impact on human health which will contribute to their nutraceuticals properties.

In this chapter, we review the diversity of bioactive compounds found in different legumes, such as flavonoids, saponins, alkaloids, carotenoids, tannins, lectins, etc. In a preventive approach, we focus on medicinal application and nutraceuticals properties of the diverse legumes bioactive compounds. All data shows that bioactive compounds and components of legumes provide beneficial effects on human health as well as help in the prevention of many chronic diseases, especially cancers, heart diseases, diabetes and neurodegenerative diseases.

Keywords Legumes · Bioactive compound · Nutraceuticals · Medicinal application · Functional foods · Healthy diet · Preventive approach · Chronic diseases · Health

Abbreviations

CHD	coronary heart disease
CRC	colorectal cancer

A. Guaadaoui (✉) · A. Hamal
Laboratory of Physiology, Genetic and Ethnopharmacology (LPGE), Department of Biology, Faculty of Sciences – Oujda (FSO), Mohammed the First University (UMP), Oujda, Morocco

M. Elyadini
Laboratory of Biochemistry, Environment and Agrifood, Department of Biology, Faculty of Sciences and technology – Mohammedia, Hassan the Second University, Mohammedia, Morocco

P. Guleria et al. (eds.), *Sustainable Agriculture Reviews 45*, Sustainable Agriculture Reviews 45, https://doi.org/10.1007/978-3-030-53017-4_6

CVD	cardiovascular disease
DNA	deoxyribonucleic acid
LDL	low density lipoprotein
L-DOPA	L-3,4-dihydroxyphenylalanine
NDD	neurodegenerative diseases

6.1 Introduction

The concept of a healthy diet is considered an aspect of good and healthy practices. Several studies have demonstrated beneficial health effects of bioactive compounds of certain eating habits and diets (Guaadaoui 2017). In fact, directions in bioactive compounds research have led to major developments in our understanding of their role in nutrition (Muzquiz et al. 2012). Identifying bioactive compounds and improving their effects on human health are the most active areas of scientific research (Guaadaoui 2017; Himri and Guaadaoui 2018).

Bioactive compounds are compounds which have the capability and/or the ability to interact with one or more component(s) of the living tissue by presenting a wide range of probable effects (Guaadaoui et al. 2014a; Guaadaoui 2017). For a nutraceutical, is defined as total or part of food that provide health benefits, including the prevention or treatment of a disease (Costa 2017). Both bioactive compounds and nutraceuticals play an extra- or non-nutritional role in the body, but the concept of bioactivity demand in addition an interaction with a living tissue.

Natural products, including foods and feeds, are a large source of various bioactive compounds. In plants, bioactive compounds are produced generally as secondary metabolites in response to the organism needs and challenges of the natural environment. Those compounds serve a wide range of functions and may present several biological activities (Guaadaoui 2017).

Bioactive compounds are a necessarily portion of the daily human intakes (Guaadaoui et al. 2015a, b, c). Food plants, including legumes, contain a plenitude of bioactive compounds that are found commonly in mixtures. Dietary bioactive compounds have ended up a quality sign of food and nourishment (Park et al. 2012). In fact, the interest in the composition of common food has extended beyond nutrients to incorporate bioactive compounds that might offer assistance to prevent many chronic diseases coexisting with inadequate consumption habits due to the impact of urbanization, industrialization and market globalization (Bovell-Benjamin 2010; Guaadaoui 2017).

As many other foods, legumes are indispensable for human diet. They are a valuable source of bioactive compounds and nutraceuticals that could prevent certain health problems. Indeed, legumes are known for the existence of various bioactive compounds, such as flavonoids, saponins, alkaloids, carotenoids, tannins, lectins, etc. Those bioactive compounds are important for legumes nutraceutical properties and provide beneficial effects on human health, as well as help in the prevention of certain diseases, such as cancers, heart diseases, diabetes, obesity, etc. (Barman et al. 2018; Kamran and Reddy 2018; Cakir et al. 2019).

Due to their remarkable bioactive compounds content and nutraceuticals properties, legumes are receiving great attention by researchers. In this chapter, we participate in the scientific community efforts by focusing on bioactive compounds found in legumes, their chemical characteristics and their medicinal application as nutraceuticals in a preventive approach, especially for some chronic diseases.

6.2 Legumes Bioactive Compounds

Legumes contain an abundance of bioactive compounds and present several biological activities that prevent many diseases. Related to the bioactive compound contents, lentils, common bean, black soybean, cowpea and peanut are the foremost-investigated legumes among many (Winda and Ignasius Radix 2018). Here, we review methodologically the chemical characteristics, sources and related studied effects on medicinal application of the different biocoumpounactives found in legumes.

6.2.1 Phenolic Compounds

Phenolic compounds are a major class of the secondary metabolites. They are pivotal to numerous critical functions of plants life, particularly in the interactions between the plant and its environment. Some plants synthesize several hundred of distinctive known phenolic compounds. The concentrations of the phenolic compounds in plants reveal several grams per kilogram, but are highly related to the environmental and physiological factors, such as stress and stage of maturity (Hansen and Wold 2010). The main phenolic compounds found in legumes are flavonoids (anthocyanin), tannins and phenolic acids (Singh and Basu 2012; Cakir et al. 2019).

Based on their chemical structure, phenolic compounds can be categorized in different ways, with more particular and detailed groups in each class (Guaadaoui 2017). The most of them are naturally occurring with functional derivatives (esters), or found associated with one or many saccharides (Garcia-Salas et al. 2010). The chemical structure of those bioactive compounds is responsible for their properties such as color, flavor and aroma (Serrano et al. 2010). Thus, the phenolic profile could be a useful parameter for the discrimination between legumes. The legume seeds, that are dark colored, are highly pigmented and so have high phenolic content (Barman et al. 2018). Lentil is reported to have the highest contents on phenolic acid (6.56 mg gallic acid equivalents/g), flavonoid (1.30 mg) and condensed tannin (5.97 mg catechin equivalents/g), followed by red kidney and black beans (Singh and Basu 2012).

As nutraceuticals, phenolic compounds potentially contribute to the maintenance of human health. Indeed, in addition to their bioactivities as anti-inflammatory, anti-bacterial, anti-viral and anti-allergenic. Phenolic compounds in legume seeds are also known to reduce the risk of chronic diseases like heart diseases, diabetes and some types of cancers (Siddhuraju and Becker 2007; Cakir et al. 2019).

Due to their ability to scavenge free radicals, chelate metal ions and inhibit lipid peroxidation, phenolic compounds can act as antioxidants (Shweta and Rana 2017; Shahwar et al. 2017), with more antioxidant potential reported for fermented legume seeds (Vedavyas et al. 2012). Studies *in vitro* and *in vivo* both conclude that legumes, particularly in the seed coat, have high antioxidant activity. Moreover, legumes contain important amounts of bioactive compounds. Therefore, they can be developed as functional food ingredients (Winda and Ignasius Radix 2018).

6.2.1.1 Phenolic Acids

Phenolic acids are a large class of polyphenols included in many foods. But, in plant foods, caffeic acid and ferulic acid are the most common (Francisco and Michael 2000). The highest content of these compounds is found in dark-colored foods such as black soybean (*Glycine max*) and kidney beans (*Phaseolus vulgaris*) (Lin and Lai 2006).

Phenolic acids are one of the key compounds responsible for most of the functional properties of many foods (Viuda-Martos et al. 2011). The most important is that many phenolic acids in legumes have shown anti-carcinogenic activity (Mathers 2002).

6.2.1.2 Anthocyanins

A large group of phenolic compounds is anthocyanins. They are naturally occurring plant metabolites belonging to the group of flavonoids and include intense color pigments that results in red, purple, blue. Anthocyanin pigments are usually found on the pericarp (the outer layers of food) (Guaadaoui 2017). Anthocyanin contents are higher in legumes that have a dark color such as black and red (Winda and Ignasius Radix 2018).

Depending on species, we may find a wide variety of anthocyanins with different predominance (Jin-Ming et al. 2003). Cyanidin glycosides were determined as the most common pigment of 40 studied species of legumes. However, malvidin glucoside was predominant in five of seven species of *Phaseolus*, and delphinidin glycosides in four of five species of *Vicia* (Nozzolillo 2011).

Thanks to the anthocyanin content, which could serve as an antioxidant, some types of legumes were thought to have health benefit properties. In the meantime, while thermal processing has been reported to reduce the anthocyanin amount and antioxidant activity, the results are not significant (Winda and Ignasius Radix 2018).

6.2.1.3 Isoflavones and Other Flavonoids

The largest group of phenolic compounds are flavonoids and their derivatives (Hollman and Arts 2000). In the biosynthesis, certain flavonoids, such as anthocyanins, flavones and flavonols, are the final products. Other groups, including chalcones, flavanones, flavanon-3-ols and flavan-3,4-diol, are intermediate and end-products at the same time (Guaadaoui 2017).

The main subclasses of flavonoids are flavonols, flavones and catechins (Hollman and Arts 2000). The structure of flavanones and flavanonols is respectively identical to flavones and flavonols, but with an addition of a 3′-hydroxyl group (Ilhami 2012). Even flavonols tend to be present at lower concentrations than anthocyanins in foods; they have the same characteristic on the correlation between color formation and food content (Liu et al. 2002; Christine and Renée 2004). The red varieties, for example, contain high levels of total flavonols than white or yellow varieties (Pérez-Gregorioa et al. 2010). The predominant flavonols are quercetin, kaempferol, myricetin and isorhamnetin. The most abundant flavones are apigenin and luteolin. The main flavan-3-ols are epi-/gallo- catechins (Kris-Etherton et al. 2004).

Isoflavones, largely reported from the *Leguminosae* family, are flavone isomers of almost the same structure. The most common forms found among many legumes are Genistein and Daidzein, with significant higher levels in soybean (74 and 47 mg/100 g, respectively) (Chang 2002; Cassidy et al. 2006).

Due to their structural similarity, isoflavones, like several other phytochemicals, can affect certain hormone modulation. They are able to act as estrogen agonists or antagonists (17β-estradiol) (Guaadaoui 2017), but isoflavones are around one thousand times less active than estradiol (Kaukovirta-Norja et al. 2004). Nevertheless, acting as phytoestrogen is one of the important bioactivities of isoflavones found in legumes.

Chalcones constitute the biosynthesis precursor of other flavonoids that are particularly abundant in various plants. Despite their presence in food to a significant concentration sometimes, chalcones are categorized as minor flavonoids. During the last decade, chalcones have been given expanding consideration because of numerous pharmacological applications and for their important bioactivities (Francisco and Michael 2000; Rahman 2011).

Apart rare exceptions in nature, flavonoids are conjugated to sugar residues. Because of their common occurrence in plants, flavonoids and their conjugates are important elements of the food and feed, and have numerous functions during plant interactions with biotic and abiotic stress conditions (Stobiecki and Piotr 2006; Guaadaoui 2017).

Flavonoids contained in legume seeds are reported to prevent several diseases such as cancer, aging, immune diseases, cardiovascular diseases and cataract (Espìn et al. 2007; Day et al. 2009). The common forms of isoflavones in legumes are natural phytoestrogens. For that, they are used to treat symptoms of menopause (Li et al. 2015; Shweta and Rana 2017). As antioxidants, they have the ability to inhibit LDL oxidation and thus reduce the risk of atherosclerosis, in addition to help in preventing osteoporosis and act as an anti-cancer agent (Shashank et al. 2015; El Bairi et al. 2016; Cheng et al. 2018).

6.2.1.4 Lignans and Phytoestrogens

Lignans are one of the main classes of polyphenols (C6-C3-C6-C3). They are distributed in different families of the plant kingdom (Aksel 2010; Adlercreutz 2010; Guaadaoui et al. 2015c). Like isoflavones, lignans are an important category of the phytoestrogens family. They have similar structural to the estradiol, but they are less estrogenic than isoflavones (Kaukovirta-Norja et al. 2004; Denny and Buttriss 2008; Guaadaoui 2017).

The cumulative amount of lignans varies between species, even in the same species, according to genetic and environmental variations (Smeds et al. 2009; Guaadaoui et al. 2015c). Matairesinol, secoisolariciresinol, pinoresinol and laricirésinol contribute significantly to total dietary intakes of lignans (Higdon and Drake 2013). In the intestinal microflora, many lignans could be transformed into "mammalian lignans" called enterolignans, such as enterolactone and enterodiol (Papadakis et al. 2008).

It has been reported that lignans induce important biological effects such as antioxidant and estrogenic/anti-estrogenic activities, as well as to prevent diabetes and coronary heart disease. Several studies suggest correlations between elevated enterolignan concentrations in biological fluids and reducing risk of chronic diseases, such as cancers (Guaadaoui 2017).

Due to their polyphenolic and non-steroidal nature quite similar to gonadal estrogen hormone, phytoestrogens in food legumes offer hormone replacement therapy alternatives, even for improving menopausal symptoms. In contrast to their other bioactivities, phytoestrogens have the ability to act as estrogenic in vascular tissue and bone, and maintain an anti-estrogenic effect on reproductive tissue and breast sites. Those bioactive compounds appear to modulate the metabolism of steroid or the detoxification of enzymes, and to interfere with the transport of calcium (Racette et al. 2009; Guaadaoui 2017).

6.2.1.5 Tannins

Tannins occur in higher plants as secondary metabolites. They are derivatives of galloyl esters (gallo-/ellagi-tannins and complex tannins) or condensed tannins (oligo-/poly-meric proanthocyanidins) which may have various interflavanyl coupling and substitution patterns (Andrés-Lacueva et al. 2010).

Tannins, notably hydrolysable tannins, have shown several pharmacological activities (Serrano et al. 2009). Due to their ability to bind proteins, tannins help to eliminate toxins from the intestine. In addition, they mark an anti-bacterial activity (Vedavyas et al. 2012; Shweta and Rana 2017). However, excessive tannin content is reported to have adverse effects as it reduces iron bioavailability and blocks enzymes in the digestive tract (Singh and Basu 2012). Most studies on legumes tannins were oriented toward biofunctional feeds used for forage and pasture.

6.2.1.6 Stilbenes

Originally found in grapes, stilbenes are polyphenols constituted by a 1,2-diphenylethylene (C6-C2-C6). Two forms of this basic structure are obtained in mixture: *cis*- and *trans*- respectively known as Z-stilbene and E-stilbene (Georg et al. 2004). The most common compound of this subclass is the resveratrol (Wang et al. 2002).

In response to environmental stress, resveratrol was found in some legumes family among other plants. Generally, resveratrol acts as phytoalexin, which gives the foods their chemopreventive activities and therapeutic properties (Guaadaoui 2017).

6.2.2 Saponins

In plants, saponins are produced as secondary metabolites. According to their aglycone, they are generally classified into steroidal saponins (less frequent) and triterpenic saponins. In rare cases, both types could be found in the same plant (Xu et al. 2007; Aksel 2010; Podolak et al. 2010). Several saponins could accumulate in a plant. The case of soybeans, for example, presents more than 20 soyasaponins (Güçlü-Ustündağ and Mazza 2007; Isanga and Zhang 2008; Kang et al. 2010).

Pulses are major source of saponins (Singh and Basu 2012). They are found in various beans, chickpeas, lentils and lupins (Cakir et al. 2019). In the human diet, soybean and chickpea constitute major sources of saponins. Broad beans, chickpea and moth bean could contain, respectively, 3.7, 3.6 and 3.4 g per kilogram dry matter of saponins. But, some saponins could be lost during processing (Singh and Basu 2012).

The bond forms with different substituents represent a significant structural diversity of saponins (Podolak et al. 2010). The molecular and structural heterogeneity gives those bioactive compounds a wide range of biochemical properties (Kim 2008; Cheok et al. 2014). Indeed, the chemopreventive role of saponins was proved in a number of studies (Guaadaoui 2017). They also have antioxidant effect that exhibits direct and selective cytotoxic action against cancer cells (Singh and Basu 2012). Saponin's chemical structure confers multiple pharmacological and biological actions, such as antitumor, hypoglycemic, vasoprotectreur immunomodulator, etc. (Afshin et al. 2014; Guaadaoui 2017; Shahwar et al. 2017). The increasing evidence of the saponins bioactivities and their health benefits has renewed their interest in recent years.

6.2.3 Carotenoids and Vitamin A

Carotenoids are an important class of terpenes found mainly in plants, with 8 isoprene units $(C_5H_8)_8$. They are responsible for the yellow, orange, red or purple pigmentation (Hanson 2003; Yahia and Ornelas-Paz 2010; Sánchez-Moreno et al. 2012; Guaadaoui 2017). Carotenoids are classified into two groups: carotenes that contain only carbon atoms and hydrogen, e.g. α-carotene, β-carotene and lycopene, and xanthophylls or oxocarotenoids, with at least one oxygen function (Sánchez-Moreno et al. 2012).

Almost all foods contain some amount of carotenoid compounds (Kim et al. 2007). Generally, the carotenoid concentration in legumes are much lower in comparison with the vegetables and fruits. However, El-Qudah (2014) identified appreciable amounts of carotenoid in faba bean, chickpea, lentil and dry beans. If these foods are consumed in higher amounts, they can contribute significantly to total vitamin A intake (EL-Qudah 2014).

Human body, as all mammals, seems to be unable to synthesize carotenoids, but can incorporate them from foods. For example, the lycopene is concentrated in the prostate; zeaxanthin and lutein are incorporated in the eye macula. Many carotenoids, particularly β-carotene, are vitamin A precursors. The conversion into vitamin A can take place in the intestinal mucosa (Blomhoff 2010; Hansen and Wold 2010). Vitamin A is reported to be abundant in winged bean (*Psophocarpus tetragonolobus*) and acts as powerful antioxidant that prevents DNA damage (Barman et al. 2018).

6.2.4 Tocols and Vitamin E

Tocols or tocochromanols are lipid-soluble molecules and are essential as nutraceuticals. They include α-, β-, γ- or δ-tocopherols and α-, β-, γ- or δ-tocotrienols (Falk and Munné-Bosch 2010; Guaadaoui et al. 2014b). The eight-tocol isoforms take the generic term of Vitamin E, where the abundant and the active form *in vivo* is α-tocopherol (Yoshida et al. 2007). In the human body, the stereoisomers are not inter-convertible (Ball 2006). Therefore, if one enantiomer has the desired beneficial effect, another might have no effect or even an adverse effect (Ryan et al. 2007; Zingg 2007).

Tocols are mainly found in peanuts and beans among other foods. The structure of tocols (vitamin E) performs an excellent antioxidant activity among various functions. Studies suggest that tocols are able to reduce the risk of some cancers, cardiovascular disease and neurodegenerative diseases such as Alzheimer's disease (Guaadaoui et al. 2014b). However, the optimal amount of vitamin E intake for health benefits is uncertain (Bramley et al. 2000).

6.2.5 *Alkaloids*

Alkaloids are a group of secondary metabolites containing nitrogen (Guaadaoui 2017). They often accumulate as a mixture with other non-nitrogen metabolites, such as polyphenols and terpenes (Fattorusso and Taglialatela-Scafati 2008).

Several alkaloids, from different sources, are well characterized pharmacologically and have discovered significant biological activities (Guaadaoui 2017). Alkaloids from edible legumes have been reported from lupins (quinolizidine), peas (trigonelline) chickpeas and lentils (Cakir et al. 2019).

6.2.6 *L-3,4-Dihydroxyphenylalanine and Betalains*

The L-3,4-dihydroxyphenylalanine (L-DOPA) is an important metabolic product of numerous plant families. In some Fabaceae/Leguminosae species, such as fava bean (*Vicia faba*), it accumulates in a large amount. Due to its neurotoxic activity, L-DOPA is used as neurotransmitter to treat effectively the Parkinson's disease (Cakir et al. 2019).

Betalains are a class of water-soluble nitrogen-containing pigments. They include betaxanthins and betacyanins with, respectively, yellow-orange and purplish red colors that replace anthocyanin pigmentations (Guaadaoui et al. 2014b). The synthesis of the betalain core moieties involves only a few enzymatic steps centred on the production and conversion of L-DOPA (Strack et al. 2003; Schwinn 2016).

Betalains have a significant bioactivity. They are reported for their antiradical properties, strong antioxidant and anti-inflammatory activities. They also have anticancer effects on human liver cells (Guaadaoui et al. 2014b; Guaadaoui 2017).

6.2.7 *Lectins*

Lectins or haemagglutinins are group of bioactive peptides, which have the ability to agglutinate cells by binding to specific carbohydrate residues on the cell surface (Cakir et al. 2019). Lectins are found in most plant foods. However, most of legume species are a good source of lectins in human food, especially beans that are considered the important source of lectins (Singh and Basu 2012; Barman et al. 2018). High level of lectins has been reported in kidney beans and very low amount in cowpea and lupin seeds (Singh and Basu 2012).

Lectins are classified as bioactive compounds for their non-nutritional role in the living tissues. Studies demonstrate a protective effect on oxidative DNA damage and cancer chemoprevention, in addition to improve an immunomodulation activity (Ruediger and Gabius 2001; Singh and Basu 2012). However, some lectins could impair the integrity of the intestinal epithelium and thus alter the absorption and utilization of nutrients, especially reduce the digestibility and biological value of dietary proteins. In addition, they might promote the development of food allergy to plants containing lectins (Singh and Basu 2012; Barman et al. 2018).

6.2.8 Fibers

Legumes are a good source of fibers. The amount of dietary fiber per total weight varies between 8.7% in black bean and 2.8% in lupins (Cakir et al. 2019). Dietary fibers include resistant starch, lignin non-starch polysaccharides and non-digestible oligosaccharides. The portion of legume fibers that is not digested by enzymes in the intestinal tract reduces cholesterol level and could improve blood glucose level for diabetic individuals (Kalogeropoulos et al. 2010; Tiwari et al. 2013; Bouchenak and Lamri-Senhadji 2013; Barman et al. 2018).

Even non-digestible legume fibers have non-nutritional role in the body, it is clear that their protective effects is due to their interaction with other chemical complex in the gastrointestinal tract, and not with a living tissue as the case for bioactive compounds like phenolic compounds. Fibers seem to be a good example for differentiate biocompounactives from nutraceuticals. Many phytochemicals, notably oligo- and poly-saccharides in addition to some lipids and proteins, could be considered to have extra-nutritional roles with health benefits, but since they not interact with one or more component(s) of a living tissue, they are not considered as bioactive compounds. Although, they could be classified as nutraceuticals.

6.3 Prevention of Chronic Diseases

As nutraceuticals, legumes provide a variety of nutritional benefits that are important to relieve global health issues (Hawana and Alraei 2019). Indeed, legumes bioactive compounds could contribute to beneficial effects on human health and help to prevent many chronic diseases, especially cancers, heart diseases, diabetes and neurodegenerative diseases.

6.3.1 Cancer Diseases

Considerable evidence links diets that are rich in plant foods, including legumes, with a lower risk of many cancers (Patterson et al. 2009; Turati et al. 2015). As mentioned above, legumes possess numerous nutrients and bioactive factors related to anti-carcinogenic activity, such as phenolic compounds, alkaloids, carotenoids, saponins, fibers, etc. Moreover, epidemiological evidence is available for diets with high legume intakes to reduce the risk of cancer (Barman et al. 2018).

6.3.1.1 Colorectal Cancer

Colorectal cancer (CRC) is one of the world's most prevalent malignancies, with a high rate of increasing in many countries. In fact, CRC is the third most commonly occurring cancer in men and the second most commonly occurring cancer in women (Patterson et al. 2009; Wild et al. 2019).

Suboptimal and unbalanced diet could be a driving risk factor for premature death (Gakidou et al. 2017). Changes in dietary habits, along with changes in related lifestyle factors such as alcohol and tobacco consumption, are projected to influence the rise in CRC (Durko and Malecka-Panas 2014). Therefore, diet has been shown to be one among the foremost vital modifiable risk factors in CRC (Garcia-Larsen et al. 2018).

Legume fibers are an important key factor to reduce the risk of CRC. Epidemiological evidence have shown that non-digestible carbohydrates, including insoluble dietary fibers in legumes, have protective effects against the CRC development. In order to explain the link between high concentrations of dietary fiber and CRC prevention, several mechanisms have been postulated (Obrador 2006; Durko and Malecka-Panas 2014). The increased fiber intake can reduce the intestinal transit time and lead to fecal carcinogens dilution. Additionally, the insoluble fibers, among other non-digestible carbohydrates, are potential prebiotics stimulating bacterial fermentation of fiber to anti-carcinogenic short-chain fatty acids (Sengupta et al. 2006), especially butyric acid that demonstrated anti-tumor and anti-inflammatory activities (Lanza et al. 2006; Martinez-Villaluenga et al. 2008). Moreover, legumes, particularly beans, contain several antioxidants and anti-mutagenic polyphenols that could restrain tumor arrangements (Boateng et al. 2007; Mahmoud et al. 2017).

6.3.1.2 Prostate Cancer

In some countries, the prevalence of prostate cancer in men exceeds all other cancers. Several studies have demonstrated positive correlations between food bioactive compounds and the prevention of prostate cancer (Song-Yi et al. 2008; Plata and Concepcion Masip 2014; Guaadaoui 2017).

Legumes, particularly soybeans, have received considerable attention for their probable contribution to lower the risk of certain cancers, including prostate cancer. Indeed, soy-based foods have important concentrations of phytoestrogens, particularly isoflavones. The (anti-)estrogenic effects of isoflavones have protective effects on hormone-related cancers, including prostate cancer (Cassidy et al. 2006; Zhang et al. 2016; Reger et al. 2017; Sivoňová et al. 2019; Kolonel 2008).

As example, epidemiologic studies have shown the reduction the prostate cancer risk among men with the highest intake of legumes, compared to men with the lowest intake. Nevertheless, for legumes excluding soy-products, results found no significant risk reduction associated with the intake of total or specific isoflavones (Song-Yi et al. 2008; Sivoňová et al. 2019). However, careful consideration should be given when isoflavones are used in the prevention and treatment of prostate cancer (Zhang et al. 2016). Nevertheless, the investigation of the correlation between some legume bioactive compounds and the reduction of prostate cancer risk demands more epidemiologic studies to be revealed more clearly.

6.3.1.3 Gastric Cancer

Globally, the fifth most common malignancy is gastric cancer. In fact, gastric cancer is the third leading cause of cancer death in the world (Siegel et al. 2019). Its prevalence often varies across countries. Overall, more than 70% of cases occur in developing countries (Colquhounet et al. 2015). Incidence rates show a high regional variation, likely due to dietary patterns, genetic differences and carcinogens exposure (Patterson et al. 2009). Therefore, there is an urgent need for effective prevention to reduce the risk of gastric cancer (Lansdorp-Vogelaar and Kuipers 2016).

To date, a poor diet constitutes one of the major key risk factors of gastric cancer. Legumes, among other fresh foods, are widely accepted to have inverse correlations with a decreased risk of gastric cancer (Xu et al. 2019). One of these beneficial effects was due, in part, to the high content of flavonoids (Petrick et al. 2015).

The human body cannot produce flavonoids, but they are present in our diet, and legumes contain an important amount of various flavonoid subclasses. Consumption of flavonoids could therefore reduce the risk of cancer development through various mechanisms, such as defense against DNA damage, blockage of particular carcinogenic pathways, initiation of apoptosis, regulation of cell proliferation, antioxidant properties, anti-inflammatory, angiogenesis inhibition, etc. (Kandaswami et al. 2005; Ramos 2007; Niedzwiecki et al. 2016; Azqueta and Collins 2016; Oh et al. 2016; Khalid et al. 2016; Terao 2017; Sznarkowska et al. 2017).

Moreover, many flavonoids have shown antimicrobial effects that inhibit the development of *H. pylori*, which is another key factor of increasing the gastric cancer risk (Harsha et al. 2017). Although, the inverse correlation between the consumption of flavonoids and the risk of gastric cancer has been examined by numerous studies. It seems that the epidemiological evidence are still unsatisfactory (Zamora-Ros et al. 2012; Petrick et al. 2015), which demands more studies to prove the biological mechanisms of legume flavonoids and other bioactive compounds in preventing gastric cancer.

6.3.2 Heart Diseases

6.3.2.1 Cardiovascular Disease

Cardiovascular disease (CVD) has the highest mortality rate worldwide (Benjamin et al. 2018). It was suggested that CVD deaths could be avoided by adequate dietary modification (Willett 2002). Of course, a healthy diet, like Mediterranean diet, in addition to other lifestyle factors (i.e. physical activity) could avoid critical CVDs (Khaw et al. 2008; Estruch et al. 2013; Guaadaoui 2017).

Legumes, particularly lentils, have been proposed as key dietary factors that can prevent CVD (Camposvega et al. 2010; Anand et al. 2015). Certain functional phytochemicals of legumes may act as antioxidants and anti-inflammatory agents, and can therefore have protective functions on pathological conditions caused by oxidative stress, including CVD (Bouchenak and Lamri-Senhadji 2013). The rich legume bioactive compounds, among which polyphenols, Vitamin E and lignans, have shown potential effects in preventing CVD (Nouri et al. 2016; Benjamin et al. 2018; Papandreou et al. 2019).

The preventive effect of legumes on the risk of CVD can be explained by different possible mechanisms. The most important is their high polyphenol content, especially flavonoids and phenolic acids (Camposvega et al. 2010), with higher concentrations reported in cooked lentils (Kalogeropoulos et al. 2010). Thanks to its bioavailability and bioaccessibility, the polyphenols absorption is an essential mechanism that may help prevent many biological processes implicated in the CVD development (Palafox-Carlos et al. 2011). However, some dietary macromolecules could interfere with polyphenols' bioavailability and bioaccessibility, which decreases the beneficial effects on preventing CVD (Patterson et al. 2009).

6.3.2.2 Coronary Heart Disease

Coronary heart disease (CHD) is among the leading causes of death in many countries (Wild et al. 2019). Lifestyle, particularly unhealthy dietary habits, is a major contributor to the development of CHD and its associated risk factors (Scarborough et al. 2011; Moran et al. 2014).

Healthy food is a nutritional strategy to prevent the risk of CHD. Indeed, it appears that intact legumes, particularly whole pulses, are one of the best carbohydrate sources, which are widely recommended in diets to reduce the risk of CHD (Mann 2007). They participate in decreasing the major risk factors for CHD: the triglycerides and the LDL-cholesterol, and affecting other risk factors such as obesity and diabetes (Anderson et al. 2000; Winham et al. 2007; Shashank et al. 2015).

6.3.3 Diabetes

The global diabetes prevalence is increasing rapidly and parallel to the outputs of the sedentary lifestyle adoption (reduced physical activity, changed diets, etc.), notably in the developing countries (Patterson et al. 2009). Lifestyle modifications, with appropriate diet (dietary therapy), in addition to exercise programs, have shown effective potentials in the prevention of diabetes (Poonam et al. 2014).

The use of legumes in a healthy diet can help in reducing related-diabetes risks and preventing diabetic factors (Rizkalla et al. 2002). Several studies have demonstrated beneficial effects of legumes consumption on diabetes complications (Poonam et al. 2014; Marventano et al. 2017).

Due to the abundance of resistant starch and non-starch oligo-/poly-saccharides, legumes are known by a low glycemic index (Atkinson et al. 2008). One of the major effect of those nutraceuticals is to slow the starch digestion in the small intestine, which decrease blood glucose (Messina 1999; Mannucci et al. 2013). Moreover, some legumes bioactive compounds may control the glucose release in diabetics and therefore ensure the health benefits of legumes, especially whole pulses (Champ 2002; Mateos-Aparicio et al. 2008; Singh et al. 2017; Clark et al. 2018). Additional studies are needed to better define anti-diabetic mechanisms responsible for modulating the cellular response, in correlation with the intake of bioactive compounds and nutraceuticals of legumes.

6.3.4 Neurodegenerative Diseases

Nowadays, neurodegenerative diseases (NDD), including Parkinson disease, Alzheimer disease, Huntington disease, etc. constitute a public health problem in many countries. The delay of the NDD evolution is a challenge for both scientific community and public health institutions. Although, increasing interest for preventive approach in halting neuronal damage/death has been shown, since the NDD treatment has revealed limited effectiveness because the majority of NDD manifest late and remain asymptomatic for most stages until leading to advanced degeneration (Joseph et al. 2009; Sofi 2009; Solanki et al. 2015; Sofi et al. 2010).

The understanding of the initiating neuronal damage mechanisms in various NDD has known many progresses. Mechanisms involved in NDD are multifactorial which restrict the benefits of therapeutic interventions. However, common pathways are presented, including inflammation and oxidative stress among others. Compared to other organs, our brain is marked by a low activity of antioxidant defense systems, so it is more susceptible to oxidative stress (Solanki et al. 2015; Javier et al. 2018).

In this context, dietary and lifestyle changes know increasing interest and constitute one of future directions in enhancing attention in NDD (Kamphuis and Scheltens 2010; Vassallo and Scerri 2013; Sharma et al. 2016). Indeed, monitoring a diet rich in bioactive compounds with proven antioxidant activity in vivo, could decrease the NDD evolution and therefore improve the life quality of patients. Moreover, such preventive strategy could reduce and/or eliminate the primary stressor, which restore the function of neurons (Javier et al. 2018).

Legume polyphenols, as other foods, are able to activate important antioxidant mechanisms. Many studies reported the role of various flavonoids in the NDD prevention. By their antioxidant properties and the interaction with cellular signaling pathways, flavonoids exhibit neuro-protective properties, such as decreasing oxidative stress and preventing inflammation and neurotoxicity, which improve cognitive function and enhance memory (Solanki et al. 2015). Resveratrol, another phenolic compound, has shown implication in various NDD pathways in human organism, more particularly for Alzheimer disease (Ahmed et al. 2017). Moreover, resveratrol was reported to treat cognitive impairment in some animal models (Marhuenda et al. 2017). More epidemiological and ethno-pathological studies are demanded for benefiting from all opportunities that give bioactive compounds and nutraceuticals of legumes.

6.4 Conclusion

Legumes have proven to be indispensable for human diet. They are a valuable source of bioactive compounds and nutraceuticals. This chapter revealed the diversity of bioactive compounds found in different legumes, such as flavonoids, saponins, alkaloids, carotenoids, tannins, lectins, etc.

Legumes bioactive compounds could help to prevent many chronic diseases coexisting with inadequate consumption habits. Indeed, scientific data show that legumes bioactive compounds provide beneficial effects on human health as well as help in the prevention of many chronic diseases, especially cancers, heart diseases, diabetes and neurodegenerative diseases.

In addition to their role in the eradication of hunger, the consolidation of healthy eating habits and the protection of the environment, legumes continue to gain attention as nutraceuticals and functional foods. Although, there are also many underutilized food legumes that might be a potential source of nutraceuticals, which demand more research in the aim to reveal the quasi-totality of the legumes potential in term of bioactive compounds and their probable beneficial effects.

References

Adlercreutz H (2010) Phytoestrogens and human health. In: Bioactive compounds in plants – benefits and risks for man and animals. The Norwegian Academy of Science and Letters, Oslo, pp 78–93. https://doi.org/10.11648/o.20100303.78

Afshin A et al (2014) Consumption of nuts and legumes and risk of incident ischemic heart disease, stroke, and diabetes: a systematic review and meta-analysis. Am J Clin Nutr 100(1):278–288. https://doi.org/10.3945/ajcn.113.076901. PMid:24898241

Ahmed T et al (2017) Resveratrol and Alzheimer's disease: mechanistic insights. Mol Neurobiol 54(4):2622–2635. https://doi.org/10.1007/s12035-016-9839-9

Aksel B (2010) A brief review on bioactive compounds in plants. In: Bioactive compounds in plants – benefits and risks for man and animals. The Norwegian Academy of Science and Letters, Oslo, pp 11–17. https://doi.org/10.11648/o.20100303.11

Anand SS et al (2015) Food consumption and its impact on cardiovascular disease: importance of solutions focused on the globalized food system: a report from the workshop convened by the world heart federation. J Am Coll Cardiol 66:1590–1614. https://doi.org/10.1016/j.jacc.2015.07.050

Anderson JW et al (2000) Whole grain foods and heart disease risk. J Am Coll Nutr 19:291. https://doi.org/10.1080/07315724.2000.10718963

Andrés-Lacueva C et al (2010) Phenolic compounds: chemistry and occurrence in fruits and vegetables. In: Fruit and vegetable phytochemicals: chemistry, nutritional value and stability. Wiley-Blackwell, Oxford, pp 53–87. https://doi.org/10.1002/9780813809397

Atkinson FS et al (2008) International tables of glycemic index and glycemic load values. Diabetes Care 31:2281. https://doi.org/10.2337/dc08-1239

Azqueta A, Collins A (2016) Polyphenols and DNA damage: a mixed blessing. Nutrients 8:785. https://doi.org/10.3390/nu8120785

Ball GFM (2006) Vitamins in foods. CRC Taylor & Francis, Boca Raton, pp 119–136. https://doi.org/10.1080/10942910600853899

Barman A et al (2018) Nutraceutical properties of legume seeds and their impact on human health (chapter 1). In: Legume seed nutraceutical research. IntechOpen, London, pp 01–24. https://doi.org/10.5772/intechopen.78799

Benjamin EJ et al (2018) Correction to: heart disease and stroke statistics 2018 update: a report from the American Heart Association. Circulation 137:493. https://doi.org/10.1161/CIR.0000000000000558

Blomhoff R (2010) Role of dietary phytochemicals in oxidative stress. In: Bioactive compounds in plants – benefits and risks for man and animals. The Norwegian Academy of Science and Letters, Oslo, pp 52–70. https://doi.org/10.11648/o.20100303.52

Boateng JA et al (2007) Inhibitory effects of selected dry beans (Phaseolus spp. L.) on azoxymethane induced formation of aberrant crypt foci in Fisher 344 male rats. Nutr Res 27:640. https://doi.org/10.1016/j.nutres.2007.07.004

Bouchenak M, Lamri-Senhadji M (2013) Nutritional quality of legumes, and their role in cardiometabolic risk prevention. J Med Food 16(3):185–198. https://doi.org/10.1089/jmf.2011.0238

Bovell-Benjamin AC (2010) Bioactivity, benefits and safety of traditional and ethnic foods (chapter 22). In: Ensuring global food safety: exploring global harmonization. Elsevier Inc, pp 363–382. https://doi.org/10.1016/B978-0-12-374845-4.00022-9

Bramley M et al (2000) Vitamin E. J Sci Food Agric 80(7):913–938. https://doi.org/10.1002/(SICI)1097-0010(20000515)80:7<913::AID-JSFA600>3.0.CO;2-3

Cakir Ö et al (2019) Nutritional and health benefits of legumes and their distinctive genomic properties. Food Sci Technol 39(1):1–12. https://doi.org/10.1590/fst.42117

Camposvega R et al (2010) Minor components of pulses and their potential impact on human health. Food Res Int 43:461–482. https://doi.org/10.1016/j.foodres.2009.09.004

Cassidy A et al (2006) Critical review of health effects of soyabean phyto-oestrogens in post-meno- pausal women. Proc Nutr Soc 65(1):76–92. https://doi.org/10.1079/PNS2005476

Champ MMJ (2002) Non-nutritive bioactive sub- stances of pulses. Br J Nutr 88:S307. https://doi.org/10.1079/BJN2002721

Chang SKC (2002) Isoflavones from soybeans and soy foods. In: Functional foods. Volume 2. Biochemical and processing aspects. CRC Press, Boca Raton, pp 22–57. https://doi.org/10.5772/18067

Cheng J et al (2018) Cyanidin chloride inhibits ovariectomy-induced osteoporosis by suppressing rankl-mediated osteoclastogenesis and associated signaling pathways. J Cell Physiol 233:2502–2512. https://doi.org/10.1002/jcp.26126

Cheok CY et al (2014) Extraction and quantification of saponins: a review. Food Res Int 59:16–40. https://doi.org/10.1016/j.foodres.2014.01.057

Christine AW, Renée JG (2004) Anthocyanins and other flavonoids. Nat Prod Rep 21(4):539–573. https://doi.org/10.1039/b311404j

Clark JL et al (2018) Rebelling against the (insulin) resistance: a review of the proposed insulin-sensitizing actions of soybeans, chickpeas, and their bioactive compounds. Nutrients 10(4):434. https://doi.org/10.3390/nu10040434

Colquhounet A et al (2015) Global patterns of cardia and non-cardia gastric cancer incidence in 2012. Gut 10:1–8. https://doi.org/10.1136/gutjnl-2014-308915

Costa JP (2017) A current look at nutraceuticals-key concepts and future prospects. Trends Food Sci Technol 62:68–78. https://doi.org/10.1016/j.tifs.2017.02.010

Day L et al (2009) Incorporation of functional ingredients into foods. Trends Food Sci Technol 20(9):388–395. https://doi.org/10.1016/j.tifs.2008.05.002
Denny A, Buttriss J (2008) Plant foods and health: focus on plant bioactives. Synthesis report no. 4. European Food Information Resource (EuroFIR). British Nutrition Foundation, Norwich, pp 1–64
Durko L, Malecka-Panas E (2014) Lifestyle modifications and colorectal cancer. Curr Colorectal Cancer Rep 10:45–54. https://doi.org/10.1007/s11888-013-0203-4
El Bairi K et al (2016) Natural products against cancer angiogenesis. Tumour Biol 37(11):14513–14536. https://doi.org/10.1007/s13277-016-5364-8
EL-Qudah JM (2014) Estimation of carotenoid contents of selected mediterranean legumes by HPLC. World J Med Sci 10(1):89–93. https://doi.org/10.5829/idosi.wjms.2014.10.1.81202
Espìn JC et al (2007) Nutraceuticals: facts and fiction. Phytochemistry 68(22–24):2986–3008. https://doi.org/10.1016/j.phytochem.2007.09.014
Estruch R et al (2013) Mediterranean diet for primary prevention of cardiovascular disease. N Engl J Med 368:1279e90. https://doi.org/10.1056/NEJMc1306659
Falk J, Munné-Bosch S (2010) Tocochromanol functions in plants: antioxidation and beyond. J Exp Bot 61(6):1549–1566. https://doi.org/10.1093/jxb/erq030
Fattorusso E, Taglialatela-Scafati O (2008) Modern alkaloids: structure, isolation, synthesis and biology. Wiley-Vch Verlag GmbH & Co. KGaA, Weinheim, pp 1–26. https://doi.org/10.1002/9783527621071
Francisco ATB, Michael NC (2000) Dietary hydroxybenzoic acid derivatives – nature, occurrence and dietary burden. J Sci Food Agric 80(7):1024–1032. https://doi.org/10.1002/(SICI)1097-0010(20000515)80:7<1024::AID-JSFA567>3.0.CO;2-S
Gakidou E et al (2017) Global, regional, and national comparative risk assessment of 84 behavioural, environmental and occupational, and metabolic risks or clusters of risks, 1990–2016: a systematic analysis for the global burden of disease study 2016. Lancet 390(10100):1345–1422. https://doi.org/10.1016/S0140-6736(17)32366-8
Garcia-Larsen V et al (2018) Diet during pregnancy and infancy and risk of allergic or autoimmune disease: a systematic review and meta-analysis. PLoS Med 15(2):e1002507. https://doi.org/10.1371/journal.pmed.1002507
Garcia-Salas P et al (2010) Phenolic-compound-extraction systems for fruit and vegetable samples. Molecules 15:8813–8826. https://doi.org/10.3390/molecules15128813
Georg W et al (2004) Synthesis and biological evaluation of stilbene-based pure estrogen antagonists. Bioorg Med Chem Lett 14:4659–4663. https://doi.org/10.1016/j.bmcl.2004.06.098
Guaadaoui A (2017) Recent advances in bioactivities of common food biocompounactives (chapter 25). In: Yahia EM (ed) Fruit and vegetable phytochemicals: chemistry and human health, vol I, 2nd edn. Wiley, Hoboken, pp 541–593. https://doi.org/10.1002/9781119158042.ch25
Guaadaoui A et al (2014a) What is a bioactive compound? A combined definition for a preliminary consensus. Int J Nutr Food Sci 3(3):174–179. https://doi.org/10.11648/j.ijnfs.20140303.16
Guaadaoui A et al (2014b) Betalains, polyacetylenes and tocols as biocompounactives: a concise review for enriching the bioactivity concept. Int J Nutr Food Sci 3(4):230–237. https://doi.org/10.11648/j.ijnfs.20140304.11
Guaadaoui A et al (2015a) The preventive approach of biocompounactives (1): a review in recent advances in common vegetables and legumes. Int J Nutr Food Sci 4(1):89–102. https://doi.org/10.11648/j.ijnfs.20150401.23
Guaadaoui A et al (2015b) The preventive approach of biocompounactives (2): a review in recent advances in common fruits. Int J Nutr Food Sci 4(2):189–207. https://doi.org/10.11648/j.ijnfs.20150402.21
Guaadaoui A et al (2015c) The preventive approach of biocompounactives (3): a review in recent advances in in cereals and some animal-based foods. Int J Nutr Food Sci 4(3):295–309. https://doi.org/10.11648/j.ijnfs.20150403.17
Güçlü-Ustündağ O, Mazza G (2007) Saponins: properties, applications and processing. Crit Rev Food Sci Nutr 47(3):231–258. https://doi.org/10.1080/10408390600698197

Hansen M, Wold AB (2010) Contents of bioactive compounds in food plants as affected bytraditional breeding and environmental factors. In: Bioactive compounds in plants – benefits and risks for man and animals. The Norwegian Academy of Science and Letters, Oslo, pp 212–222

Hanson JR (2003.: 147p) Natural products: the secondary metabolites. The Royal Society of Chemistry, London. https://doi.org/10.1039/9781847551535

Harsha C et al (2017) Antiulcer properties of fruits and vegetables: a mechanism based perspective. Food Chem Toxicol 108:104–119. https://doi.org/10.1016/j.fct.2017.07.023

Hawana S, Alraei R (2019) Benefits and preparation effects on polyphenol-rich fruits and vegetables towards cardiovascular disease prevention. J Nutr Food Sci 9:1. https://doi.org/10.4172/2155-9600.1000749

Higdon JJ, Drake VJ (2013) Evidence-based approach to phytochemicals and other dietary factors, 2nd edn. Georg Thieme Verlag KG, Stuttgart, pp 126–132. https://doi.org/10.1055/b-002-76301

Himri I, Guaadaoui A (2018) Cell and organ drug targeting: types of drug delivery systems and advanced targeting strategies (chapter I). In: Grumezescu AM (ed) Design of nanostructures for cells, organs and tissue engineering applications, 1st edn. Elsevier, Oxford. ISBN: 9780128136652, pp 01–66. https://doi.org/10.1016/B978-0-12-813665-2.00001-6

Hollman PCH, Arts ICW (2000) Flavonols, flavones and flavanols-nature, occurrence and dietary burden. J Sci Food Agric 80:1081–1093. https://doi.org/10.1002/(SICI)1097-0010(20000515)80:7<1081::AID-JSFA566>3.0.CO;2-G

Ilhami G (2012) Antioxidant activity of food constituents: an overview. Arch Toxicol 86:345–391. https://doi.org/10.1007/s00204-011-0774-2

Isanga J, Zhang GN (2008) Soybean bioactive components and their implications to health – a review. Food Rev Int 24:252–276. https://doi.org/10.1080/87559120801926351

Javier MH et al (2018) Bioactive compounds contained in Mediterranean diet and their effects on neurodegenerative diseases. In: Current topics on superfoods. IntechOpen. https://doi.org/10.5772/intechopen.74084

Jin-Ming K et al (2003) Analysis and biological activities of anthocyanins. Phytochemistry 64(5):923–933. https://doi.org/10.1016/S0031-9422(03)00438-2

Joseph J et al (2009) Nutrition, brain aging, and neurodegeneration. J Neurosci 29(41):12795–12801. https://doi.org/10.1523/JNEUROSCI.3520-09.2009

Kalogeropoulos N et al (2010) Nutritional evaluation and bioactive microconstituents (phytosterols, tocopherols, polyphenols, triterpenic acids) in cooked dry legumes usually consumed in the Mediterranean countries. Food Chem 121(3):682–690. https://doi.org/10.1016/j.foodchem.2010.01.005

Kamphuis PJ, Scheltens P (2010) Can nutrients prevent or delay onset of Alzheimer's disease? J Alzheimers Dis 20(3):765–775. https://doi.org/10.3233/JAD-2010-091558

Kamran F, Reddy N (2018) Bioactive peptides from legumes: functional and nutraceutical potential. Recent Adv Food Saf 1(3):134–139. https://doi.org/10.1002/9780813811048

Kandaswami C et al (2005) The antitumor activities of flavonoids. In Vivo 19:895–909

Kang J et al (2010) Non-isoflavone phytochemicals in soy and their health effects. J Agric Food Chem 58:8119–8133. https://doi.org/10.1021/jf100901b

Kaukovirta-Norja A et al (2004) Germination: a means to improve the functionality of oat. Agric Food Sci 13(1–2):100–112. https://doi.org/10.2137/1239099041838049

Khalid EB et al (2016) Natural products against cancer angiogenesis. Tumour Biol 37:14513–14536. https://doi.org/10.1007/s13277-016-5364-8

Khaw KT et al (2008) Combined impact of health behaviours and mortality in men and women: the EPIC-Norfolk prospective population study. PLoS Med 5(1):12. https://doi.org/10.1371/journal.pmed.0050012

Kim J (2008) Protective effects of Asian dietary items on cancers – soy and ginseng. Asian Pac J Cancer Prev 9:543–548

Kim Y et al (2007) Tocopherol and carotenoid contents of selected Korean fruits and vegetables. J Food Compos Anal 20(6):458–465. https://doi.org/10.1016/j.jfca.2007.02.001

Kolonel LN (2008) Legume and isoflavone intake and prostate cancer risk: the multiethnic cohort study. Int J Cancer 123(4):927–932. https://doi.org/10.1002/ijc.23594

Kris-Etherton PM et al (2004) Bioactive compounds in nutrition and health-research methodologies for establishing biological function: the antioxidant and anti-inflammatory effects of flavonoids on atherosclerosis. Annu Rev Nutr 24:511–538. https://doi.org/10.1146/annurev.nutr.23.011702.073237

Lansdorp-Vogelaar I, Kuipers EJ (2016) Screening for gastric cancer in Western countries. Gut: Int J Gastroenterol Hepatol 65:543–544. https://doi.org/10.1136/gutjnl-2014-308588

Lanza E et al (2006) High dry bean intake and re- duced risk of advanced colorectal adenoa re-currence among participants in the polyp prevention trial. J Nutr 136:1896. https://doi.org/10.1093/jn/136.7.1896

Li LJ et al (2015) Quantitative efficacy of soy isoflavones on menopausal hot flashes. Br J Clin Pharmacol 79(4):593–604. https://doi.org/10.1111/bcp.12533. PMid:25316502

Lin PY, Lai HM (2006) Bioactive compounds in legumes and their germinated products. J Agric Food Chem 54:3807–3814. https://doi.org/10.1021/jf060002o

Liu M et al (2002) Antioxidant and antiproliferative activities of raspberries. J Agric Food Chem 50:2926–2930. https://doi.org/10.1021/jf0111209

Mahmoud S et al (2017) Crop genetic engineering: an approach to improve fungal resistance in plant system (chapter 23). In: Dhananjaya PS et al (eds) Plant-microbe interactions in agro-ecological perspectives: volume 2. Springer, Singapore, pp 581–592. https://doi.org/10.1007/978-981-10-6593-4_23

Mann J (2007) Dietary carbohydrate: relationship to cardiovascular disease and disorders of carbohydrate metabolism. Eur J Clin Nutr 61:100. https://doi.org/10.1038/sj.ejcn.1602940

Mannucci E et al (2013) Is glucose control important for prevention of cardiovascular disease in diabetes. Diabetes Care 36(Suppl 2):S259–S263. https://doi.org/10.2337/dcS13-2018

Marhuenda J et al (2017) Melatonin and hydroxytyrosol protect against oxidative stress related to the central nervous system after the ingestion of three types of wine by healthy volunteers. Food Funct 8(1):64–74. https://doi.org/10.1039/c6fo01328g

Martinez-Villaluenga C et al (2008) Alpha-galactosides: antinutri- tional factors or functional ingredients? Crit Rev Food Sci Nutr 48:301. https://doi.org/10.1080/10408390701326243

Marventano S et al (2017) Legume consumption and CVD risk: a systematic review and meta-analysis. Public Health Nutr 20(2):245–254. https://doi.org/10.1017/S1368980016002299

Mateos-Aparicio I et al (2008) Soybean, a promising health source. Nutr Hosp 23(4):305–312

Mathers JC (2002) Pulses and carcinogenesis: potential for the prevention of colon, breast and other cancers. Br J Nutr 88(3):S273–S279. https://doi.org/10.1079/BJN2002717. PMid:12498627

Messina MJ (1999) Legumes and soybeans: overview of their nutritional profiles and health effects. Am J Clin Nutr 70:439S–450S. https://doi.org/10.1093/ajcn/70.3.439s

Moran A et al (2014) 1990–2010 global cardiovascular disease atlas. Glob Heart 9:3–16. https://doi.org/10.1016/j.gheart.2014.03.1220

Muzquiz M, Varela A, Burbano C, Cuadrado C, Guillamón E, Pedrosa MM (2012) Bioactive compounds in legumes: Pronutritive and antinutritive actions. Implications for nutrition and health. Phytochem Rev 11(2-3):227–244. https://doi.org/10.1007/s11101-012-9233-9

Niedzwiecki A, Roomi MW, Kalinovsky T, Rath M (2016) Anticancer efficacy of polyphenols and their combinations. Nutrients 8:552. https://doi.org/10.3390/nu8090552

Nouri F et al (2016) Intake of legumes and the risk of cardiovascular disease: frailty modeling of a prospective cohort study in the Iranian middle-aged and older population. Eur J Clin Nutr 70:217–221. https://doi.org/10.1038/ejcn.2015.153

Nozzolillo C (2011) A survey of anthocyanin pigments in seedling legumes. Can J Bot 51:911–915. https://doi.org/10.1139/b73-113

Obrador A (2006) Fibre and colorectal cancer: A controversial question. Br J Nutr 96:S46. https://doi.org/10.1079/BJN20061700

Oh J et al (2016) Therapeutic effectiveness of anticancer phytochemicals on cancer stem cells. Toxins 8:199. https://doi.org/10.3390/toxins8070199

Palafox-Carlos H et al (2011) The role of dietary fiber in the bioaccessibility and bioavailability of fruit and vegetable antioxidants. J Food Sci 76:R6–R15. https://doi.org/10.1111/j.1750-3841.2010.01957.x

Papadakis EN et al (2008) Effect of the form of the sesame-based diet on the absorption of lignans. Br J Nutr 100:1213–1219. https://doi.org/10.1017/S0007114508978272

Papandreou C et al (2019) Legume consumption and risk of all-cause, cardiovascular, and cancer mortality in the PREDIMED study. Clin Nutr 38:348–356. https://doi.org/10.1016/j.clnu.2017.12.019

Park YS et al (2012) Analytical determination of bioactive compounds as an indication of fruit quality. J AOAC Int 95(6):1725–1732. https://doi.org/10.5740/jaoacint.12-130

Patterson CA, Maskus H, Dupasquier C (2009) Pulse crops for health. Cereal Foods World 54(3):109–112. https://doi.org/10.1094/CFW-54-3-0108

Pérez-Gregorioa RM et al (2010) Identification and quantification of flavonoids in traditional cultivars of red and white onions at harvest. J Food Compos Anal 23:592–598. https://doi.org/10.1016/j.jfca.2009.08.013

Petrick JL et al (2015) Dietary intake of flavonoids and oesophageal and gastric cancer: incidence and survival in the United States of America (USA). Br J Cancer 112:1291–1300. https://doi.org/10.1038/bjc.2015.25

Plata BA, Concepcion Masip T (2014) Prostate cancer epidemiology. Arch Esp Urol 67:373–382

Podolak I et al (2010) Saponins as cytotoxic agents: a review. Phytochem Rev (9):425–474. https://doi.org/10.1007/s11101-010-9183-z

Poonam S et al (2014) Antidiabetic potential of commonly consumed legumes: a review. Crit Rev Food Sci Nutr 54(5):655–672. https://doi.org/10.1080/10408398.2011.604141

Racette SB et al (2009) Phytosterol-deficient and high-phytosterol diets developed for controlled feeding studies. J Am Diet Assoc 109(12):2043–2051. https://doi.org/10.1016/j.jada.2009.09.009

Rahman MA (2011) Chalcone: a valuable insight into the recent advances and potential pharmacological activities. Chem Sci J CSJ-29:1–16. https://doi.org/10.1002/chin.201132239

Ramos S (2007) Effects of dietary flavonoids on apoptotic pathways related to cancer chemoprevention. J Nutr Biochem 18:427–442. https://doi.org/10.1016/j.jnutbio.2006

Reger MK et al (2017) Dietary intake of isoflavones and coumestrol and the risk of prostate cancer in the prostate, lung, colorectal and ovarian Cancer screening trial. Int J Cancer 142:719–728. https://doi.org/10.1002/ijc.31095

Rizkalla SW et al (2002) Health benefits of low glycaemic index foods, such as pulses, in diabetic patients and healthy individuals. Br J Nutr 88:S255. https://doi.org/10.1079/BJN2002715

Ruediger H, Gabius HJ (2001) Plant lectins: occurrence, biochemistry, functions and applications. Glycoconj J 18:589–613. https://doi.org/10.1023/A:1020687518999

Ryan D et al (2007) Bioactivity of oats as it relates to cardiovascular disease. Nutr Res Rev 20(2):147–162. https://doi.org/10.1017/S0954422407782884

Sánchez-Moreno C et al (2012) Nutritional quality of fruits (chapter 5). In: Handbook of fruits and fruit processing (2nd edition), part 1: biology, biochemistry, nutrition, and microbiology. Wiley, Ames, pp 73–84. https://doi.org/10.1002/9781118352533.ch5

Scarborough P et al (2011) Differences in coronary heart disease, stroke and cancer mortality rates between England, Wales, Scotland and Northern Ireland: the role of diet and nutrition. BMJ Open 1:e000263. https://doi.org/10.1136/bmjopen-2011-000263

Schwinn KE (2016) The dope on l-DOPA formation for betalain pigments. New Phytol 210(1):06–09. https://doi.org/10.1111/nph.13901

Sengupta S, Muir JG, Gibson PR (2006) Does butyrate protect from colorectal cancer? J Gastroenterol Hepatol (Australia) 21(1):209–218. https://doi.org/10.1111/j.1440-1746.2006.04213.x

Serrano J et al (2009) Tannins: current knowledge of food sources, intake, bioavailability and biological effects. Mol Nutr Food Res 53:S310–S329. https://doi.org/10.1002/mnfr.200900039

Serrano M et al (2010) Antioxidant and nutritive constituents during sweet pepper development and ripening are enhanced by nitrophenolate treatments. Food Chem 118(3):497–503. https://doi.org/10.1016/j.foodchem.2009.05.006

Shahwar D et al (2017) Health functional compounds of lentil (Lens culinaris Medik): a review. Int J Food Prop 20:1–15. https://doi.org/10.1080/10942912.2017.1287192

Sharma K et al (2016) Enhancing attention in neurodegenerative diseases: current therapies and future directions. Transl Neurosci 7(1):98–109. https://doi.org/10.1515/tnsci-2016-0016

Shashank A et al (2015) Nutraceutical potential of soybean: review. Asian J Clin Nutr 7:22–32. https://doi.org/10.3923/ajcn.2015.22.32

Shweta KM, Rana A (2017) Bioactive components of Vigna species: current prospective. Bull Environ Pharmacol Life Sci 6:1–13. Online ISSN 2277-1808

Siddhuraju P, Becker K (2007) The antioxidant and free radical scavenging activities of processed cowpea (*Vigna unguiculata* (L.) Walp.) seed extracts. Food Chem 101:10–19. https://doi.org/10.1016/j.foodchem.2006.01.004

Siegel RL, Miller KD, Jemal A (2019) Cancer statistics, 2019. CA Cancer J Clin 69:7–34. https://doi.org/10.3322/caac.21551

Singh J, Basu PS (2012) Non-nutritive bioactive compounds in pulses and their impact on human health: an overview. Food Nutr Sci 3:1664–1672. https://doi.org/10.4236/fns.2012.312218

Singh B et al (2017) Bioactive constituents in pulses and their health benefits. J Food Sci Technol 54(4):858–870. https://doi.org/10.1007/s13197-016-2391-9

Sivoňová MK et al (2019) Androgen receptor and soy isoflavones in prostate cancer (review). Mol Clin Oncol 10(2):191–204. https://doi.org/10.3892/mco.2018.1792

Smeds AI et al (2009) Characterization of variation in the lignan content and composition of winter rye, spring wheat, and spring oat. J Agric Food Chem 57:5837–5842. https://doi.org/10.1021/jf9004274

Sofi F (2009) The Mediterranean diet revisited: evidence of its effectiveness grows. Curr Opin Cardiol 24(5):442–446. https://doi.org/10.1097/HCO.0b013e32832f056e

Sofi F et al (2010) Effectiveness of the Mediterranean diet: can it help delay or prevent Alzheimer's disease? J Alzheimers Dis 20(3):795–801. https://doi.org/10.3233/JAD-2010-1418

Solanki I et al (2015) Flavonoid-based therapies in the early management of neurodegenerative diseases. Adv Nutr 6:64–72. https://doi.org/10.3945/an.114.007500

Song-Yi P et al (2008) Legume and isoflavone intake and prostate cancer risk: the multiethnic cohort study. Int J Cancer 123(4):927–932. https://doi.org/10.1002/ijc.23594

Stobiecki M, Piotr K (2006) Isolation and identification of flavonoids. In: The science of flavonoids. Springer Science+Business Media, Inc., New York, pp 47–70. https://doi.org/10.1007/978-0-387-28822-2_2

Strack D, Vogt T, Schliemann W (2003) Recent advances in betalain research. Phytochemistry 62:247–269. https://doi.org/10.1016/S0031-9422(02)00564-2

Sznarkowska A et al (2017) Inhibition of cancer antioxidant defense by natural compounds. Oncotarget 8:15996–16016. https://doi.org/10.18632/oncotarget.13723

Terao J (2017) Factors modulating bioavailability of quercetin-related flavonoids and the consequences of their vascular function. Biochem Pharmacol 139:15–23. https://doi.org/10.1016/j.bcp.2017.03.021

Tiwari AT et al (2013) Pigeon pea seed husks as potent natural resource of antioxidant and anti hyperglycaemic activity. Int J Green Pharm 7:252–257. https://doi.org/10.4103/0973-8258.120247

Turati F et al (2015) Fruit and vegetables and cancer risk: a review of southern european studies. Br J Nutr 113:S102–S110. https://doi.org/10.1017/S0007114515000148

Vassallo N, Scerri C (2013) Mediterranean diet and dementia of the Alzheimer type. Curr Aging Sci 6(2):150–162. https://doi.org/10.2174/1874609811306020003

Vedavyas R et al (2012) Antioxidant activity of raw, cooked and Rhizopus oligosporus fermented beans of Canavalia of coastal sand dunes of Southwest India. J Food Sci Technol 51(11):3253–3260. https://doi.org/10.1007/s13197-012-0830-9

Viuda-Martos M et al (2011) Spices as functional foods. Crit Rev Food Sci Nutr 51(1):13–28. https://doi.org/10.1080/10408390903044271

Wang Y et al (2002) An LC-MS method for analyzing total resveratrol in grape juice, cranberry juice, and in wine. J Agric Food Chem 50:431–435. https://doi.org/10.1021/jf010812u
Wild CP, Espina C, Bauld L, Bonanni B, Brenner H, Brown K, Dillner J, Forman D, Kampman E, Nilbert M, Steindorf K, Storm H, Vineis P, Baumann M, Schüz J (2019) Cancer prevention Europe. Mol Oncol 13(3):528–534. https://doi.org/10.1002/1878-0261.12455
Willett WC (2002) Balancing life-style and genomics research for disease prevention. Science 8:296–695. https://doi.org/10.1126/science.1071055
Winda CH, Ignasius Radix APJ (2018) Antioxidant activity of anthocyanins in common legume grains (chapter 8). In: Ronald RW et al (eds) Polyphenols: mechanisms of action in human health and disease, 2nd edn. Academic, Amsterdam, pp 81–92., ISBN 9780128130063. https://doi.org/10.1016/B978-0-12-813006-3.00008-8
Winham DM et al (2007) Pinto bean consumption reduces biomarkers for heart disease risk. J Am Coll Nutr 26:243. https://doi.org/10.1080/07315724.2007.10719607
Xu FQ et al (2007) Steroidal saponins from Lysimachia Paridiformis. J Asian Nat Prod Res 9:493–497. https://doi.org/10.1080/10286020600727434
Xu Y et al (2019) Association of whole grain, refined grain, and cereal consumption with gastric cancer risk: a meta-analysis of observational studies. Food Sci Nutr 7:256–265. https://doi.org/10.1002/fsn3.878
Yahia EM, Ornelas-Paz JJ (2010) Chemistry, stability, and biological actions of carotenoids. In: Fruit and vegetable phytochemiclas: chemistry, nutritional value and stability, 1st edn. Wiley-Blackwell, Oxford, pp 177–222. https://doi.org/10.1002/9780813809397
Yoshida Y et al (2007) Chemical reactivities and physical effects in comparison between tocopherols and tocotrienols: physiological significance and prospects as antioxidants. J Biosci Bioeng 104(6):439–444. https://doi.org/10.1263/jbb.104.439
Zamora-Ros R et al (2012) Dietary flavonoid and lignan intake and gastric adenocarcinoma risk in the european prospective investigation into cancer and nutrition (epic) study. Am J Clin Nutr 96:1398–1408. https://doi.org/10.3945/ajcn.112.037358
Zhang H-Y et al (2016) Isoflavones and prostate cancer: a review of some critical issues. Chin Med J 129(3):341–347. https://doi.org/10.4103/0366-6999.174488
Zingg JM (2007) Molecular and cellular activities of vitamin E analogues. Mini-Rev Med Chem 7:543–558. https://doi.org/10.2174/138955707780619608

Chapter 7
Legume Symbiotic Interaction from Gene to Whole Plant

Kaouthar Feki, Faiçal Brini, Moncef Mrabet, and Haythem Mhadhbi

Abstract Bacterial and arbuscular mycorrhizal symbiosis are both beneficial for bacteria or fungus and host plants. The bacterial symbiosis enhances plant growth and productivity and it is beneficial for sustainable agriculture. In fact, rhizobia are soil bacteria that form symbiosis with legumes and help in fixing atmospheric nitrogen by converting into ammonia inside the root nodules. Legumes-rhizobia interaction is specificity controlled by various genetic and molecular mechanisms. Here we review the different components and molecules implicated in this interaction. Moreover, we detailed the genes coding for Nod factor secreted by bacteria and their receptors present on the cell membrane of plant host. The perception of Nod factors by their receptors leads to activate various signaling pathways that involve different transcription factors depending on the symbiotic stage. Nodule inception NIN proteins are crucial for nodulation organogenesis through the activation of target proteins. Finally, we reviewed the response of rhizobia and plants to environmental stress and their strategy to adapt stress.

Keywords Legumes · Rhizobia · Nitrogen · Nodulation · Agriculture · Mutualistic · Interaction · Symbiosis · Stress · Defense

Abbreviations

AMF	arbuscular mycorrhizal fungi
AMT	ammonium transporter
ARID	AT-rich interaction domain
ERF	ethylene response factor

K. Feki · M. Mrabet · H. Mhadhbi
Laboratory of Legumes, Center of Biotechnology of Borj-Cédria, Hammam Lif, Tunisia

F. Brini (✉)
Biotechnology and Plant Improvement Laboratory, Center of Biotechnology of Sfax (CBS), University of Sfax, Sfax, Tunisia
e-mail: faical.brini@cbs.rnrt.tn

P. Guleria et al. (eds.), *Sustainable Agriculture Reviews 45*, Sustainable Agriculture Reviews 45, https://doi.org/10.1007/978-3-030-53017-4_7

ERN	ERF required for nodulation
LYK	Lysin motif receptor-like kinase
MAPK	mitogen-activated protein kinase
NCR	nodule cysteine-rich
NFP	Nod factor perception
NFR	Nod factor receptor
NPL	nodulation pectate lyase
NSP	nodulation signaling pathway
PGPR	plant growth promoting rhizobacteria
PT	phosphate transporter
ROS	reactive oxygen species.

7.1 Introduction

Symbiotic interaction between legumes and soil bacteria supports plant growth and development through nitrogen fixation process. This process is essential for sustainable agriculture and it reduces the requirement of nitrogen fertilizers. In fact, agriculture productivity is restricted by the levels of nitrogen and phosphorus that are crucial for plant growth.

So far, soil microorganisms are frequently used to ameliorate crops production. These beneficial microorganisms are plant growth promoting rhizobacteria (PGPR), arbuscular mycorrhizal fungi (AMF) and nitrogen-fixing rhizobia. They implicate various mechanisms to enhance plant growth either by direct or indirect ways. Rhizobia act directly by facilitating nitrogen and phosphate absorption, iron sequestration and inducing osmolytes and exopolysaccharides production, or act indirectly by stimulating plant defense to biotic stress and producing compounds like hydrogen cyanide, siderophore and antibiotics (Fig. 7.1) (Grover et al. 2011; Ojuederie et al. 2019).

In general, the symbiotic systems include the arbuscular mycorrhizal symbiosis and root nodules symbiosis in legumes. The first symbiosis implicates fungi and plants, but the second one involves soil bacteria and plants. In agriculture, co-inoculation plants with arbuscular mycorrhizal fungi and plant growth promoting rhizobacteria (PGPR) increases plants productivity through the amelioration of plant growth, nutrient acquisition, phytoremediation and abiotic stress tolerance (Gamalero and Glick 2011; Ramasamy et al. 2011).

The phosphate and nitrogen absorption requires several genes in arbuscular mycorrhizal fungus. Subsequently, these minerals were transported to the host through ammonium (AMT) and phosphate transporters (PT) like GmAMT4 in soybean and MtPT4 in *Medicago*, respectively (Kobae et al. 2010; Javot et al. 2007). In rhizobium-legume symbiosis, rhizobia genes *nif* and *fix* are responsible for nitrogen fixation, while *nod* and *nol* genes control nodules formation (Freiberg et al. 1997; Barnett et al. 2001).

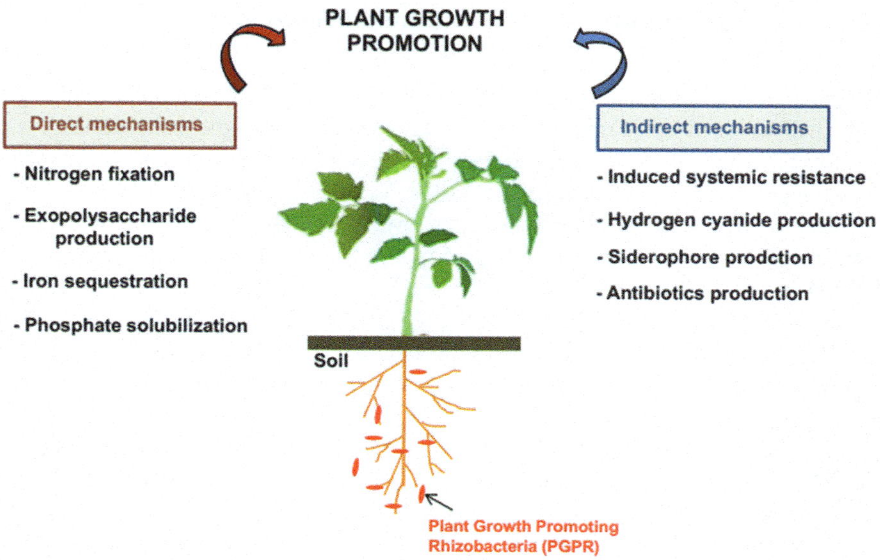

Fig. 7.1 Plant growth promoting rhizobacteria promotes plant growth under water deficit by direct and indirect mechanisms. Directly, the beneficial bacteria enhance nitrogen fixation, the production of some compounds like exopolysaccharide iron sequestration and mobilization of phosphorus in the soils. Indirectly, these microbes stimulate plant defense to biotic stress, the production of some compounds like hydrogen cyanide, siderophore and antibiotics

Nodulation process is controlled by plants feedback systems. This autoregulation of nodulation implicates a signal between shoots and roots to control nodule number. It was shown that many kinases receptors in different legumes species control the nodule number like leucine-rich repeat receptor-like kinase protein from *L. japonicas* and *Glycine max* designed HAR1 and NARK, respectively (Oka-Kira and Kawaguchi 2006; Ferguson et al. 2010).

The development of symbiosis requires specificity of both partners. This specificity is the consequence of some signals from host and bacteria at early and later stage of interaction. At early stage, R proteins involved in plants resistance are implicated in host specificity during nodulation process, like the two R proteins in soybean encoded by the two alleles *Rj2* and *Rfg1* (Yang et al. 2010). Moreover, the soybean *Rfg1* and *Rj4* genes controlled the nodulation to specific bacteria strains (Tang et al. 2016; Fan et al. 2017). At later stage, antimicrobial peptides secreted by host regulate nitrogen-fixing efficiency (Wang et al. 2017, 2018).

In this chapter, we present an overview of our current understanding of the mechanisms contributing to symbiosis from host-bacteria interaction to nodulation and nitrogen fixation, plant immunity under symbiosis and the response of plants to environmental stress during symbiosis.

7.2 Legumes Symbiotic Interaction Mechanism

In the soil, bacteria termed as rhizobia are capable of fixing atmospheric nitrogen that will be used by plants. In this symbiotic interaction, the atmospheric N_2 is reduced to ammonium by large group of rhizobial microorganisms and then used by many legumes species (Herridge et al. 2008; Van Hameren et al. 2013).

When nitrogen is present in soil at low level, plants liberate into the rhizosphere phenolic compounds like flavonoids and isoflavonoids, leading to the activation of the nodulation protein nodD of the rhizobia. In turn, the regulator nodD triggers the genes responsible for Nod factors production (Haeze and Holsters 2002). After their secretion by the rhizobia, they are perceived by a specific Nod factors receptors-like kinases localized at the root hair cells membrane. As consequence, some molecular and physiological responses are enhanced in plant cells, which are root hair deformation, cortical cell division and increased flavonoid production (Tikhonovich and Provorov 2007) (Fig. 7.2).

After Nod factors-host receptors recognition, rhizobia degraded the cell wall forming an infection thread by invagination of the plasma membrane of the root hair cells. Moreover, bacteria secrete polysaccharides that are crucial for infection thread formation, like succinoglycan secreted by *S. meliloti* (Brewin 2004). Then, bacteria enter to host cells by endocytosis-like process forming intracellular structures called symbiosomes, within bacteria are differentiated into symbiotic forms called bacteroids. This occurs due to the bacteria BacA protein which protects against the antimicrobial activity of nodule cysteine-rich (NCR) peptides present in some plants species like *M. truncatula* and *P. sativum* (Marlow et al. 2009; Haag et al. 2013).

Once transformed to bacteroids, bacteria reduce atmospheric N_2 to ammonium in nodules using the enzymatic complex of nitrogenase. Then, ammonium is transported to plants and the carbon source is supplied to bacteroids from the macrosymbiont (Gibson et al. 2008; Haag et al. 2013). In general, the nitrogen and carbon metabolism differ between the determinate and the indeterminate nodules. The first type is formed in some plants like *Glycine max* and *Vicia faba*, and the nodules have a round shape caused by the absence of persistent meristem and developmental zones. However, the indeterminate nodules have a persistent meristem and many developmental zones, resulting in nodules with elongated shape (White et al. 2007; Gibson et al. 2008; Haag et al. 2013).

7.2.1 Signal Molecules Involved in Symbiosis Interaction

Symbiosis process is very complex and implicates a coordinated exchange of different signals between plants and the microbial symbiosis. Among them flavonoids are secreted by plants and lipochitooligosaccharides known as Nod factor that are secreted by bacteria (Janczarek et al. 2015).

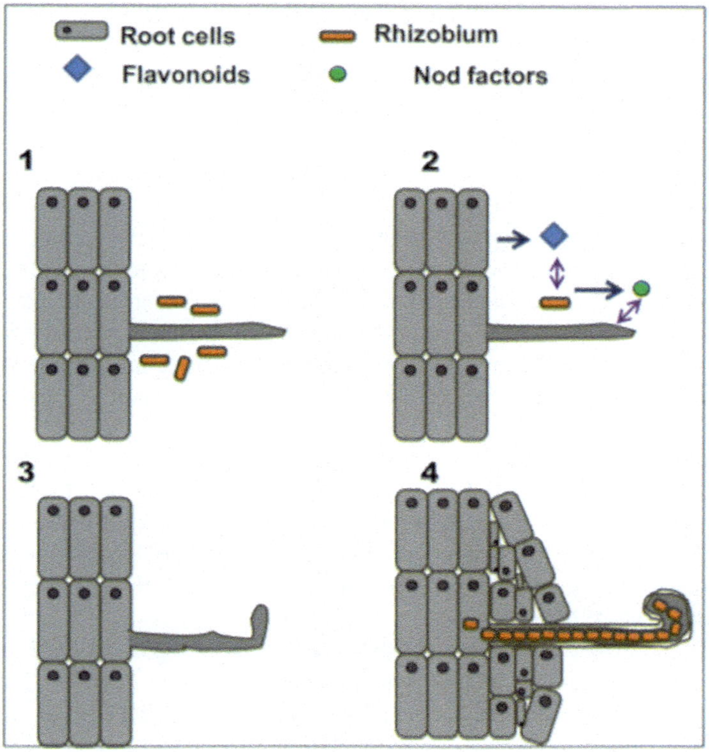

Fig. 7.2 The process of bacterial symbiosis and the exchange of signal molecules from bacterium to plant cells. In the presence of rhizobia (1), plant cells excrete flavonoids and others compounds that will be recognized by soil bacteria. This lead to the production of the Nod factors coded by nod genes in bacteria, and they are perceived by specific Nod factors receptors present at the plasma membrane of root hair cells (2). This recognition produces some physiological response like root hair curling (3) and the formation of thread infection, resulting bacteria penetration in root hair cells (4)

In the rhizosphere, there are various flavonoid compounds secreted by root cells. So far, it is not clear which flavonoids is more essential for nodulation. However, it was demonstrated that methoxychalcone and genistein are excellent signal molecules in *Medicago* and soybeans, respectively, which form determinate nodules (Liu and Murray 2016). Nevertheless, in some case genistein acts as repressor of nodulation formation by its inhibition of *nod* gene expression of *S. meliloti* (Peck et al. 2006).

In general, flavonoids are secreted in the rhizobia infection zone at the root hair level. These compounds attract specific bacterial species and they are also signal molecules for *nod* genes induction (Cooper 2007; Subramanian et al. 2007; Janczarek et al. 2015). The combination of flavonoids enhances significantly *nod* gene expression compared to one flavonoids type. This is the case of the *S. meliloti* which is attracted by four different flavonoids from alfalfa roots (Cooper 2004; Janczarek et al. 2015).

In some case, plants secrete non-flavonoids compounds, like betaines that affect the expression of *nod* genes in rhizobia. Betaines are identified from *Medicago* and activate the expression of *nod* genes of *S. meliloti* (Phillips et al. 1992). In addition, some phytohormones like jasmonic acid efficiently activate *nod* genes (Mabood et al. 2006; Poustini et al. 2007).

After flavonoids secretion by host plants, the microorganisms secrete signal molecules depending upon the stage of symbiotic interaction from the adhesion to root cells to nodules invasion (Janczarek et al. 2015). Rhizobia produce many types of surface polysaccharides involving in the attachment to root hair in order to assure the nitrogen fixation. Among these polysaccharides, there are exopolysacharides that are excluded in the environment, lipopolysaccharide and capsular polysaccharide.

Exopolysacharides are crucial for rhizobial infection in various symbiotic interactions (Wang et al. 2018). The exopolysacharides receptor 3 (EPR3) identified from *L. japonicas* has three extracellular LysM domains and a kinase domain (Kawaharada et al. 2015). Recently, it was shown that exopolysacharides receptors are involved in bacteria entry after the activation of Nod factors receptors (Kawaharada et al. 2017).

Once fixed to host plants, rhizobia produce lipochitin oligosaccharide called Nod factors, coded by *nodABC* genes, which play a crucial role in nodules formation. However, some bradyrhizobia bacteria, like *B. elkanii*, lack *nodABC* genes and consequently implicate the type III secretion system (T3SS) for nodule organogenesis (Giraud et al. 2007; Okazaki et al. 2013, 2016).

7.2.2 Nod Factor Receptors

The perception of rhizobia by host receptors is essential for development of infection thread and nodulation, because the inactivation of nod factor receptor abolishes plant-bacteria interaction and consequently inhibits thread formation (Arrighi et al. 2006).

In general, there are two types of Nod factor receptors that contain three domains. The first domain is the kinase domain present in the cytosol. The second one is the transmembrane domain and the third domain is localized in the extracellular region. This latter domain is composed by three lysine motifs (LysM) in the case of the Nod factor receptor (NFR) or composed by leucine rich repeats (LRR) domain in the case of the symbiosis receptor-like kinase (SYMRK) (Madsen et al. 2003; Radutoiu et al. 2003; Arrighi et al. 2006; Smit et al. 2007). These receptors are primordial for nodulation (Stracke et al. 2002; Esseling et al. 2004).

So far, several Nod factor receptors were characterized in different legumes species like Nod factors receptors NFR1 and NFR5 in *L. japonicus*. Their genes orthologous were identified in *P. sativum* Sym37 and Sym10 and lysin motif receptor-like kinase LYK3 and Nod factor perception NFP in *M. truncatula*. Thus, one bacterial strain has many Nod factor receptors. The Nod factor receptors that lack the kinase activity are implicated in early response by recognition of Nod factors like Nod factor receptor NFR5, Nod factor perception (NFP) and Sym10. Interestingly, these receptors are functional only in complex with Nod factors receptor possessing kinase activity like NFR1, LYK3 and Sym37 (Ardourel et al. 1994; Limpens et al. 2003; Madsen et al. 2003; Radutoiu et al. 2003; Bensmihen et al. 2011; Pietraszewska et al. 2013).

7.2.3 Nodulation Genes

In rhizobia, Nod factors synthesis involves various *nod* genes with different function during nodulation process. *NodD* genes code for the transcription regulators that consequently activate *nod* genes expression (Egelhoff et al. 1985; Honma et al. 1990). It occurs when the N-terminal DNA binding domain of NodD protein binds to the nod boxes of the target *nodABC* genes involved in the synthesis of the basic Nod factors (Brencic and Winans 2005; Hassan and Mathesius 2012).

The two forms NodD1 and NodD2 are induced by flavonoids compounds, while NodD3 is constitutively active (Hartwig et al. 1990; Phillips et al. 1992). NodD proteins belong to the family of Lys-R type of transcriptional activator. It was shown that NodD1 is activator of *nod* genes in contrast to NodD2. The mechanism of NodD1 action was determined by the identification of the crucial residues in four types of nodD1 *S. meliloti* mutants implicated in binding to *nod* genes promoters, multimerization and interaction with RNA polymerase (Gottfert et al. 1992: Peck et al. 2013).

Similarly to nodD, regulatory protein SyrM of *Rhizobiurn meliloti* is also involved for optimal nodules formation (Barnett and Long 1990; Kondorosi et al. 1991). In addition, there are others regulators of *nod* genes expression which are nolR and nodVW. NolR plays a pivotal role for optimal nodulation and bacterial growth because NolR repression mutants generated a delayed nodulation phenotype. In fact, nolR binds to nod boxes and consequently competes with nodD1 or represses the expression of *nodA* gene expression of (Kondorosi et al. 1989; Chen et al. 2005).

In *B. japonicum*, nodVW regulatory system is crucial for nodulation in the mutant *NodD1* (Gottfert et al. 1990). This system involves the kinase nodV, which autophosphorylates and phosphorylates the regulator protein NodW at the residue Asp70. This phosphorylation is crucial for nodulation (Loh et al. 1997). Similarly to NodW, NwsB controls the expression of *nod* genes in *B. japonicum* after its phosphorylation by the kinase NwsA (Groß et al. 1993; Loh et al. 2002).

7.2.4 Nod Factor Signaling Pathway

The recognition between legumes and rhizobia leads to the induction of nodulation signaling pathway and the inactivation of this pathway inhibits the nodulation (Broghammer et al. 2012). This is the case of the *Medicago* mutants *Mtnin* which were unable to form nodules after inoculation with *S. meliloti* (Marsh et al. 2007). *So far, a multitude of transcription factors were characterized by gene knockout in different plants species, and they target nodule inception NIN genes expression* (Soyano and Hayashi 2014).

The oscillations of calcium concentration in the nucleus activate the signaling pathway cascade, and Ca^{2+} ions bind to the Ca^{2+}/calmodulin-dependent protein kinase (CCaMK). The deletion of the autoinhibition domain of CCaMK protein leads to activation of downstream genes and consequently nodulation even in the absence of rhizobia (Gleason et al. 2006; Swainsbury et al. 2012). This kinase is involved in symbiosis through its association and phosphorylation of CYCLOPS protein that trans-activate the *NIN* gene (Lévy et al. 2004; Singh et al. 2014). Moreover, the complex formed by CCaMK, CYCLOPS and DELLA activates the expression of the Reduced Arbuscular Mycorrhiza 1 (*RAM1* gene) which is primordial for arbuscular development in the arbuscular mycorrhizal symbiosis (Fig. 7.3) (Pimprikar et al. 2016).

During nodulation signaling, the two nodulation signaling pathway proteins NSP1 and NSP2 combine together and activate *Early Nodulin 11* (*ENOD11*) gene expression. Furthermore, this complex is positive regulator of ethylene response factor (ERF) required for nodulation ERN1 that also activates *ENOD11* gene. However, this different *ENOD11* gene induction is through the binding of NSP1 or ERN1 to different promoter regulatory regions, which is separated by 26 bp, and is related to different symbiotic stages (Kalo et al. 2005; Smit et al. 2005; Hirsch et al. 2009; Cerri et al. 2012) (Fig. 7.4).

In addition to NSP1, NSP2 associates with another interacting protein IPN2 through the GRAS domain and the coiled-coil domain of the MYB transcription factor IPN2. This transcription factor is involved directly with nodulation development during rhizobial infection, but it is not related to arbuscular mycorrhizal development (Kang et al. 2014) (Fig. 7.4).

In *Lotus japonicas*, SYMRK interacting protein 1 (SIP1) contains a conserved AT-rich interaction domain (ARID) that interacts with the nod factor receptor SYMRK through its kinase domain, leading to the induction of *NIN* gene during early plant-rhizobia interaction (Zhu et al. 2008). It was also demonstrated that this SymRK interacting protein 1 (SIP1) is primordial not only in bacterial symbiosis but also in arbuscular mycorrhizal symbiosis (Wang et al. 2013).

Nodulation signaling pathway is controlled by phytohormones, especially by cytokinin that promotes nodulation. In fact, the inactivation of cytokinin receptor LHK1 in *snf2* mutant *Lotus japonicus* plant caused a spontaneous nodulation in the absence of rhizobia. On the other hand, cytokinin triggers the GRAS protein NSP2 to activate *NIN* gene (Murray et al. 2007; Tirichine et al. 2007). The cytokinin

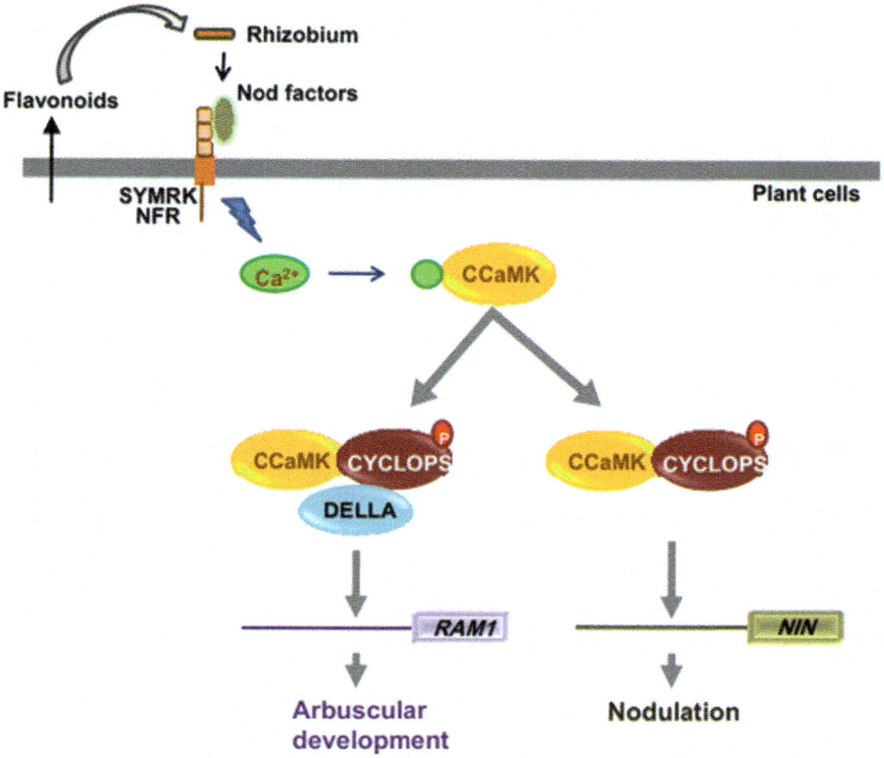

Fig. 7.3 Nod factors signaling pathway involving the *Ca^{2+}/calmodulin-dependent protein kinase. At low level of nitrogen in the soil, plants secrete flavonoids and as consequence Nod factors are produced by rhizobia. Once the Nod factors are recognized by Nod factors receptors (NFR) or by* symbiosis receptor-like kinase (SYMRK), *calcium ions bind to Ca^{2+}/calmodulin-dependent protein kinase CCaMK, which in turn they bind to CYCLOPS protein and its phosphorylate. As consequence, this complex activates the nodule inception NIN gene in the case of bacterial symbiosis. However, in the case of arbuscular mycorrhizal symbiosis, this complex is associated with DELLA and* activates the expression of reduced arbuscular mycorrhiza *RAM1* gene

receptor MtCRE1 in *Medicago truncatula* triggers the nod factor signaling cascade by the activation of the two transcription factors MtERN1 and MtNSP2 resulting in *MtNIN* genes induction (Plet et al. 2011). Recently, it was shown that cytokinin pathway interplays with the gibberellins pathway through the DELLA1 protein to control early nodulation (Farde et al. 2017). Moreover, basic helix-loop-helix transcription factor bHLH476 and the oxidase/dehydrogenase CKX1 are two targets of cytokinin signaling pathway. It was shown that this pathway is also regulated by the response regulator type B MtRR4 (Ariel et al. 2012).

Contrary to cytokinin, abscisic acid inhibits Nod factor signaling pathway and consequently abolish *NIN* gene expression in *Medicago* (Ding et al. 2008). Similarly, gibberellin negatively regulates the nodulation (Maekawa et al. 2009). Moreover,

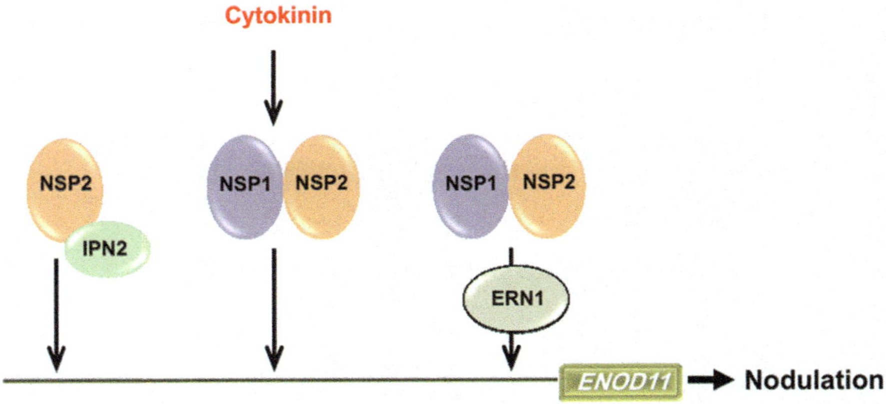

Fig. 7.4 The role of the two nodulation signaling pathway NSP1 and NSP2 in nodulation signaling. NSP1 associates with NSP2 protein and activates nodule inception *NIN* gene expression. This complex is positive regulator of ethylene response factor (ERF) required for nodulation ERN1 that also activates *ENOD11* gene at different promoter regulatory regions. Moreover, NSP2 binds to the MYB transcription factor IPN2 and this complex activates also *NIN* genes

ethylene and jasmonic acid reduce nodules number on roots by regulation of various nodulation process (Heidstra et al. 1997; Sun et al. 2006).

7.3 Regulation of Nodule Organogenesis

The perception of Nod factors secreted by rhizobia generates an induction of the signaling pathways that enhances the *NIN* gene. However, the overexpression of nodule inception (NIN) proteins produces structures like nodules in root cortex even in absence of rhizobia (Soyano et al. 2013).

In addition to legumes, other angiosperms plants have NIN-like proteins (NLP) divided in three homologous subfamilies in monocots (Schauser et al. 2005). *Arabidopsis* has nine NLP proteins that are implicated in response to nitrate through their binding to nitrate responsive elements (NRE) present in nitrate-inducible genes like nitrite reductase *NIR1* gene (Konishi and Yanagisawa 2013). During nodulation process, nitrate inhibits nodule development by the repression of *NIN* genes expression (Barbulova et al. 2007). Soyano et al. (2015) proposed that NIN and nitrate have antagonistic effect to repress nitrate-inducible genes and NIN target genes, respectively.

During nodule organogenesis, NIN regulates several genes essential for activation of cell division like the nuclear factor-Y *NF-YA1/B1* and the nodulation pectate lyase *NPL* gene (Xie et al. 2012; Soyano et al. 2013). In *Lotus japonicus*, LjNPL protein has pectate lyase activity essential for rhizobia penetration to epidermis cells and the formation of thread infection (Xie et al. 2012).

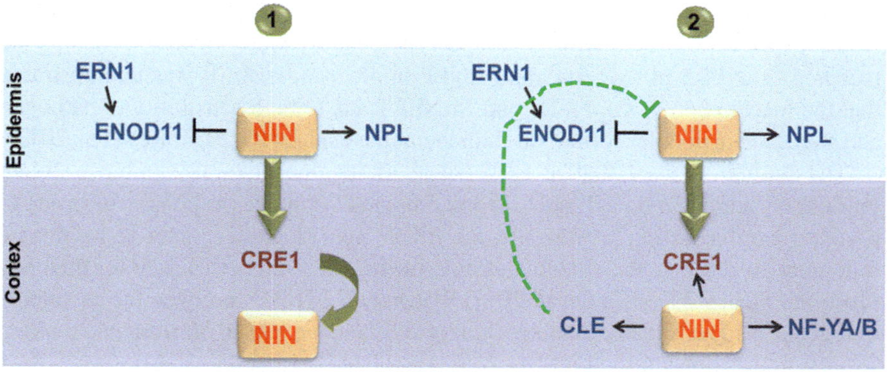

Fig. 7.5 Regulation of nodule organogenesis by nodule inception proteins. The perception of Nod factors enhances the expression of nodule inception NIN by the activation with the complex formed by *Ca^{2+}/calmodulin-dependent protein kinase* CCaMK and *CYCLOPS*. (*1*) *Within epidermis, NIN induces* nodulation pectate lyase *NPL expression but it inhibits* early nodulin *ENOD11 expression. Moreover, the* cytokinin signaling is enhanced leading to activation of NIN in root cortex. (*2*) Within root cortex, nodule inception (NIN) proteins activate nuclear factor NF-YA/B and CLE peptides to regulate nodulation organogenesis. CLE peptides are implicated in a negative regulation of nodule organ formation

Moreover, when NIN is expressed in root epidermis, it enhances the expression of the cytokinin receptor *CRE1* while it inhibits the expression of *ENOD11*. However, NIN can promote nodulation independently of cytokinin signaling in root cortex. Thus, it was proposed that NIN activates nodulation process in root cortex but it suppress further induction of Nod factors response in the root epidermis (Fig. 7.5) (Vernié et al. 2015).

Nodulation process is regulated through the expression of various CLAVATA3/embryo-surrounding region (CLE) peptides that are also induced by nodule inception NIN present in cortex (Soyano et al. 2014). Thus, NIN plays a crucial role in nodulation organogenesis by the coordination of the epidermis and cortex responses (Vernié et al. 2015).

7.4 Plants Immunity During Legume-Rhizobium Symbiosis

At the cell surface, pattern recognition receptors (PRRs) recognize microbe-associated molecular pattern, generating elevation of calcium and reactive oxygen species (ROS) levels in cytosol and activation of mitogen-activated protein kinase (MAPK) (Boller and Felix, 2009). Surprisingly, rhizobia suppress ROS production, salicylic acid accumulation and down-regulate pathogenesis-related 2 (*PR2*) gene through Nod factors signal (Shaw and Long 2003; Mitra and Long 2004).

The PRRs include receptor-like kinases and they are involved not only in microbe detection but also in symbiosis interaction through Nod factors perception (Deakin and Broughton 2009; Zipfel 2014; Zipfel and Oldroyd 2017). So far, in response to Nod factors, various lysine motif receptors like kinase were identified such as the receptor GmNFR5 in soybean and MtNFP in *M. truncatula*. It was demonstrated that the inactivation of GmNFR5 and MtNFP leads to both sensibility to pathogen and alteration in nodules formation (Indrasumunar et al. 2010, 2011; Rey et al. 2013).

The finding showed that at early stage of legume-rhizobia symbiosis, plant immune system is induced and rhizobia suppress immune responses in order to promote symbiosis interaction. On the other hand, legume plants active innate immunity to exclude the rhizobia strains having less nitrogen fixation efficiency (Toth and Stacey 2015; Cao et al. 2017; Wang et al. 2018). Moreover, the advantage of *R* genes is to restrict nodulation of rhizobia, like gene *NS1* in *M. truncatula* which reduces nodulation caused by *S. meliloti* strain Rm41 (Liu et al. 2014).

The nodule restriction was also demonstrated in soybean through several dominant genes such as *Rj2*, *Rfg1* and *Rj4* (Yang et al. 2010; Tang et al. 2016; Fan et al. 2017). Interestingly, these plants *R* genes were controlled by the bacterial type III secretion system generating the induction of effector-triggered immunity ETI (Krishnan et al. 2003; Boller and Felix 2009; Okazaki et al. 2009).

7.5 Symbiosis Interaction Under Abiotic Stress

Desiccation and salt stress are considered as major abiotic stresses that alter crop production. As we described above, the symbiotic nitrogen fixation is essential for sustainable agriculture. The response of legumes to abiotic stress can be ameliorated by rhizobia and symbiotic processes (Yang et al. 2009). The latter can survive under severe conditions like salinity and desiccation, resulting in the activation of the nitrogen fixation process (Vriezen et al. 2007; Mhadhbi et al. 2011).

In addition, plant growth promoting rhizobacterium (PGPR) enhance drought and salinity adaptation of the host through the stimulation of lateral root formation and shoot growth (Gopalakrishnan et al. 2015; Rolli et al. 2015; Mondal et al. 2017). So far, various bacteria imparting some degree of tolerance to legumes under various abiotic stresses have been known, indicating the application of microorganisms in agriculture (Grover et al. 2011; Kunert et al. 2016).

Under stress conditions, rhizobia modulate their cytoplasmic osmolarity by producing high levels of osmoprotectants like proline and potassium. Moreover, the high production of exopolysaccharides helps *Pseudomonas putida* strain to protect from dessication by enhancing water retention (Sandhya et al. 2009). Interestingly, the thermotolerant strains enhance heat shock proteins HSPs or cryoprotective proteins during their exposition to high or low temperatures, respectively (Koda et al. 2001; Ali et al. 2009). Moreover, the arbuscular mycorrhizal symbiosis imparts water deficit tolerance to plants by the alteration of their genes expression, decreasing the content of malondialdehyde in leaves and the enhancement of antioxidant enzymes (Lozano and Azcon 2000; Wu et al. 2006). Likewise, salt tolerance was observed in various plants colonized by arbuscular mycorrhizal fungi (Feng et al. 2002; Ben Khaled et al. 2003).

So far, many reports demonstrated that drought changes roots at physiological and molecular levels (Kunert et al. 2016). In fact, drought influences root architecture and biomass (Tanaka et al. 2014; Vadez 2014; Fenta et al. 2014). Moreover, drought up-regulates various genes coding for proteins involved in many biochemical pathways, and depending on the duration and the intensity of stress (Stolf-Moreira et al. 2011; Tripathi et al. 2016). Consequently, tolerant legume plants have high nodules number compared to the drought-sensitive plants (Sulieman et al. 2015). Nevertheless, the nodule biomass is low in soybean roots exposed to drought, resulting in the impairment of nitrogenase activity (Márquez-García et al. 2015; Kunert et al. 2016).

Further studies showed that the nitrogen fixation rate declines under severe drought conditions. This decrease is related to the down-regulation of proteins involved in ethylene biosynthesis (Larrainzar et al. 2009, 2014). However, it has been shown recently that osmotic stress and salinity up-regulate the expression of the regulatory protein NodD1 that controls the expression of nodulation genes and genes involved not only in nitrogen fixation but also in mobility and synthesis of polysaccharides (Cerro et al. 2019).

The response of plant to salt stress implicates some ionic transporter like high affinity K^+ transporter called HKT proteins. The expression of *HKT1* in plant can be modulated by volatile compounds emitted from some PGPRs. In fact, under saline conditions, *HKT1* expression is up-regulated in shoots leading to better recirculation of Na^+ in whole plant (Zhan et al. 2008). Similarly, the production of the volatile butanediol by *Pseudomonas chlororaphis* leads to drought tolerance in *Arabidopsis* (Cho et al. 2008).

Thus, rhizobia are not only implicated in nitrogen fixation but also promotes host tolerance to multiple abiotic stresses, through the involvement of different bacterial strategies (Grover et al. 2011; Kunert et al. 2016; Staudinger et al. 2016; Defez et al. 2017).

7.6 Conclusions

Legumes-rhizobia interaction occurs due to the specific recognition between hosts and soil bacteria. Various studies have demonstrated that specific interaction is due to exchange of multiple signals between plants and its microsymbionts. Despite the identification of various signal molecules and their receptors, little is known about the interaction of Nod factors receptors to each others to perceive rhizobial signals. As a consequence of Nod factors recognition, signaling pathways involving a multitude of transcription factors are induced leading to activate NIN protein, which is the key of nodules organogenesis and rhizobial infection. So far, various genes involved in nodulation process are identified and characterized in various legumes. This knowledge opens scope for developing genetic tools to improve symbiotic nitrogen fixation in legumes.

References

Ali SZ, Sandhya V, Grover M, Kishore N, Rao LV, Venkateswarlu B (2009) *Pseudomonas* sp. strain AKM-P6 enhances tolerance of sorghum seedlings to elevated temperatures. Biol Fertil Soil 46:45–55. https://doi.org/10.1007/s00374-009-0404-9

Ardourel M, Demont N, Debelle F, Maillet F, de Billy F, Prome J-C, Denarie J (1994) Rhizhobium meliloty lipooligosaccharide nodulation factors. Plant Cell 6:1357–1374. https://doi.org/10.1105/tpc.6.10.1357

Ariel F, Brault-Hernandez M, Laffont C, Huault E, Brault M, Plet J, Moison M, Blanchet S, Ichanté JL, Chabaud M et al (2012) Two direct targets of cytokinin signaling regulate symbiotic nodulation in *Medicago truncatula*. Plant Cell 24:3838–3852. https://doi.org/10.1105/tpc.112.103267

Arrighi JF, Barre A, Ben Amor B, Bersoult A, Soriano LC, Mirabella R, de Carvalho-Niebel F, Journet EP, Gherardi M, Huguet T et al (2006) The *Medicago truncatula* lysin motif-receptor-like kinase gene family includes NFP and new nodule-expressed genes. Plant Physiol 142:265–279. https://doi.org/10.1104/pp.106.084657

Barbulova A, Rogato A, D'Apuzzo E, Omrane S, Chiurazzi M (2007) Differential effects of combined N sources on early steps of the Nod Factor-dependent transduction pathway in *Lotus japonicus*. Mol Plant-Microbe Interact 20:994–1003. https://doi.org/10.1094/MPMI-20-8-0994

Barnett MJ, Long SR (1990) DNA sequence and translational product of a new nodulation-regulatory locus: SyrM has sequence similarity to NodD proteins. J Bacteriol 172:3695–3700. https://doi.org/10.1128/jb.172.7.3695-3700

Barnett MJ, Fisher RF, Jones T, Komp C, Abola AP, Barloy-Hubler F, Bowser L, Capela D, Galibert F, Gouzy J et al (2001) Nucleotide sequence and predicted functions of the entire Sinorhizobium meliloti pSymA megaplasmid. Proc Natl Acad Sci USA 98:9883–9888. https://doi.org/10.1073/pnas.161294798

Ben Khaled L, Gomez AM, Ourraqi EM, Oihabi A (2003) Physiological and biochemical responses to salt stress of mycorrhized and/or nodulated clover seedlings (Trifolium alexandrinum L.). Agronomie 23:571–580

Bensmihen S, de Billy F, Gough C (2011) Contribution of NFP LysM domains to the recognition of Nod factors during the *Medicago truncatula/Sinorhizobium meliloti* symbiosis. PLoS One 6(11):e26114. https://doi.org/10.1371/journal.pone.0026114

Boller T, Felix G (2009) A renaissance of elicitors: perception of microbe-associated molecular patterns and danger signals by pattern-recognitionreceptors. Annu Rev Plant Biol 60:379–406. https://doi.org/10.1146/annurev.arplant.57.032905.105346

Brencic A, Winans SC (2005) Detection of and response to signals involved in host microbe interactions by plant-associated bacteria. Microbiol Mol Biol Rev 69:155–194. https://doi.org/10.1128/MMBR.69.1.155-194.2005

Brewin NJ (2004) Plant cell wall remodelling in the Rhizobium legume symbiosis. Crit Rev Plant Sci 23:293–316. https://doi.org/10.1080/07352680490480734

Broghammer A, Krusell L, Blaise M, Sauer J, Sullivan JT, Maolanon N et al (2012) Legume receptors perceive the rhizobial lipochitin oligosaccharide signal molecules by direct binding. Proc Natl Acad Sci 109:13859–13864. https://doi.org/10.1073/pnas.1205171109

Cao Y, Halane MK, Gassmann W, Stacey G (2017) The role of plant innate immunity in the legume-rhizobium symbiosis. Annu Rev Plant Biol 68:535–561. https://doi.org/10.1146/annurev-arplant-042916-041030

Cerri MR, Frances L, Laloum T, Auriac MC, Niebel A, Oldroyd GED, Barker DG, Fournier J, de Carvalho-Niebel F (2012) *Medicago truncatula* ERN transcription factors: regulatory interplay with NSP1/NSP2 GRAS factors and expression dynamics throughout rhizobial infection. Plant Physiol 160:2155–2172. https://doi.org/10.1104/pp.112.203190

Cerro DP, Megías M, López-Baena FJ, Gil-Serrano A, Pérez-Montaño F, Ollero FJ (2019) Osmotic stress activates nif and fix genes and induces the *Rhizobium tropici* CIAT 899 Nod factor production via NodD2 by up-regulation of the nodA2 operon and the nodA3 gene. PLoS One 14(3):e0213298. https://doi.org/10.1371/journal.pone.0213298

Chen H, Gao K, Kondorosi E, Kondorosi A, Rolfe BG (2005) Functional genomic analysis of global regulator NolR in Sinorhizobium meliloti. Mol Plant-Microbe Interact 18(12):1340–1352. https://doi.org/10.1094/MPMI-18-1340

Cho Sm KBR, Han SH, Anderson AJ, Park JY, Lee YH, Cho BH, Yang KY, Ryu CM, Kim YC (2008) 2R, 3r-butanediol, a bacterial volatile produced by Pseudomonas chlororaphis O6, is invoplved in induction of systemic tolerance to drought in *Arabdopsis thaliana*. Mol Plant-Microbe Interact 21(8):1067–1075. https://doi.org/10.1094/MPMI-21-8-1067

Cooper JE (2004) Multiple responses of rhizobia to flavonoids during legume root infection. Adv Bot Res 41:1–62. https://doi.org/10.1016/S0065-2296(04)41001-5

Cooper JE (2007) Early interactions between legumes and rhizobia: disclosing complexity in a molecular dialogue. J Appl Microbiol 103:1355–1365. https://doi.org/10.1111/j.1365-2672.2007.03366.x

Deakin WJ, Broughton WJ (2009) Symbiotic use of pathogenic strategies: rhizobial protein secretion systems. Nat Rev Microbiol 7:312–321. https://doi.org/10.1038/nrmicro2091

Defez R, Andreozzi A, Dickinson M, Charlton A, Tadini L, Pesaresi P, Bianco C (2017) Improved drought stress response in alfalfa plants nodulated by an IAA over-producing rhizobium strain. Front Microbiol 8:2466. https://doi.org/10.3389/fmicb.2017.02466

Ding Y, Kalo P, Yendrek C, Sun J, Liang Y, Marsh JF, Harris JM, Oldroyd GE (2008) Abscisic acid coordinates nod factor and cytokinin signaling during the regulation of nodulation in *Medicago truncatula*. Plant Cell 20:2681–2695. https://doi.org/10.1105/tpc.108.061739

Egelhoff TT, Fisher RF, Jacobs TW, Mulligan JT, Long SR (1985) Nucleotide sequence of *Rhizobium meliloti* 1021 nodulation genes: nod is read divergently from nodABC. DNA 4:241–248. https://doi.org/10.1089/dna.1985.4.241

Esseling JJ, Lhuissier FG, Emons AM (2004) A nonsymbiotic root hair tip growth phenotype in NORK-mutated legumes: implications for nodulation factor-induced signaling and formation of a multifaceted root hair pocket for bacteria. Plant Cell 16:933–944. https://doi.org/10.1105/tpc.019653

Fan Y, Liu J, Lyu S, Wang Q, Yang S, Zhu H (2017) The soybean Rfg1 gene restricts nodulation by *Sinorhizobium fredii* USDA193. Front Plant Sci 8:1548. https://doi.org/10.3389/fpls.2017.01548

Farde CF, Kisiala A, Brault M, Emery RJN, Diet A, Frugier F (2017) DELLA1-mediated gibberellin signaling regulates cytokinin-dependent symbiotic nodulation. Plant Physiol 175(4):1795–1806. https://doi.org/10.1104/pp.17.00919

Feng G, Zhang FS, Li XL, Tian CY, Tang C, Renegal Z (2002) Improved tolerance of maize plants to salt stress by arbuscular mycorrhiza is related to higher accumulation of leaf P-concentration of soluble sugars in roots. Mycorrhiza 12:185–190. https://doi.org/10.1007/s00572-002-0170-0

Fenta BA, Beebe SE, Kunert KJ, Burridge JD, Barlow KM, Lynch PJ et al (2014) Field phenotyping of soybean roots for drought stress tolerance. Agronomy 4:418–435. https://doi.org/10.3390/agronomy4030418

Ferguson BJ, Indrasumunar A, Hayashi S, Lin MH, Lin YH, Reid DE et al (2010) Molecular analysis of legume nodule development and autoregulation. J Integr Plant Biol 52:61–76. https://doi.org/10.1111/j.1744-7909.2010.00899.x

Freiberg C, Fellay R, Bairoch A, Broughton WJ, Rosenthal A, Perret X (1997) Molecular basis of symbiosis between Rhizobium and legumes. Nature 387:394–401. https://doi.org/10.1038/387394a0

Gamalero E, Glick BR (2011) Mechanisms used by plant growth-promoting bacteria. In: Bacteria in agrobiology: plant nutrient management, pp 17–46. https://doi.org/10.1007/978-3-642-21061-7_2

Gibson KE, Kobayashi H, Walker GC (2008) Molecular determinants symbiotic chronic infection. Annu Rev Genet 42:413–441. https://doi.org/10.1146/annurev.genet.42.110807.091427

Giraud E, Moulin L, Vallenet D, Barbe V, Cytryn E, Avarre JC et al (2007) Legumes symbioses: absence of Nod genes in photosynthetic bradyrhizobia. Science 316:1307–1312. https://doi.org/10.1126/science.1139548

Gleason C, Chaudhuri S, Yang T, Munoz A, Poovaiah BW, Oldroyd GE (2006) Nodulation independent of rhizobia induced by a calcium-activated kinase lacking autoinhibition. Nature 441(7097):1149–1152. https://doi.org/10.1038/nature04812

Gopalakrishnan S, Srinivas V, Alekhya G, Prakash B, Kudapa H, Rathore A, Varshney RK (2015) The extent of grain yield and plant growth enhancement by plant growth-promoting broad-spectrum *Streptomyces* sp. in chickpea. Springer Plus 4:31. https://doi.org/10.1186/s40064-015-0811-3

Gottfert M, Groß P, Hennecke H (1990) Proposed regulatory pathway encoded by the nodV and nodW genes, determinants of host specificity in Bradyrhizobium japonicum. Proc Natl Acad Sci USA 87:2680–2684. https://doi.org/10.1073/pnas.87.7.2680

Gottfert M, Holzhauser D, Bani D, Hennecke H (1992) Structural and functional analysis of two different *nodD* genes in *Bradyrhizobium japonicum* USDA110. Mol Plant-Microbe Interact 5:257–265

Groß P, Michel P, Hennecke H, Gottfert M (1993) A novel responseregulator is able to suppress the nodulation defect of a Bradyrhizobium japonicum nodW mutant. Mol Gen Genet 241:531–541. https://doi.org/10.1007/bf00279895

Grover M, Ali SZ, Sandhya V, Rasul A, Venkateswarlu B (2011) Role of microorganisms in adaptation of agriculture crops to abiotic stresses. World J Microbiol Biotechnol 27:1231–1240. https://doi.org/10.1007/s11274-010-0572-7

Haag AF, Arnold MFF, Myka KK, Kerscher B, Dall'Angelo S, Zanda M, Mergaert P, Ferguson GP (2013) Molecular insights into bacteroid development during Rhizobium-legume symbiosis. FEMS Microbiol Rev 37:364–383. https://doi.org/10.1111/1574-6976.12003

Haeze DW, Holsters M (2002) Nod factor structures, responses, and perception during initiation of nodule development. Glycobiology 12:79–105. https://doi.org/10.1093/glycob/12.6.79r

Hartwig UA, Maxwell CA, Joseph CM, Phillips DA (1990) Chrysoeriol and luteolin released from alfalfa seeds induce nod genes in Rhizobium meliloti. Plant Physiol 92:116–122. https://doi.org/10.1104/pp.92.1.116

Hassan S, Mathesius U (2012) The role of flavonoids in root-rhizosphere signalling: opportunities and challenges for improving plant–microbe interactions. J Exp Bot 63:3429–3444. https://doi.org/10.1093/jxb/err430

Heidstra R, Yang WC, Yalcin Y, Peck S, Emons AM, van Kammen A, Bisseling T (1997) Ethylene provides positional information on cortical cell division but is not involved in Nod factor-induced root hair tip growth in Rhizobium-legume interaction. Development 124:1781–1787

Herridge DF, Peoples MB, Boddey RM (2008) Global inputs of biological nitrogen fixation in agricultural systems. Plant Soil 311:1–18. https://doi.org/10.1007/s11104-008-9668-3

Hirsch S, Kim J, Munoz A, Heckmann AB, Downie JA, Oldroyd GE (2009) GRAS proteins form a DNA binding complex to induce gene expression during nodulation signaling in *Medicago truncatula*. Plant Cell 21:545–557. https://doi.org/10.1105/tpc.108.064501

Honma MA, Asomaning M, Ausubel FM (1990) Rhizobium meliloti nodD genes mediate host-specific activation of nodABC. J Bacteriol 172:901–911. https://doi.org/10.1128/jb.172.2.901-911.1990

Indrasumunar A, Kereszt A, Searle I, Miyagi M, Li D, Nguyen CD et al (2010) Inactivation of duplicated nod factor receptor 5(NFR5) genes in recessive loss-of-function non-nodulation mutants of allotetraploid soybean (*Glycinemax* L. Merr.). Plant Cell Physiol 51:201–214. https://doi.org/10.1093/pcp/pcp178

Indrasumunar A, Searle I, Lin MH, Kereszt A, Men A, Carroll BJ et al (2011) Nodulation factor receptor kinase 1 alpha controls nodule organ number in soybean (*Glycinemax* L.Merr). Plant J 65:39–50. https://doi.org/10.1111/j.1365-313X.2010.04398.x

Janczarek M, Rachwal K, Marzec A, Grzadziel J (2015) Signal molecules and cell-surface components involved in early stages of the legume-rhizobium interactions. Appl Soil Ecol 85:94–113

Javot H, Pumplin N, Harrison MJ (2007) Phosphate in the arbuscular mycorrhizal symbiosis: transport properties and regulatory roles. Plant Cell Environ 30:310–322. https://doi.org/10.1111/j.1365-3040.2006.01617.x

Kalo P, Gleason C, Edwards A, Marsh J, Mitra RM, Hirsch S, Jakab J, Sims S, Long SR, Rogers J et al (2005) Nodulation signaling in legumes requires NSP2, a member of the GRAS family of transcriptional regulators. Science 308:1786–1789. https://doi.org/10.1126/science.1110951

Kang H, Chu X, Wang C, Xiao A, Zhu H, Yuan S, Yang Z, Ke D, Xiao S, Hong Z, Zhang Z (2014) A MYB coiled-coil transcription factor interacts with NSP2 and is involved in nodulation in *Lotus japonicus*. New Phytol 201:837–849. https://doi.org/10.1111/nph.12593

Kawaharada Y, Kelly S, Nielsen MW, Hjuler CT, Gysel K, Muszynski A et al (2015) Receptor-mediated exopolysaccharide perception controls bacterial infection. Nature 523:308–312. https://doi.org/10.1038/nature14611

Kawaharada Y, Nielsen MW, Kelly S, James EK, Andersen KR, Rasmussen SR et al (2017) Differential regulation of the Epr3 receptor coordinates membrane-restricted rhizobial colonization of root nodule primordia. Nat Commun 23:14534. https://doi.org/10.1038/ncomms14534

Kobae Y, Tamura Y, Takai S, Banba M, Hata S (2010) Localized expression of arbuscular mycorrhiza-inducible ammonium transporters in soybean. Plant Cell Physiol 51:1411–1415. https://doi.org/10.1093/pcp/pcq099

Koda N, Asaeda Y, Yamade K, Kawahara H, Obata H (2001) A novel cryoprotective protein (CRP) with high activity from the icenucleating bacterium, Pantoea agglomerans IFO12686. Biosci Biotechnol Biochem 65:888–894. https://doi.org/10.1271/bbb.65.888

Kondorosi E, Gyuris J, Schmidt J, John M, Duda E, Hoffmann B, Schell J, Kondorosi A (1989) Positive and negative control of *nod* gene expression in *Rhizobiurn meliloti* is required for optimal nodulation. EMBO J 8:1331–1340. https://doi.org/10.1002/j.1460-2075.1989.tb03513.x

Kondorosi E, Buiré M, Cren M, Iyer N, Hoffmann B, Kondorosi A (1991) Involvement of the *syrM* and *nodD3* genes of *Rhizobiurn meliloti* in *nod* gene activation and in optimal nodulation of the plant host. Mol Microbiol 5:3035–3048. https://doi.org/10.1111/j.1365-2958.1991.tb01863.x

Konishi M, Yanagisawa S (2013) *Arabidopsis* NIN-like transcription factors have a central role in nitrate signalling. Nat Commun 4:1617. https://doi.org/10.1038/ncomms2621

Krishnan HB, Lorio J, Kim WS, Jiang G, Kim KY, DeBoer M et al (2003) Extracellular proteins involved in soybean cultivar-specific nodulation are associated with pilus-like surface appendages and exported by a type III protein secretion system in Sinorhizobium fredii USDA257. Mol Plant-Microbe Interact 16:617–625. https://doi.org/10.1094/MPMI.2003.16.7.617

Kunert KJ, Vorster BJ, Fenta BA, Kibido T, Dionisio G, Foyer CH (2016) Drought stress responses in soybean roots and nodules. Front Plant Sci 7:1015. https://doi.org/10.3389/fpls.2016.01015

Larrainzar E, Wienkoop S, Scherling C, Kempa S, Ladrera R, Arrese-Igor C et al (2009) Carbon metabolism and bacteroid functioning are involved in the regulation of nitrogen fixation in Medicago truncatula under drought and recovery. Mol Plant-Microbe Interact 22:1565–1576. https://doi.org/10.1094/MPMI-22-12-1565

Larrainzar E, Molenaar JA, Wienkoop S, Gil-Quintana E, Alibert B, Limami AM et al (2014) Drought stress provokes the down-regulation of methionine and ethylene biosynthesis pathways in *Medicago truncatula* roots and nodules. Plant Cell Environ 37:2051–2063. https://doi.org/10.1111/pce.12285

Lévy J, Bres C, Geurts R, Chalhoub B, Kulikova O, Duc G, Journet EP, Ané JM Lauber E, Bisseling T et al (2004) A putative Ca^{2+} and calmodulin-dependent protein kinase required for bacterial and fungal symbioses. Science 303:1361–1364. https://doi.org/10.1126/science.1093038

Limpens E, Franken C, Smit P, Willemse J, Bisseling T, Geurts R (2003) LysM domain receptor kinases regulating rhizobial Nod factor-induced infection. Science 302:630–633. https://doi.org/10.1126/science.1090074

Liu CW, Murray JD (2016) The role of flavonoids in nodulation host-range specificity: an update. Plan Theory 5:E33. https://doi.org/10.3390/plants5030033

Liu J, Yang S, Zheng Q, Zhu H (2014) Identification of a dominant gene in *Medicago truncatula* that restricts nodulation by Sinorhizobium meliloti strain Rm41. BMC Plant Biol 14:167. https://doi.org/10.1186/1471-2229-14-167

Loh J, Garcia M, Stacey G (1997) NodV and NodW, a second flavonoid recognition system regulating nod gene expression in *Bradyrhizobium japonicum.* J Bacteriol 179:3013–3020. https://doi.org/10.1128/jb.179.9.3013-3020.1997

Loh J, Lohar D, Andersen B, Stacey G (2002) A two-component regulator mediates population density dependent expression of the *Bradyrhizobium japonicum* nodulation genes. J Bacteriol 184:1759–1766. https://doi.org/10.1128/jb.184.6.1759-1766.2002

Lozano JMR, Azcon R (2000) Symbiotic efficiency and infectivity of an autochthonous arbuscular mycorrhizal *Glomus* sp. from saline soils and *Glomus deserticola* under salinity. Mycorrhiza 10:137–143. https://doi.org/10.1007/s005720000075

Mabood F, Souleimanov A, Khan W, Smith DL (2006) Jasmonates induce Nod factor production by *Bradyrhizobium japonicum.* Plant Physiol Biochem 44:759–765. https://doi.org/10.1016/j.plaphy.2006.10.025

Madsen EB, Madsen LH, Radutoiu S, Olbryt M, Rakwalska M, Szczyglowski K, Sato S, Kaneko T, Tabata S, Sandal N et al (2003) A receptor kinase gene of the LysM type is involved in legume perception of rhizobial signals. Nature 425:637–640. https://doi.org/10.1038/nature02045

Maekawa T, Maekawa-Yoshikawa M, Takeda N, Imaizumi-Anraku H, Murooka Y, Hayashi M (2009) Gibberellin controls the nodulation signaling pathway in *Lotus japonicus.* Plant J 58:183–194. https://doi.org/10.1111/j.1365-313X.2008.03774.x

Marlow VL, Haag AF, Kobayashi H, Fletcher V, Scocchi M, Walker GC, Ferguson GP (2009) Essential role for the BacA protein in the uptake of a truncated eukaryotic peptide in *Sinorhizobium meliloti.* J Bacteriol 191:1519–1527. https://doi.org/10.1128/JB.01661-08

Márquez-García B, Shaw D, Cooper JW, Karpinska B, Quain MD, Makgopa EM et al (2015) Redox markers for drought-induced nodule senescence, a process occurring after drought-induced senescence of the lowest leaves in soybean (Glycine max Merr.). Ann Bot 116:497–510. https://doi.org/10.1093/aob/mcv030

Marsh JF, Rakocevic A, Mitra RM, Brocard L, Sun J, Eschstruth A, Long SR, Schultze M, Ratet P, Oldroyd GED (2007) *Medicago truncatula* NIN is essential for Rhizobial-independent nodule organogenesis induced by autoactive calcium/calmodulin-dependent protein kinase. Plant Physiol 144:324–335. https://doi.org/10.1104/pp.106.093021

Mhadhbi H, Chihaoui S, Mhamdi R, Mnasri B, Jebara M, Mhamdi R (2011) A highly osmotolerant rhizobial strain confers a better tolerance of nitrogen fixationand enhanced protective activities to nodules of phaseolus vulgaris under drought stress. Afr Biotechnol 22:4555–4563. https://doi.org/10.5897/AJB10.1991

Mitra RM, Long SR (2004) Plant and bacterial symbiotic mutants define three transcriptionally distinct stages in the development of the *Medicago truncatula/Sinorhizobiummeliloti* symbiosis. Plant Physiol 134:595–604. https://doi.org/10.1104/pp.103.031518

Mondal HK, Mehta S, Kaur H, Gera R (2017) Characterization of abiotic stress tolerant rhizobia as PGPR of mothbean, clusterbean and mungbean grown in hyper-arid zone of Rajasthan. Int J Bio-Resour Stress Manag 8:309–315. https://doi.org/10.23910/IJBSM/2017.8.2.1793

Murray JD, Karas BJ, Sato S, Tabata S, Amyot L, Szczyglowski K (2007) A cytokinin perception mutant colonized by Rhizobium in the absence of nodule organogenesis. Science 315:101–104. https://doi.org/10.1126/science.1132514

Ojuederie OB, Olanrewaju OS, Babalola OO (2019) Plant growth promoting rhizobacterial mitigation of drought stress in crop plants: implications for sustainable agriculture. Agronomy 9:712. https://doi.org/10.3390/agronomy9110712

Oka-Kira E, Kawaguchi M (2006) Long-distance signaling to control root nodule number. Curr Opin Plant Biol 9:496–502. https://doi.org/10.1016/j.pbi.2006.07.012

Okazaki S, Zehner S, Hempel J, Lang K, Göttfert M (2009) Genetic organization and functional analysis of the type III secretion system of *Bradyrhizobium elkanii.* FEMS Microbiol Lett 295:88–95. https://doi.org/10.1111/j.1574-6968.2009.01593.x

Okazaki S, Kaneko T, Sato S, Saeki K (2013) Hijacking of leguminous nodulation signaling by the rhizobial type III secretion system. Proc Natl Acad Sci U S A 110:17131–17136. https://doi.org/10.1073/pnas.1302360110

Okazaki S, Tittabutr P, Teulet A, Thouin J, Fardoux J, Chaintreuil C et al (2016) Rhizobium–legume symbiosis in the absence of Nod factors: two possible scenarios with or without the T3SS. ISME J 10:64–74. https://doi.org/10.1038/ismej.2015.103

Peck MC, Fisher RF, Long SR (2006) Diverse flavonoids stimulate NodD1 binding to nod gene promoters in *Sinorhizobium meliloti*. J Bacteriol 188:5417–5427. https://doi.org/10.1128/JB.00376-06

Peck MC, Fisher RF, Bliss R, Long SR (2013) Isolation and characterization of mutant Sinorhizobium meliloti NodD1 proteins with altered responses to luteolin. J Bacteriol 195:3714–3723. https://doi.org/10.1128/JB.00309-13

Phillips D, Joseph C, Maxwell C (1992) Trigonelline and stachydrine released from alfalfa seeds activate NodD2 protein in Rhizobium *meliloti*. Plant Physiol 99:1526–1531. https://doi.org/10.1104/pp.99.4.1526

Pietraszewska-Bogiel A, Lefebvre B, Koini MA, Klaus-Heisen D, Takken FL, Geurts R et al (2013) Interaction of Medicago truncatula lysin motif receptor-like kinases, NFP and LYK3, produce-din Nicotiana benthamiana induces defence-like responses. PLoSONE 8:e65055. https://doi.org/10.1371/journal.pone.0065055

Pimprikar P, Carbonnel S, Paries M, Katzer K, Klingl V, Bohmer MJ, Karl L, Floss DS, Harrison MJ, Parniske M, Gutjahr C (2016) A CCaMK-CYCLOPS-DELLA complex activates transcription of RAM1 to regulate arbuscule branching. Curr Biol 25:987–998. https://doi.org/10.1016/j.cub.2016.01.069

Plet J, Wasson A, Ariel F, Le Signor C, Baker D, Mathesius U, Crespi M, Frugier F (2011) MtCRE1-dependent cytokinin signaling integrates bacterial and plant cues to coordinate symbiotic nodule organogenesis in *Medicago truncatula*. Plant J 65:622–633. https://doi.org/10.1111/j.1365-313X.2010.04447.x

Poustini K, Mabood F, Smith DL (2007) Preincubation of Rhizobium leguminosarum bv. phaseoli with jasmonate and genistein signal molecules increases bean (*Phaseolus vulgaris* L.) nodulation, nitrogen fixation and biomass production. J Agric Sci Technol 9:107–117

Radutoiu S, Madsen LH, Madsen EB, Felle HH, Umehara Y, Gronlund M, Sato S, Nakamura Y, Tabata S, Sandal N et al (2003) Plant recognition of symbiotic bacteria requires two LysM receptor-like kinases. Nature 425:585–592. https://doi.org/10.1038/nature02039

Ramasamy K, Joe M, Kim K, Lee S et al (2011) Synergistic effects of arbuscular mycorrhizal fungi and plant growth promoting rhizobacteria for sustainable agricultural production. Korean J Soil Sci Fertil 44:637–649. https://doi.org/10.7745/KJSSF.2011.44.4.637

Rey T, Nars A, Bonhomme M, Bottin A, Huguet S, Balzergue S et al (2013) NFP, a Lys M protein controlling Nod factor perception, also intervenes in *Medicago truncatula* resistance to pathogens. New Phytol 198:875–886. https://doi.org/10.1111/nph.12198

Rolli E, Marasco R, Vigani G, Ettoumi B, Mapelli F, Deangelis ML et al (2015) Improved plant resistance to drought is promoted by the root-associated microbiome as a water stress-dependent trait. Environ Microbiol 17:316–331. https://doi.org/10.1111/1462-2920.12439

Sandhya V, Ali SZ, Grover M, Reddy G, Venkateswarlu B (2009) Alleviation of drought stress effects in sunflower seedlings by exopolysaccharides producing *Pseudomonas putida* strain P45. Biol Fertil Soil 46:17–26. https://doi.org/10.1007/s00374-009-0401-z

Schauser L, Wieloch W, Stougaard J (2005) Evolution of NIN-like proteins in Arabidopsis, rice, and *Lotus japonicus*. J Mol Evol 60:229–327. https://doi.org/10.1007/s00239-004-0144-2

Shaw SL, Long SR (2003) Nod factor inhibition of reactive oxygen efflux in a host legume. Plant Physiol 132:2196–2204. https://doi.org/10.1104/pp.103.021113

Singh S, Katzer K, Lambert J, Cerri M, Parniske M (2014) Cyclops, a DNA-binding transcriptional activator, orchestrates symbiotic root nodule development. Cell Host Microbe 15:139–152. https://doi.org/10.1016/j.chom.2014.01.011

Smit P, Raedts J, Portyanko V, Debellé F, Gough C, Bisseling T, Geurts R (2005) NSP1 of the GRAS protein family is essential for rhizobial Nod factor-induced transcription. Science 308:1789–1791. https://doi.org/10.1126/science.1111025

Smit P, Limpens E, Geurts R, Fedorova E, Dolgikh E, Gough C, Bisseling T (2007) *Medicago* LYK3, an entry receptor in rhizobial nodulation factor signaling. Plant Physiol 145:183–191. https://doi.org/10.1104/pp.107.100495

Soyano T, Hayashi M (2014) Transcriptional networks leading to symbiotic nodule organogenesis. Curr Opin Plant Biol 20:146–154. https://doi.org/10.1016/j.pbi.2014.07.010

Soyano T, Kouchi H, Hirota A, Hayashi M (2013) Nodule inception directly targets NF-Y subunit genes to regulate essential processes of root nodule development in *Lotus japonicus*. PLoS Genet 9:e1003352. https://doi.org/10.1371/journal.pgen.1003352

Soyano T, Hirakawa H, Sato S, Hayashi M, Kawaguchi M (2014) Nodule inception creates a long-distance negative feedback loop involved in homeostatic regulation of nodule organ production. Proc Natl Acad Sci U S A 111:14607–14612. https://doi.org/10.1073/pnas.1412716111

Soyano T, Shimoda Y, Hayashi M (2015) NODULE INCEPTION antagonistically regulates gene expression with nitrate in *Lotus japonicas*. Plant Cell Physiol 56:368–376. https://doi.org/10.1093/pcp/pcu168

Staudinger C, Mehmeti-Tershani V, Gil-Quintana E, Gonzalez EM, Hofhansl F, Bachmann G, Wienkoop S (2016) Evidence for a rhizobia-induced drought stress response strategy in *Medicago truncatula*. J Proteome 136:202–213. https://doi.org/10.1016/j.jprot.2016.01.006

Stolf-Moreira R, Lemos EGM, Carareto-Alves L, Marcondes J, Pereira SS (2011) Transcriptional profiles of roots of different soybean genotypes subjected to drought stress. Plant Mol Biol Rep 29:19–34. https://doi.org/10.1186/1471-2164-14-687

Stracke S, Kistner C, Yoshida S, Mulder L, Sato S, Kaneko T, Tabata S, Sandal N, Stougaard J, Szczyglowski K et al (2002) A plant receptor-like kinase required for both bacterial and fungal symbiosis. Nature 417:959–962. https://doi.org/10.1038/nature00841

Subramanian S, Stacey G, Yu O (2007) Distinct, crucial roles for flavonoids during legume nodulation. Trends Plant Sci 12:282–285. https://doi.org/10.1016/j.tplants.2007.06.006

Sulieman S, Van Ha C, Nasr Esfahani M, Watanabe Y, Nishiyama R, Pham CT et al (2015) DT2008: a promising new genetic resource for improved drought tolerance in soybean when solely dependent on symbiotic N2 fixation. Biomed Res Int 2015:1–7. https://doi.org/10.1155/2015/687213

Sun J, Cardoza V, Mitchell DM, Bright L, Oldroyd G, Harris JM (2006) Crosstalk between jasmonic acid, ethylene and Nod factor signaling allows integration of diverse inputs for regulation of nodulation. Plant J 46:961–970. https://doi.org/10.1111/j.1365-313X.2006.02751.x

Swainsbury DJ, Zhou L, Oldroyd GE, Bornemann S (2012) Calcium ion binding properties of *Medicago truncatula* calcium/calmodulin-dependent protein kinase. Biochemistry 51:6895–6907. https://doi.org/10.1021/bi300826m

Tanaka N, Kato M, Tomioka R, Kurata R, Fukao Y, Aoyama T et al (2014) Characteristics of a root hair-less line of *Arabidopsis thaliana* under physiological stresses. J Exp Bot 65:1497–1512. https://doi.org/10.1093/jxb/eru014

Tang F, Yang S, Liu J, Zhu H (2016) Rj4, a gene controlling nodulation specificity in soybeans, encodes a thaumatin-like protein, but not the one previously reported. Plant Physiol 170:26–32. https://doi.org/10.1104/pp.15.01661

Tikhonovich IA, Provorov N (2007) Beneficial plant–microbe interactions. Comprehensive Mol Phytopathol:365–420. https://doi.org/10.1016/B978-044452132-3/50018-3

Tirichine L, Sandal N, Madsen LH, Radutoiu S, Albrektsen AS, Sato S, Asamizu E, Tabata S, Stougaard J (2007) A gain-of-function mutation in a cytokinin receptor triggers spontaneous root nodule organogenesis. Science 315:104–107. https://doi.org/10.1126/science.1132397

Toth K, Stacey G (2015) Does plant immunity play a critical role during initiation of the legume-rhizobium symbiosis? Front Plant Sci 6:401. https://doi.org/10.3389/fpls.2015.00401

Tripathi P, Rabara RC, Reese RN, Miller MA, Rohila JS, Subramanian S et al (2016) A tool box of genes, proteins, metabolites and promoters for improving drought tolerance in soybean includes the metabolite coumestrol and stomatal development genes. BMC Genomics 17:102. https://doi.org/10.1186/s12864-016-2420-0

Vadez V (2014) Root hydraulics: the forgotten side of roots in drought adaptation. Field Crop Res 165:15–24. https://doi.org/10.1016/j.fcr.2014.03.017

Van Hameren B, Hayashi S, Gresshoff PM, Ferguson BJ (2013) Advances in the identification of novel factors required in soybean nodulation, a process critical to sustainable agriculture and food security. J Plant Biol Soil Health 1:1–6

Vernié T, Kim J, Frances L, Ding Y, Sun J, Guan D, Niebel A, Gifford ML, de Carvalho-Niebel F, Oldroyd GED (2015) The NIN transcription factor coordinates diverse nodulation programs in different tissues of the *Medicago truncatula* root. Plant Cell 27:3410–3424. https://doi.org/10.1105/tpc.15.00461

Vriezen JAC, Bruijn FJ, Nusslein K (2007) Responses of rhizobia to desiccation in relation to osmotic stress, oxygen, and temperature. Appl Environ Microbiol 73:3451–3459. https://doi.org/10.1128/AEM.02991-06

Wang C, Zhu H, Jin L, Chen T, Wang L, Kang H, Hong Z, Zhang Z (2013) Splice variants of the SIP1 transcripts play a role in nodule organogenesis in *Lotus japonicus*. Plant Mol Biol 82:97–111. https://doi.org/10.1007/s11103-013-0042-3

Wang Q, Yang S, Liu J, Terecskei K, Ábrahám E, Gombár A et al (2017) Host-secreted antimicrobial peptide enforces symbiotic selectivity in *Medicago truncatula*. Proc Natl Acad Sci 114:6854–6859. https://doi.org/10.1073/pnas.1700715114

Wang Q, Liu J, Li H, Yang S, Körmöczi P, Kereszt A et al (2018) Nodule specific cysteine-rich peptides negatively regulate nitrogen-fixing symbiosis in a strain-specific manner in *Medicago truncatula*. Mol Plant-Microbe Interact 31:240–248. https://doi.org/10.1094/MPMI-08-17-0207-R

White J, Prell J, James EK, Poole P (2007) Nutrient sharing between symbionts. Plant Physiol 144:604–614. https://doi.org/10.1104/pp.107.097741

Wu QS, Xia RX, Zou YN (2006) Reactive oxygen metabolism in mycorrhizal and non-mycorrhizal citrus (Poncirus trifoliata) seedlings subjected to water stress. J Plant Physiol 163:1101–1110. https://doi.org/10.1016/j.jplph.2005.09.001

Xie F, Murray JD, Kim J, Heckmann AB, Edwards A, Oldroyd GE, Downie JA (2012) Legume pectate lyase required for root infection by rhizobia. Proc Natl Acad Sci U S A 109:633–638. https://doi.org/10.1073/pnas.1113992109

Yang J, Kloepper JW, Ryu CM (2009) Rhizosphere bacteria help plants tolerate abiotic stress. Trends Plant Sci 14:1–4. https://doi.org/10.1016/j.tplants.2008.10.004

Yang S, Tang F, Gao M, Krishnan HB, Zhu H (2010) R gene-controlled host specificity in the legume–rhizobia symbiosis. Proc Natl Acad Sci 107:18735–18740. https://doi.org/10.1073/pnas.1011957107

Zhang H, Kim MS, Sun Y, Dowd SE, Shi H, Paré PW (2008) Soil bacteria confer plant salt tolerance by tissue-specific regulation of sodium transporter HKT1. Mol Plant-Microbe Interact 21:737–744. https://doi.org/10.1094/MPMI-21-6-0737

Zhu H, Chen T, Zhu M, Fang Q, Kang H, Hong Z, Zhang Z (2008) A novel ARID DNA-binding protein interacts with SymRK and is expressed during early nodule development in Lotus japonicus. Plant Physiol 148:337–347. https://doi.org/10.1104/pp.108.119164

Zipfel C (2014) Plant pattern-recognition receptors. Trends Immunol 35:345–351. https://doi.org/10.1016/j.it.2014.05.004

Zipfel C, Oldroyd GE (2017) Plant signalling in symbiosis and immunity. Nature 543:328–336. https://doi.org/10.1038/nature22009

Chapter 8
Optimizing *Rhizobium*-Legume Symbiosis in Smallholder Agroecosystems

Morris Muthini, Richard Awino, Kibet Charles Kirui, Kipkorir Koech, Abdul A. Jalloh, and Ezekiel Mugendi Njeru

Abstract Legumes represent some of the most important crops in the world that are very useful in smallholder and large-scale farming systems. They are commonly referred to as the poor man's meat due to their high dietary proteins and other micronutrients, including iron, zinc, folate, and thiamine that are important for human and livestock nutrition. The role of legumes in boosting food security and generating income in rural households across the globe is indispensable. Other than boosting healthy diet choices in rural households, legumes have also been used to promote soil health and fertility as they collectively account for 80% of biological nitrogen fixation. Industrial application of legumes is also gaining popularity with grain legumes currently used in synthesis of bioplastics. Legume production commands 27% of global crop production with a significant proportion coming from low input rural smallholder farms. However, legume production has declined over the years as a result of poor soil health management strategies and climate change. These have led to utilization of intensive agriculture practices that dictates the use of inorganic fertilizers and other agrochemicals to boost crop production. Intensive agricultural practices are unsustainable, costly and not eco-friendly. The plant growth-promoting microorganisms like rhizobia, mycorrhiza, and phosphate solubilizing bacteria are eco-friendly and can sustainably be used to improve soil fertility and improve legume production. Besides, co-inoculation of legume plants with rhizobia and phosphate solubilizing microorganisms can sustainably improve legume production and soil fertility. Here, we review the potential benefits of utilization of rhizobia, and other plant growth promoting microorganisms for sustainable improvement of legume production in smallholder agroecosystems.

Keywords Rhizobia · Legumes · Smallholder agroecosystems · Plant growth-promoting microorganisms

---

M. Muthini · R. Awino · K. C. Kirui · K. Koech · A. A. Jalloh · E. M. Njeru (✉)
Department of Biochemistry, Microbiology and Biotechnology, Kenyatta University, Nairobi, Kenya
e-mail: njeru.ezekiel@ku.ac.ke

P. Guleria et al. (eds.), *Sustainable Agriculture Reviews 45*, Sustainable Agriculture Reviews 45, https://doi.org/10.1007/978-3-030-53017-4_8

Abbreviations

ACC	1-aminocyclopropane-1-carboxylic Acid
AMF	Arbuscular Mycorrhizal Fungi
BNF	Biological Nitrogen Fixation
HAP	Heme Activator Protein
IAA	Indole Acetic Acid
miRNA	micro Ribonucleic Acid
NCR	Nodule Cysteine Rich
NCRPs	Nodule Cysteine Rich Peptides
PGPR	Plant Growth Promoting Rhizobacteria
sRNA	small Ribonucleic Acid

8.1 Introduction

Legumes are a significant part of diet in many cultural settings in the world and are some of the most ecologically essential crops (Magrini et al. 2016). Most legumes have high dietary protein and oil content making them pertinent in human and animal diets as well as a promising source of biofuels (Checcucci et al. 2017). They are also important sources of essential micronutrients in the human diet including, iron, zinc, niacin, calcium, folate, and thiamine. These nutrients vary depending on the legume variety and the environmental conditions (De Jager et al. 2019). Food legumes ensure food and nutritional security to family farming systems that produce most of the food consumed in the world. Legumes are also a suitable source of income to smallholder farmers, for they have better prices as compared to other cereal crops (Venance et al. 2016; Muoni et al. 2019). Despite being essential crops for human and livestock population, legume production has registered decline over the years owing to climate change in sub-Saharan Africa, increasing human population and shrinking farmland (Belmain et al. (2013). Continual soil fertility depletion has worsened the scenario making it difficult for the resource strained smallholder agroecosystems to cope and maintain high productivity of these vital crops (Dwivedi et al. 2015).

Most of the smallholder farmers are characterized by lack of financial capital; poor produce marketing strategies, low farm inputs as well as limited size of their farmland (Kuivanen et al. 2016). To improve crop production in smallholder farms, intensive agricultural practices which include increased use of agrochemicals has been on the rise (Gopalakrishnan et al. 2015). With continuing increase in population growth, efforts are being made to get alternative ecofriendly and sustainable techniques of soil health management (Mungai et al. 2016). Soil health can be sustainably managed using organic treatments including; farm yard manure, crop residues, compost and cropping methods like crop rotation, natural fallow, intercropping, legumes and cover crops (Gopalakrishnan et al. 2015; Komarek et al. 2017).

Rhizospheric microorganisms including plant growth promoting rhizobacteria, biological nitrogen fixing bacteria and mycorrhizal fungi play a significant role in sustainable soil fertility management (Mendes et al. 2013). Eighty percent of biologically fixed nitrogen in the soil is attributed to the relationship between nitrogen fixing rhizobia and leguminous plants (Mabrouk et al. 2018). Legume symbiotic interactions with rhizobia have been hailed as a fundamental interaction not only because they boost legume production, but they also improve soil fertility status, significantly boosting the production of other associated crops through biological nitrogen fixation (BNF). These underpin the significance of promoting these interactions in smallholder agroecosystems where intensive and continual farming coupled with poor agronomic practices, have considerably led to decline in soil fertility status (Diep et al. 2016).

8.2 Significance of Legumes and Production Constraints

Other than the nutritional value, legumes are utilized in soil management. They have a unique interaction with rhizobia, which leads to biological nitrogen fixation and hence improvement of soil fertility (Muoni et al. 2019). According to Jeger et al. (2019), farmers grow legumes as cover crops which lessen surface runoff and in turn limit soil erosion. This is particularly significant in the developing nations where the legumes significantly supplement the cereals produced (Temba et al. 2016). Legumes have also been exploited for use in oil production owing to the high oil content of the seeds. Oil production from soybean is second to palm oil production worldwide (Lim et al. 2015). Currently, the grain legumes have also been exploited for a wide range of industrial applications such as synthesis of bioplastics and bread industries. Alfalfa, one of the most economically important forage legumes, is important in dairy and beef production farms (Kulkarni et al. 2018).

Despite their significance, the production of legumes has slightly improved in comparison to the cereals whose world production has seen tremendous improvement. Poor soil fertility management practices, biotic and abiotic stresses have severely strained legume production. Moreover, drought and heat stress due to climate change have made it difficult for the smallholder agroecosystems to cope. Besides, global warming has favored the proliferation of new bacterial, fungal and viral pathogens leading to increased losses in legume production (Rubiales and Mikic 2015). Limited by resources, many smallholding systems are unable to adequately mitigate these challenges (Daryanto et al. 2015). In particular, the farmers cannot afford to install efficient irrigation systems, and knowledge gaps have made it difficult for the farmers to seek eco-friendly and effective pest and soil management techniques (Tsion and Steven 2019). Moreover, application of inorganic fertilizers and pesticides has led to further deterioration of farms through increased acidity and eco-toxicological effects on the critical soil microflora negatively affecting rhizospheric micro-ecology (Mahmood et al. 2016). The bactericidal effect of the chemicals on the plant growth-promoting rhizobacteria has a far-reaching impact on legume production. These beneficial microorganisms through their physiological activities solubilize phosphorous, another critical legume yield-limiting element. The complexing capability of the Al^{3+} is also responsible for leaching of vital nutrients from the subsurface. Overuse and suboptimal use of the pesticides occasioned by inadequate knowledge has exacerbated incidences of fungal resistance, increasingly making more soils less productive (Brown et al. 2017).

8.3 Legume Production in Smallholder Agroecosystems

Smallholder agroecosystems play a significant role in ensuring the eco-genetic stability of many biosphere resources (Morel et al. (2012). Smallholder farmers form the main bulk of agricultural producers, and even though their products are not often market-oriented, collectively they are responsible for the largest share of the global market value (Kuivanen et al. 2016). While many households have always preferred cultivating staple crops to ensure food security, crop diversification has been infused in many smallholder agroecosystems for purposes of ecological and economic needs (Kuivanen et al. 2016). With agroecosystems in eco-climatic zones of abundant precipitation able to go for high-value crops such as coffee and tea, the smallholder farmers in the arid and semi-arid lands who are the vast majority have legumes as their most important crop for economic purposes (Chibarabada et al. 2017). This preference is down to the fact that most legumes are quick maturing and able to do well with moderately low rainfall. Most smallholder farmers intercrop legumes, especially common bean and cowpea with other staple cereals like maize to meet the food demands of their households (Muoni et al. 2019). The crops planted in each farm is to a great extend determined by household needs including nutritional requirements and feed for livestock (Kuivanen et al. 2016). Commonly grown food legumes in smallholder farming systems include cowpea, green grams, pigeon pea, common beans, soy beans, and ground nuts. However, Jack beans as well velvet beans are among non-edible legumes planted as a source of food for livestock. Legume production in smallholder farming systems has declined over the past years owing to inadequate capital to purchase farm inputs and, knowledge gaps on the effective use of the same (Pannell et al. 2014). Continual land use has also led to soil deterioration evidenced by increased demand for nitrogen fertilizers in both developed and developing nations calling for the need to upscale remedial regiments. The ecological effects and health hazards imposed by the chemical residues of farm inputs are well documented and calls for natural and environmentally friendly farm practices.

Legume-rhizobia interactions remain one of the most well studied symbiosis that has been well exploited for use in the improvement of legume production (Karmakar et al. 2015). It is pertinent to review the gains made, challenges and future perspectives that can be employed to improve legume production to levels that will not only be resilient to the climate changes but also meet the burgeoning population needs in terms of food, biofuels, and fodder for animals.

8.4 Potential of Rhizobia to Improve Legume Production in Smallholder Agroecosystems

Nitrogen is undoubtedly one of the most important macronutrients required by plants for suitable development and productivity. It is an essential mineral for the synthesis of vital macromolecules like amino acids and chlorophyll (Martin 2017). Nitrogen deficiency impairs plant growth and development, leading to a considerable decline in productivity (Kalaji et al. 2016). The potential of rhizobia to improve legume production is a well-studied area. Nearly all the agriculturally significant legume species have been observed to have the capacity of forming symbiosis with rhizobia (Gupta et al. 2015). These symbioses with agriculturally important legumes are not only significant for the host legumes development but are also responsible for more than half of all the nitrogen that gets into the soil ecosystems through biological nitrogen fixation (Youseif et al. 2017). This has the capability of greatly promoting the productivity of the subsequent non-legume plants making *Rhizobium*-legume symbiosis of immense value in smallholder agroecosystems where crop diversification is particularly integral, (Checcucci et al. 2017). Increased biological nitrogen fixation activity in the soil greatly decreases the rate of nitrogen depletion in the soil (Giller et al. 2013).

Rhizobium-legume symbiosis has been shown to promote legume production in several ways, directly and indirectly (Fig. 8.1). Directly, *Rhizobium* bacteria promote legume production through nitrogen fixation enabling the plants to obtain the

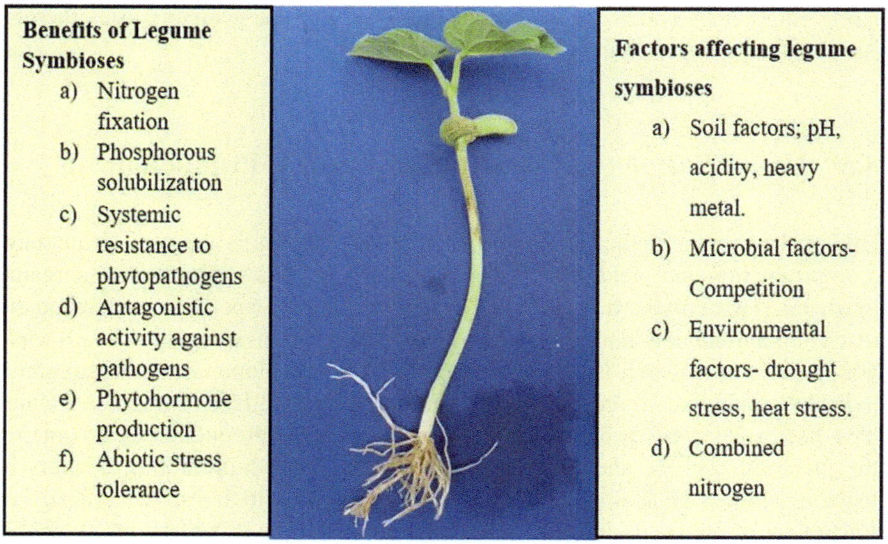

Fig. 8.1 Legume symbioses; Legume symbioses with microbial population confers to the legume a host of benefits ranging from nitrogen fixation, mineral mobilization, abiotic stress tolerance and antagonistic activity. Environmental factors and soil factors impact strongly on the efficacy of the legume symbioses

important yield determining mineral (Gopalakrishnan et al. 2015). In addition, most rhizobia isolates are particularly effective in phosphate solubilization, especially in acidic soils where phosphorus has been immobilized (Collavino et al. 2010). The significance of rhizobia in these soil ecosystems is underpinned by the fact that phosphorous immobilization by heavy metal ions like Al^{3+} makes it expensive for farmers to ameliorate through phosphate fertilizers that in higher concentrations in the soil present more harm than good (Sulieman and Tran 2015). Phosphorous is the second most important nutrient in legume production, and its deficiency in many soils is significantly yield-limiting (Tak et al. 2012). In addition, the potential of the rhizobia to promote the production of phytohormones is a yield boosting mechanism. Some rhizobia strains can modify the hormone levels of the host through the production of enzymes such as indole 3 acetic acid (IAA) and 1-aminocyclopropan e-1-carboxylic acid (ACC) deaminase (Carlos et al. 2016).

Rhizobia also promote legume production indirectly through a number of ways. Some strains of the bacteria have been shown to have important capabilities of inhibiting the growth of pathogens and other unwanted organisms thereby acting as biocontrol. This ability of rhizobia to act as biocontrol is mainly linked to production of antibiotics, cell-wall degrading enzymes, hydrocyanic acid, and siderophore (Das et al. 2017). A number of rhizobia strains have been documented to exhibit antagonistic activity against the common fungal pathogens and other harmful bacteria, thus providing vital bio-protection to the crop symbionts (Kucuk and Cevheri 2015; Checcucci et al. 2017). *Pseudomonas fluorescens* and *Paenibacillus polymyxa* were shown to reduce the incidence of damping off disease in alfalfa (Sarhan and Shehata 2014). Besides, some rhizobia strains produce siderophores, iron chelating compounds which act as biocontrol by limiting the available irons thereby starving the pathogen (Saha et al. 2016).

8.5 Use of *Rhizobium* Inoculants in Legume Production

One of the current challenges facing smallholder farmers is the need to increase crop production and yield quality with limited externalities. The use of inorganic fertilizers is not only harmful to the environment but also expensive and beyond the reach of many smallholder farmers in the developing nations (Doganova and Karnoe 2015). A lot of research is presently focused on the development of effective biological formulations to replace the inorganic fertilizers. Effective rhizobia isolates have been used to boost biological nitrogen fixation and production by colonizing the plant rhizosphere and confer numerous plant growth-promoting benefits to legumes (Fig. 8.2). Co-inoculation has been explored to boost the efficacy of *Rhizobium* bacteria in improving legume performance (Qureshi et al. 2012; Mweetwa et al. 2016). Co-inoculation of *Rhizobium* with *Bacillus megaterium*, Plant Growth Promoting Rhizobacteria (PGPR) was observed to significantly improve nodule weight and overall legume performance (Korir et al. 2017; Atieno et al. 2012). Rhizobia isolates introduced, by co-inoculation, into the legume

A B

Fig. 8.2 (**a**) Healthy common bean plants inoculated with an effective *Rhizobium* isolate, (**b**) Well developed root system with nodules of a common bean plant, inoculated with an effective native *Rhizobium* isolate. In both controlled and field conditions, inoculation of legumes with effective *Rhizobium* isolates have been shown to promote growth through increased biological nitrogen fixation, siderophore production, phytohormone production and systemic pathogen resistance

rhizosphere promote legume production by creating additional infection sites, increasing the release of plant growth-promoting substances such as secondary metabolites, direct nitrogen fixation, solubilization and uptake of nutrients, and siderophore production (Mishra et al. 2009; Vargas et al. 2017; de Souza et al. 2015, 2016). Different *Rhizobium* isolates have been shown to have differing ability to promote legume production in different mechanisms, especially when co-inoculated with other isolates. For example, *Pseudomonas fluorescens* produces growth-promoting compound and subsequently enhance nitrogen fixation in soybeans when co-inoculated with *B. japonicum*, *Pseudomonas striata* increase nodulation when co-inoculated with *Bradyrhizobium sp* and clover production is improved by the inoculation of *P. fluorescens* and *Rhizobium leguminosarum*. Moreover, legumes can form a dual symbiotic relationship with rhizobia and arbuscular mycorrhizal fungi (AMF) (Kaschuk et al. 2009). According to Stajkovic et al. (2009) AMF and mesorhizobia co-inoculation increase legume yields and growth under controlled and field conditions. AMF increases plant N and P uptake and subsequently promote biomass production and legume photosynthesis. A co-inoculation of *Glomus mosseae* and *R. leguminosarum* can increase plant biomass, rate of photosynthesis, nodule formation, and nitrogen fixation (Bhowmik and Das 2018). Subsequently, legume yield is bound to increase.

The efficacy of the commercial *Rhizobium* inoculants has been observed to vary with edaphic and agroclimatic factors. Notably, there are complex microbial interactions in the soil ranging from positive and negative that affects the performance of the introduced commercial rhizobia inoculants. Occasionally, the commercial inoculants fail to significantly establish themselves in the native soil due to their allochthonous nature and lack of the capacity to establish in foreign conditions (Batista et al. 2015). Therefore, research on isolation and use of autochthonous rhizobia strains has gained a lot of ground. The native rhizobia from any ecological zone are significant as they are well adapted to the local edaphic factors and easily establish symbiosis with the legumes (Gronemeyer et al. 2014). Furthermore, indigenous rhizobia isolates are more important to smallholder farmers because of their low cost, abiotic stress resistance and wide host range (Gyogluu et al. 2018), and can sustainably be used to promote legume production in smallholder farms (Fig. 8.3).

However, screening is important since not all native rhizobia have the desired efficacy in nodulation and effective symbiosis establishment. In a study by Thuita et al. (2012), commercial rhizobia inoculants proved more effective in nodulation and improving nitrogen content of select soybean varieties. Similar results have also been obtained by Muleta (2017). It is also imperative to isolate the native rhizobia from the crops of interest for which the symbiosis is desired. Isolation of *Rhizobium* and or field tests should always be done on the target farms for effective strain

Fig. 8.3 A smallholder farming unit in Embu County, Kenya showing cowpeas (*Vigna unguiculata*) inoculated with indigenous *Rhizobium* isolates. The indigenous isolates have the ability to form beneficial symbiosis with the legumes as they are able to favorably compete with the native microflora and establish symbiosis with the cowpea plants in comparison to the exotic strains. (Image courtesy of the The Future Leaders – African Independent Researchers (FLAIR) project titled; using root-associated microorganisms to enhance sustainable crop production and resilience of smallholder agroecosystems to climate change)

isolation. Previous studies have demonstrated that native, naturalized rhizobia isolates have a competitive advantage in nodule occupancy and effectiveness as compared to commercial rhizobia inoculants (Irisarri et al. 2019). For instance, higher nitrogen fixation by clover plants inoculated with native naturalized rhizobia isolates has been reported as compared to commercial rhizobia inoculant (Irisarri et al. 2019). According to Chibeba et al. (2017), indigenous rhizobia isolates have also been observed to be very effective in nitrogen fixation and enhancement of soya bean production. Similar effects have also been reported on *Phaseolus vulgaris* (Karaca and Uyanöz 2012). The fast-growing rhizobia isolates are more competitive as compared to slow-growing rhizobia strains (Gyogluu et al. 2018).

8.6 The Regulatory Role of Small Ribonucleic Acid (sRNA) and Micro-Ribonucleic Acid (miRNA) on Nodule Development and Nitrogen Fixation

Small Ribonucleic acids (sRNAs) and the micro- Ribonucleic acids (miRNAs) are non coding Ribonucleic acids playing pertinent roles in gene expression regulation. The symbiotic nodule development and functionality are regulated by the genetic system of both legume plants and the nitrogen-fixing microorganism. Through experimental and bioinformatics approaches, several legume specific miRNAs have been identified (Hussain et al. 2018). Small RNAs are particularly crucial in plant gene regulation, consequently affecting plant growth and development (Simon et al. 2009; Formey et al. 2016). Besides, most of the miRNAs code for transcription factors important for plant development and physiology. These essential molecules have mostly been identified through conventional computational approaches that most often couple prediction of the secondary structures, including but not limited to hairpin forming pri-miRNA precursor sequences with the presence of a conserved mature miRNA sequence (Subramanian et al. 2008). Other than computational approaches, experimental protocols are also used in miRNA identification. These include cloning and sRNA library sequencing, with the former having been successfully used in legumes and the study of the nitrogen-fixing nodules. Besides playing pertinent roles in plant gene expression, the micro RNAs also regulate legume-microbe interactions. A number of the micro- RNAs have been well documented to promote nodule development directly and indirectly. In particular, micro RNA, miR167, which is a CCAAT-Transcription factor MtHAP2–1 has been reported to be integral in nodule formation in *Medicago truncatula* (Formey et al. 2014). Gifford et al. (2008) established that miR167 plays a crucial phenotypic role in plants, especially in roots and shoots, by targeting the auxin response factor8. Other miRNAs such as miR164 and miR396 indirectly affect legume nodule development as they influence auxin response in roots. This effect on nodule development is through repression of HAP2, which encodes nodulation response translation factor (Simon et al. 2009). Other studies have also shown a strong linkage between nodule meristems in most legumes and the miR397, which is common in most legumes and is associated with not only symbiotic but also pathogenic interactions in the roots. Notably, gma-miR482, gma-miR1515 increased the nodule numbers in soybean without any notable impact on the soybean root development (Formey et al. 2014). The miRNAs are essential in the regulation of most of the cellular activities since they play a significant role in post-transcriptional regulation of mRNA expression (Jayaswal et al. 2011).

miRNAs also function in plant nutrient management. miR395, for instance, has been demonstrated to show increased expression with sulfate starvation. miR395 targets the adenosine triphosphate sulfurylases, thereby playing a critical role in sulfate homeostasis. In other studies, PvmiR399, which is a common bean miR399, also plays a vital role in phosphorous remobilization and homeostasis (Simon et al. 2009).

The eukaryotic sRNAs are critical in the post-transcriptional regulation of gene expression. Mostly, this is achieved by targeting 3′ end of cognate mRNAs. Most of the trans encoded RNAs often exert the gene regulatory role through base pairing with the complementary mRNA target. Chaperone protein Hfq facilitates this sRNA-mRNA interaction (Zhan et al. 2016). Numerous studies have shown Hfq genes to regulate nitrogen fixation. This is supported by the fact that inactivation of Hfq is reported to significantly reduces nitrogenase activity in many legume symbioses. The non-protein coding RNAs; NfiS from *P. stutzeri* strains control nitrogenase activity by the post-transcriptional regulation of nifK-mRNA through an unknown mechanism, thus, controlling nitrogen fixation (Zhan et al. 2016). There is a need for intensified studies to determine the exact expression patterns for most of the sRNAs, miRNAs and their targets. This may reveal significant insights into the processes of nutrient delivery in the plants, enhanced legume-rhizobia interaction and this can be exploited for improved legume crop productivity in smallholder agroecosystems.

8.7 Challenges in Legume-*Rhizobium* Symbiosis

8.7.1 Environmental Stress

Despite the isolation and identification of many effective rhizobia strains, a number of challenges on improving legume production using rhizobia as biofertilizer still exist. Like any other living system, legume-rhizobia symbiosis is affected by a host of biotic and abiotic stresses. Salt stress, for instance, has been well documented to affect *Rhizobium*- legume symbiosis (Karmakar et al. 2015). Nodulation of legumes is severely suppressed at high salinity (Dong et al. 2017). However, different strains of rhizobia have been reported to show varying degrees to salt and osmotic stress (Gulati 2018). Currently, emphases have been laid on the isolation and determination of nitrogen fixation efficacy of salt and water stress resistant rhizobia. It is, however, important to note that, the tolerance of the legume host to salt stress is more significant in ensuring symbiotic success and nodulation efficacy. Most legumes have been established to thrive well in neutral or slightly acidic soil conditions (Burns et al. 2017).

Most symbiotic rhizobia have been documented to show varying degree of tolerance to acidity with most of the rhizobia survival and persistence significantly constrained by soil acidity in the tropical soils (Aliverdi and Ahmadvand 2018). In particular, *R. loti* has been observed to have the capacity to proliferate and nodulate the host legumes at pH 4.5 while many *Bradyrhizobium* cannot multiply at the stated pH (Dong et al. 2017). Some crops acidify the soils, and this has a significant impact on the establishment of effective *Rhizobium*- legume symbiosis (Karmakar et al. 2015). While it is important for both legumes and the rhizobia to be pH tolerant, in some legumes such as the mung bean it has been established that only one of the symbionts need to be pH tolerant for effective nodulation (Mekonnen 2018).

Combined nitrogen available in the soil also has a strong effect on the success of legume symbioses. High amounts of nitrate ions in the soil have been observed to reduce the rate of biological nitrogen fixation (BNF) and nodulation. At high nitrate ion concentration, most plants conserve energy as BNF is more energy demanding than nitrate ion assimilation (Mabrouk et al. 2018). Similar to nitrogen, phosphorus is another significant nutrient that determines the legume yields. Phosphorous deficiency is well documented to affect nodulation and nitrogen-fixing capability of most strains of rhizobia (Fageria et al. 2017). Several studies have indicated that the addition of phosphate formulations in the fields significantly increases the nodulation capacity of many legumes (Bashir et al. 2011). Also, rhizobia strains have been observed to be sensitive to acidic soils containing heavy metal toxicity. Aluminium toxicity has been previously reported to impact on the potential of rhizobia to fix nitrogen (Artigas Ramírez et al. 2018). The heavy metals in addition to persisting in the soil for too long, they have been revealed as eco-toxicologic to both *Rhizobium* and legume crops (Jaiswal et al. 2018).

Salinity and pH effects go beyond the physiological effect on the legumes and the rhizobia. These parameters, when supplied below or above the optimum triggers the existence or inexistence of some important nutrients in the soil. For instance, salt stress and pH have demonstrated the potential to influence the levels of vital minerals like calcium required for effective attachment of *Rhizobium* to the roots of legumes (Argaw 2016; Aliverdi and Ahmadvand 2018). The deficiency in calcium affects the nodulation capacity of the legumes.

Heat stress below or above the optimum is a potential factor that affects *Rhizobium*- legume symbiosis. Most rhizobia strains have been profiled to be more effective at an optimum temperature of between 28 °C and 31 °C. However, others like rhizobia isolated from *Acacia senegal* shows activity at temperatures of up to 44 °C. High temperatures above the optimum have been reported to affect the nodule structure as well as delay nodulation. For instance, temperature above 45 °C, has been associated with plasmid curing which causes the loss of effectiveness of many rhizobia strains (Aranjuelo et al. 2015).

8.7.2 Soil Amendment and Other Agronomic Practices

Owing to continually diminishing agricultural lands productivity, there is a greater focus shift towards soil amendment techniques aimed at renewing soil health and enhancing crop productivity. With many soils becoming poorer, many agroecosystems have resorted to the use of inorganic fertilizers to boost soil productivity (Atieno and Lesueur 2019). Addition of nitrogenous fertilizers has been established to reduce the rate of nodulation in most legumes. In addition, the use of sewage fertilizers also has associated side effects. Untreated sewage sludge fertilizers can be contaminated with heavy metals that are well known to affect microbial physiology severely (Iheanacho et al. 2017). *Bradyrhizobium* strains are documented to be more resistant to the heavy metals in comparison to the rhizobia strains (Dong et al. 2017). However, resistance to tellurite and selenite has not been reported in *Bradyrhizobium* strains but a few *Rhizobium* strains such as *R. fredii*. Not all heavy metal infestation is disadvantageous since nickel has been observed to increase nodulation capacity by *Bradyrhizobium japonicum* on soybean plant (Gulati 2018). Intensive use of chemicals such as fungicides and pesticides to control phytopathogens in the farms has also negatively impacted on legume microbial symbioses with the chemicals having toxicological effects on the rhizobia (Karmakar et al. 2015). While these chemicals are effective in controlling the pathogens and pests, their non-specific action harms other beneficial soil microflora, including important legume symbionts. Herbicide bentazone in particular, when administered at the rightly specified amounts, affects the root hair structures and nodulation capacity of some legumes (Lengai and Muthomi 2018).

8.7.3 *Rhizobium-Rhizobium* and *Rhizobium*-Legume Interactions

Interactions between the rhizobia and the legume plants can impact negatively on the effectiveness of *Rhizobium*-legume symbiosis. In particular, the evolution of the Nodule Cysteine Rich peptides (NCRPs) in most of the legume plant species has been observed to antagonistically act against the symbiosis (Checcucci et al. 2017). While they promote nitrogen fixation, the NCRPs reduce bacterial fitness. Most rhizobia have evolved NCR peptidases. These improve bacterial proliferation and fitness but limit nitrogen fixation efficiency. To establish and promote plant growth, the Rhizobia must be able to survive and outcompete the native microbes (Ferguson et al. 2019). Most often, the effective *Rhizobium* are outcompeted by the ineffective native rhizobia impairing nitrogen fixation efficiency in legumes.

Indigenous rhizobia isolate also form a competitive barrier leading to failure of foreign inoculant strains It has also been observed that the success of inoculant strains in enhancing legume yield is limited in high numbers of infective indigenous rhizobia isolates in the soil (Remigi et al. 2016). Availability of effective indigenous rhizobia isolates in the soil also makes it difficult if not impossible to determine the effectiveness of inoculant rhizobia strains. It is therefore imperative for inoculant strains to be antagonistic to native rhizobia for its effectiveness to be realized.

8.8 Future Prospects

Legumes are a centerpiece in any green agricultural revolution. *Rhizobium*-legume-symbiosis not only boosts legume production but significantly contribute to the bio-stimulation of other beneficial non-rhizobia microorganisms that can boost production of non-legume crops. This underscores the need to promote this symbiosis. With environmental stress unquestionably impacting significantly on the symbiosis, need to develop more abiotic resistant *Rhizobium* strains, and legume genotypes is still a key focus area for the future. This can be achieved through crop breeding with wild legumes showing promising potential in regards to heat tolerance, acid tolerance, and drought tolerance. Research on competitiveness of most of the rhizobia strains has only been cursorily done. Through genetic engineering there is need to improve the competitive attributes of the effective rhizobia. For select rhizobia, the secondary replicons can be manipulated and be used to develop highly effective strains for competitiveness and abiotic stress tolerance. These can be achieved through combination with accessory replicons containing gene functions for these desired abilities. With mycorrhizal fungi inoculation well established to promote plant endurance to drought stress, upscaling synergistic inoculation of mycorrhizal fungi and rhizobia will be beneficial to the legumes. The ability of the symbiotic rhizobia to work closely and synergistically with other plant growth-promoting rhizobacteria should be explored. Other than mycorrhizal fungi and plant growth promoting rhizobacteria, bio-prospecting *Trichoderma* species that can confer advantages to the legume symbioses can also be explored since much of the current studies have only been cursorily done (Table 8.1).

With microbial interactions well established to have an impact on the effectiveness of rhizobia strains, it is imperative to seek a deeper understanding of the interactions among the commercial *Rhizobium* inoculants, native rhizobia and the non-*Rhizobium* microbial population found in the rhizospheric soils. Interactions between these organisms will directly improve *Rhizobium*-legume symbioses and improve crop production in smallholder agroecosystems.

Table 8.1 A summary of the strategies for upscaling *Rhizobium*-legume symbiosis. The approaches need to be multi-faceted and not just aimed at the legumes and rhizobia only

Target	Strategies	References
Soil management factors	Reducing reliance on nitrogen fertilizers Bioremediation of heavy metal polluted soil ecosystems.	Ojuederie and Babalola (2017) and Iheanacho et al. (2017)
	Intercropping of legumes and cereals.	Stagnari et al. (2017)
	Conservation tillage.	Piazza et al. (2019)
	Biofertilizer and biopesticide application.	Zaim et al. (2017) and Kalia and Gosal (2011)
	Determination of soil characteristics and effects on legume-*Rhizobium* symbiosis.	de Castro Pires et al. (2018)
Legume factors	Determination of high yielding and stress resistant legumes genotypes.	Darkwa et al. (2016)
	Genetic modification.	Lupwayi et al. (2006)
Rhizobium factors	Determination of effective cross nodulation groups.	Mahmood and Athar (2008)
	Determination of stress resistant isolates	Benjelloun et al. (2019)
	Isolation and screening for effective hyper nodulating autochthonous strains.	Koskey et al. (2017, 2018)
	Co-inoculation of rhizobia with other beneficial microbes.	Gopalakrishnan et al. (2015) and Musyoka et al. (2020)
	Use of synergistic soil microbes.	Larimer et al. (2014) and Mweetwa et al. (2016)

Acknowledgements This work was supported by The Future Leaders – African Independent Researchers (FLAIR) Fellowship Programme, which is a partnership between the African Academy of Sciences and the Royal Society funded by the UK Government's Global Challenges Research Fund, Research.

References

Aliverdi A, Ahmadvand G (2018) Herbicide toxicity to soybean–*Rhizobium* symbiosis as affected by soil pH. Bull Environ Contam Toxicol 101(4):434–438. https://doi.org/10.1007/s00128-018-2417-2

Aranjuelo I, Aldasoro J, Arrese-Igor C, Erice G, Sanz-Sáez Á (2015) How does high temperature affect legume nodule symbiotic activity? In: Legume nitrogen fixation in a changing environment, pp 67–87. https://doi.org/10.1007/978-3-319-06212-9_4

Argaw A (2016) Effectiveness of *Rhizobium* inoculation on common bean productivity as determined by inherent soil fertility status. J Crop Sci Biotechnol 19(4):311–322. https://doi.org/10.1007/s12892-016-0074-8

Artigas Ramírez MD, Silva JD, Ohkama-Ohtsu N, Yokoyama T (2018) *In vitro* rhizobia response and symbiosis process under aluminium stress. Can J Microbiol 64(8):511–526. https://doi.org/10.1139/cjm-2018-0019

Atieno M, Lesueur D (2019) Opportunities for improved legume inoculants: enhanced stress tolerance of rhizobia and benefits to agroecosystems. Symbiosis 77(3):191–205. https://doi.org/10.1007/s13199-018-0585-9

Atieno M, Hermann L, Okalebo R, Lesueur D (2012) Efficiency of different formulations of *Bradyrhizobium japonicum* and effect of co-inoculation of *Bacillus subtilis* with two different strains of *Bradyrhizobium japonicum*. World J Microbiol Biotechnol 28:2541–2550. https://doi.org/10.1007/s11274-012-1062-x

Bashir K, Ali S, Umair A (2011) Effect of different phosphorus levels on xylem sap components and their correlation with growth variables of mash bean. Sarhad J Agric 27:1–6

Batista L, Irisarri P, Rebuffo M, Cuitiño MJ, Sanjuan J, Monza J (2015) Nodulation competitiveness as a requisite for improved rhizobial inoculants of *Trifolium pratense*. Biol Fertil Soils 51(1):11–20. https://doi.org/10.1007/s00374-014-0946-3

Belmain SR, Haggar J, Holt J, Stevenson PC (2013) Managing legume pests in sub-Saharan Africa: challenges and prospects for improving food security and nutrition through agro-ecological intensification

Benjelloun I, Thami Alami I, Douira A, Udupa SM (2019) Phenotypic and genotypic diversity among symbiotic and non-symbiotic Bacteria present in chickpea nodules in Morocco. Front Microbiol 10:1885. https://doi.org/10.3389/fmicb.2019.01885

Bhowmik SN, Das A (2018) Biofertilizers: a sustainable approach for pulse production. In: Legumes for soil health and sustainable management. Springer, Singapore, pp 445–485

Brown DM, Okoro S, van Gils J, van Spanning R, Bonte M, Hutchings T, Linden O, Egbuche U, Bruun KB, Smith JWN (2017) Comparison of landfarming amendments to improve bioremediation of petroleum hydrocarbons in Niger Delta soils. Sci Total Environ 596-597:284–292. https://doi.org/10.1016/j.scitotenv.2017.04.072

Burns H, Norton M, Tyndall P (2017) Improving yield potential of legumes on acidic soils. Australian government grain research and development program corporation

Carlos MHJ, Stefani PVY, Janette AM, Melani MSS, Gabriela PO (2016) Assessing the effects of heavy metals in ACC deaminase and IAA production on plant growth-promoting bacteria. Microbiol Res 188:53–61

Checcucci A, DiCenzo GC, Bazzicalupo M, Mengoni A (2017) Trade, diplomacy, and warfare: the quest for elite rhizobia inoculant strains. Front Microbiol 8:2207. https://doi.org/10.3389/fmicb.2017.02207

Chibarabada T, Modi A, Mabhaudhi T (2017) Expounding the value of grain legumes in the semi- and arid tropics. Sustainability 9(1):60. https://doi.org/10.3390/su9010060

Chibeba AM, Kyei-Boahen S, de Fátima Guimarães M, Nogueira MA, Hungria M (2017) Isolation, characterization and selection of indigenous *Bradyrhizobium* strains with outstanding symbiotic performance to increase soybean yields in Mozambique. Agric Ecosyst Environ 246:291–305. https://doi.org/10.1016/j.agee.2017.06.017

Collavino MM, Sansberro PA, Mroginski LA, Aguilar OM (2010) Comparison of in vitro solubilization activity of diverse phosphate-solubilizing bacteria native to acid soil and their ability to promote *Phaseolus vulgaris* growth. Biol Fertil Soils 46:727–738. https://doi.org/10.1007/s00374-010-0480-x

Darkwa K, Ambachew D, Mohammed H, Asfaw A, Blair MW (2016) Evaluation of common bean (Phaseolus vulgaris L.) genotypes for drought stress adaptation in Ethiopia. Crop J 4(5):367–376

Daryanto S, Wang L, Jacinthe PA (2015) Global synthesis of drought effects on food legume production. PLoS One 10(6):e0127401. https://doi.org/10.1371/journal.pone.0127401

Das K, Prasanna R, Saxena AK (2017) Rhizobia: a potential biocontrol agent for soilborne fungal pathogens. Folia Microbiol 62(5):425–435

de Castro Pires R, dos Reis Junior FB, Zilli JE, Fischer D, Hofmann A, James EK, Simon MF (2018) Soil characteristics determine the rhizobia in association with different species of Mimosa in Central Brazil. Plant Soil 423:411–428. https://doi.org/10.1007/s11104-017-3521-5

De Jager I, Borgonjen-Van Den Berg KJ, Giller KE, Brouwer ID (2019) Current and potential role of grain legumes on protein and micronutrient adequacy of the diet of rural Ghanaian infants and young children: using linear programming. Nutr J 18(1):1–16. https://doi.org/10.1186/s12937-019-0435-5

de Souza R, Ambrosini A, Passaglia LMP (2015) Plant growth-promoting bacteria as inoculants in agricultural soils. Genet Mol Biol 38(4):401–419. https://doi.org/10.1590/S1415-475738420150053

de Souza EM, Bassani VL, Sperotto RA, Granada CE (2016) Inoculation of new rhizobial isolates improve nutrient uptake and growth of bean (Phaseolus vulgaris) and arugula (Eruca sativa). J Sci Food Agric 96(10):3446–3453

Diep CN, My NTX, Nhu VTP (2016) Isolation and characterization of endophytic bacteria in soybean root nodules. World J Pharm Pharm Sci 5:222–241

Doganova L, Karnoe P (2015) Building markets for clean technologies: controversies, environmental concerns and economic worth. Ind Mark Manag 44:22–31. https://doi.org/10.1016/j.indmarman.2014.10.004

Dong R, Zhang J, Huan H, Bai C, Chen Z, Liu G (2017) High salt tolerance of a *Bradyrhizobium* strain and its promotion of the growth of *Stylosanthes guianensis*. Int J Mol Sci 18(8):1625. https://doi.org/10.3390/ijms18081625

Dwivedi A, Dev I, Kumar V, Yadav RS, Yadav M, Gupta D, Tomar SS (2015) Potential role of maize-legume intercropping systems to improve soil fertility status under smallholder farming systems for sustainable agriculture in India. Int J Life Sci Biotechnol Pharma Res 4(3):145

Fageria NK, He Z, Baligar VC (2017) Phosphorus management in crop production. CRC Press, Boca Raton. https://doi.org/10.1201/9781315162096

Ferguson BJ, Mens C, Hastwell AH, Zhang M, Su H, Jones CH, Gresshoff PM (2019) Legume nodulation: the host controls the party. Plant Cell Environ 42(1):41–51. https://doi.org/10.1111/pce.13348

Formey D, Sallet E, Lelandais-Brière C, Ben C, Bustos-Sanmamed P, Niebel A et al (2014) The small RNA diversity from *Medicago truncatula* roots under biotic interactions evidences the environmental plasticity of the miRNAome. Genome Biol 15(9):1–21. https://doi.org/10.1186/s13059-014-0457-4

Formey D, Martín-Rodríguez JÁ, Leija A, Santana O, Quinto C, Cárdenas L, Hernández G (2016) Regulation of small RNAs and corresponding targets in nod factor-induced *Phaseolus vulgaris* root hair cells. Int J Mol Sci 17(6):1–15. https://doi.org/10.3390/ijms17060887

Gifford ML, Dean A, Gutierrez RA, Coruzzi GM, Birnbaum KD (2008) Cell-specific nitrogen responses mediate developmental plasticity. Proc Natl Acad Sci 105(2):803–808

Giller KE, Franke AC, Abaidoo R, Baijukya F, Bala A, Boahen S et al (2013) N_2 Africa: putting nitrogen fixation to work for smallholder farmers in Africa. In: Agro-ecological Intensification of Agricultural Systems in the African Highlands. Routledge, London, pp 156–174

Gopalakrishnan S, Sathya A, Vijayabharathi R, Varshney RK, Gowda CLL, Krishnamurthy L (2015) Plant growth promoting rhizobia: challenges and opportunities. 3 Biotechnology 5(4):355–377. https://doi.org/10.1007/s13205-014-0241-x

Gronemeyer JL, Kulkarni A, Berkelmann D, Hurek T, Reinhold-Hurek B (2014) Identification and characterization of rhizobia indigenous to the Okavango region in Sub-Saharan Africa. Appl Environ Microbiol:AEM-02417. https://doi.org/10.1128/AEM.02417-14

Gulati R (2018) Legume symbiosis under abiotic stresses-a review. Mol Physiol Abiotic Stress Plant Prod:186

Gupta G, Parihar SS, Ahirwar NK, Snehi SK, Singh V (2015) Plant growth promoting rhizobacteria (PGPR): current and future prospects for development of sustainable agriculture. J Microb Biochem Technol 7(2):096–102. https://doi.org/10.4172/1948-5948.1000188

Gyogluu C, Mohammed M, Jaiswal SK, Kyei-Boahen S, Dakora FD (2018) Assessing host range, symbiotic effectiveness, and photosynthetic rates induced by native soybean rhizobia isolated from Mozambican and South African soils. Symbiosis 75(3):257–266. https://doi.org/10.1007/s13199-017-0520-5

Hussain SS, Hussain M, Irfan M, Siddique KHM (2018) Legume, microbiome, and regulatory functions of miRNAs in systematic regulation of symbiosis. In: Egamberdieva D, Ahmad P (eds) *Plant Microbiome: Stress Response*, Microorganisms for sustainability, vol 5. Springer, Singapore. https://doi.org/10.1007/978-981-10-5514-0_12

Iheanacho EU, Ndulaka JC, Onuh CF (2017) Environmental pollution and heavy metals. Environ Pollut 5(5):2321–9122

Irisarri P, Cardozo G, Tartaglia C, Reyno R, Gutiérrez P, Lattanzi FA et al (2019) Selection of competitive and efficient rhizobia strains for white clover. Front Microbiol 10:768. https://doi.org/10.3389/fmicb.2019.00768

Jaiswal SK, Naamala J, Dakora FD (2018) Nature and mechanisms of aluminium toxicity, tolerance and amelioration in symbiotic legumes and rhizobia. Biol Fertil Soils 54(3):309–318. https://doi.org/10.1007/s00374-018-1262-0

Jayaswal V, Lutherborrow M, Ma DDF, Yang YH (2011) Identification of microRNA-mRNA modules using microarray data. BMC Genomics 12(*1*):138. https://doi.org/10.1186/1471-2164-12-138

Kalaji HM, Jajoo A, Oukarroum A, Brestic M, Zivcak M, Samborska IA et al (2016) Chlorophyll a fluorescence as a tool to monitor physiological status of plants under abiotic stress conditions. Acta Physiol Plant 38(4):102. https://doi.org/10.1007/s11738-016-2113-y

Kalia A, Gosal SK (2011) Effect of pesticide application on soil microorganisms. Arch Agron Soil Sci 57:569–596. https://doi.org/10.1080/03650341003787582

Karaca Ü, Uyanöz R (2012) Effectiveness of native rhizobium on nodulation and growth properties of dry bean (Phaseolus vulgaris L.). Afr J Biotechnol 11(37):8986–8991. https://doi.org/10.5897/AJB11.4087

Karmakar K, Rana A, Rajwar A, Sahgal M, Johri BN (2015) Legume-rhizobia symbiosis under stress. In: Arora NK (ed) Plant microbes Symbiosis: applied facets. Springer, New Delhi, pp 241–258

Kaschuk G, Kuyper TW, Leffelaar PA, Hungria M, Giller KE (2009) Are the rates of photosynthesis stimulated by the carbon sink strength of rhizobial and arbuscular mycorrhizal symbioses? Soil Biol Biochem 41(6):1233–1244

Komarek AM, Drogue S, Chenoune R, Hawkins J, Msangi S, Belhouchette H, Flichman G (2017) Agricultural household effects of fertilizer price changes for smallholder farmers in Central Malawi. Agric Syst 154(April):168–178. https://doi.org/10.1016/j.agsy.2017.03.016

Korir H, Mungai NW, Thuita M, Hamba Y, Masso C (2017) Co-inoculation effect of rhizobia and plant growth promoting rhizobacteria on common bean growth in a Low Phosphorous Soil. Front Plant Sci 8:141

Koskey G, Mburu SW, Njeru EM, Kimiti JM, Ombori O, Maingi JM (2017) Potential of native rhizobia in enhancing nitrogen fixation and yields of climbing beans (*Phaseolus vulgaris L.*) in contrasting environments of eastern Kenya. Front Plant Sci 8:443. https://doi.org/10.3389/fpls.2017.00443

Koskey G, Mburu SW, Kimiti JM, Ombori O, Maingi JM, Njeru EM (2018) Genetic characterization and diversity of *Rhizobium* isolated from root nodules of mid-altitude climbing bean (*Phaseolus vulgaris* L.) varieties. Front Microbiol 9:968. https://doi.org/10.3389/fmicb.2018.00968

Kucuk C, Cevheri C (2015) In vitro antagonism of *Rhizobium* strains isolated from various legumes. J Pure Appl Microbiol 9(1):503–511

Kuivanen KS, Alvarez S, Michalscheck M, Adjei-Nsiah S, Descheemaeker K, Mellon-Bedi S, Groot JCJ (2016) Characterising the diversity of smallholder farming systems and their constraints and opportunities for innovation: a case study from the northern region, Ghana. NJAS – Wageningen J Life Sci 78:153–166. https://doi.org/10.1016/j.njas.2016.04.003

Kulkarni KP, Tayade R, Asekova S, Song JT, Shannon JG, Lee JD (2018) Harnessing the potential of forage legumes, alfalfa, soybean, and cowpea for sustainable agriculture and global food security. Front Plant Sci 9:1314. https://doi.org/10.3389/fpls.2018.01314

Larimer A, Clay K, Bever J (2014) Synergism and context dependeny of interactions between arbuscular mycorrhizal fungi and rhizobia with a prairie legume. Ecology 95:1045–1054. https://doi.org/10.1890/13-0025.1

Lengai GM, Muthomi JW (2018) Biopesticides and their role in sustainable agricultural production. J Biosci Med 6(06):7. https://doi.org/10.4236/jbm.2018.66002

Lim CI, Biswas W, Samyudia Y (2015) Review of existing sustainability assessment methods for Malaysian palm oil production. Procedia CIRP 26:13–18. https://doi.org/10.1016/j.procir.2014.08.020

Lupwayi NZ, Clayton GW, Rice WA (2006) Rhizobial inoculants for legume crops. J Crop Improv 15:289–321. https://doi.org/10.1300/J411v15n02_09

Mabrouk Y, Hemissi I, Salem IB, Mejri S, Saidi M, Belhadj O (2018) Potential of rhizobia in improving nitrogen fixation and yields of legumes. Symbiosis 107.

Magrini MB, Anton M, Cholez C, Corre-Hellou G, Duc G, Jeuffroy MH et al (2016) Why are grain-legumes rarely present in cropping systems despite their environmental and nutritional benefits? Analyzing lock-in in the French agrifood system. Ecol Econ 126:152–162. https://doi.org/10.1016/j.ecolecon.2016.03.024

Mahmood A, Athar M (2008) Cross inoculation studies: response of Vigna mungo to inoculation with rhizobia from tree legumes growing under arid environment. Int J Sci Environ Technol 5:135–139. https://doi.org/10.1007/bf03326006

Mahmood I, Imadi SR, Shazadi K, Gul A, Hakeem KR (2016) Effects of pesticides on environment. In: Hakeem K, Akhtar M, Abdullah S (eds) Plant, soil and microbes. Springer, Cham

Martin B (2017) The nitrogen cycle. The Rosen Publishing Group, Inc

Mekonnen M (2018) Evaluating the symbiotic effectiveness of common bean (Phaseolus vulgaris L.) Nodulating Rhizobial mutants on extreme pH and high salt soil conditions. J Plant Pathol Microbiol 9(11). https://doi.org/10.4172/2157-7471.1000462

Mendes R, Garbeva P, Raaijmakers JM (2013) The rhizosphere microbiome: significance of plant beneficial, plant pathogenic, and human pathogenic microorganisms. FEMS Microbiol Rev 37(5):634–663. https://doi.org/10.1111/1574-6976.12028

Mishra PK, Mishra S, Selvakumar G, Bisht JK, Kundu S, Gupta HS (2009) Coinoculation of Bacillus thuringeinsis-KR1 with Rhizobium leguminosarum enhances plant growth and nodulation of pea (Pisum sativum L.) and lentil (Lens culinaris L.). World J Microbiol Biotechnol 25(5):753–761

Morel MA, Braña V, Castro-Sowinski S (2012) Legume crops, importance and use of bacterial inoculation to increase production. Crop Plant 12:218–240

Muleta D (2017) Legume response to arbuscular mycorrhizal fungi inoculation in sustainable agriculture. In: Microbes for legume improvement, pp 227–260. https://doi.org/10.1007/978-3-319-59174-2_10

Mungai LM, Snapp S, Messina JP, Chikowo R, Smith A, Anders E et al (2016) Smallholder farms and the potential for sustainable intensification. Front Plant Sci:1–17. https://doi.org/10.3389/fpls.2016.01720

Muoni T, Barnes AP, Öborn I, Watson CA, Bergkvist G, Shiluli M, Duncan AJ (2019) Farmer perceptions of legumes and their functions in smallholder farming systems in East Africa. Int J Agric Sustain 17(3):205–218. https://doi.org/10.1080/14735903.2019.1609166

Musyoka DM, Njeru EM, Nyamwange MMe, Maingi JM (2020) Arbuscular mycorrhizal fungi and Bradyrhizobium co-inoculation enhances nitrogen fixation and growth of green grams (*Vigna radiata* L.) under water stress. J Plant Nutr:1–12. https://doi.org/10.1080/01904167.2020.1711940

Mweetwa AM, Chilombo G, Gondwe BM (2016) Nodulation, nutrient uptake and yield of common bean inoculated with *Rhizobia* and *Trichoderma* in an acid soil. J Agric Sci 8:61–71. https://doi.org/10.5539/jas.v8n12p61

Ojuederie OB, Babalola OO (2017) Microbial and plant-assisted bioremediation of heavy metal polluted environments: a review. Int J Environ Res Public Health 14:1504. https://doi.org/10.3390/ijerph14121504

Pannell DJ, Llewellyn RS, Corbeels M (2014) The farm-level economics of conservation agriculture for resource-poor farmers. Agric Ecosyst Environ 187:52–64. https://doi.org/10.1016/j.agee.2013.10.014

Piazza G, Ercoli L, Nuti M, Pellegrino E (2019) Interaction between conservation tillage and nitrogen fertilization shapes prokaryotic and fungal diversity at different soil depths: evidence from a 23-year field experiment in the mediterranean area. Front Microbiol 19(1):133. https://doi.org/10.3389/fmicb.2019.02047

Qureshi MA, Shakir MA, Iqbal A, Akhtar N, Khan A (2012) Co-inoculation of phosphate solubilizing bacteria and rhizobia for improving growth and yield of mung bean (*Vigna radiata* L.). J. Anim. Plant Sci. 21:491–497

Remigi P, Zhu J, Young JPW, Masson-Boivin C (2016) Symbiosis within symbiosis: evolving nitrogen-fixing legume symbionts. Trends Microbiol 24(1):63–75. https://doi.org/10.1016/j.tim.2015.10.007

Rubiales D, Mikic A (2015) Introduction: legumes in sustainable agriculture. Crit Rev Plant Sci 34: 2–3. (December 2014). https://doi.org/10.1080/07352689.2014.897896

Saha M, Sarkar S, Sarkar B, Sharma BK, Bhattacharjee S, Tribedi P (2016) Microbial siderophores and their potential applications: a review. Environ Sci Pollut Res 23(5):3984–3999

Sarhan EA, Shehata HS (2014) Potential plant growth-promoting activity of Pseudomonas spp. and Bacillus spp. as biocontrol agents against damping-off in alfalfa. Plant Pathol J (Faisalabad) 13(1):8–17

Simon SA, Meyers BC, Sherrier DJ (2009) MicroRNAs in the rhizobia legume symbiosis. Plant Physiol 151(3):1002–1008. https://doi.org/10.1104/pp.109.144345

Stagnari F, Maggio A, Galieni A, Pisante M (2017) Multiple benefits of legumes for agriculture sustainability: an overview. Chem Biol Technol Agric 4:2. https://doi.org/10.1186/s40538-016-0085-1

Stajković O, De Meyer S, Miličić B, Willems A (2009) Isolation and characterization of endophytic non-rhizobial bacteria from root nodules of alfalfa (Medicago sativa L.). Botanica Serbica 33(1):107–114

Subramanian S, Fu Y, Sunkar R, Barbazuk WB, Zhu JK, Yu O (2008) Novel and nodulation-regulated microRNAs in soybean roots. BMC Genomics 9:1–14. https://doi.org/10.1186/1471-2164-9-160

Sulieman S, Tran LSP (2015) Phosphorus homeostasis in legume nodules as an adaptive strategy to phosphorus deficiency. Plant Sci 239:36–43

Tak HI, Ahmad F, Babalola OO, Inam A (2012) Growth, photosynthesis and yield of chickpea as influenced by urban waste water and different levels of phosphorus. Int. J. Plant Res. 2:6–13

Temba MC, Njobeh PB, Adebo OA, Olugbile AO, Kayitesi E (2016) The role of compositing cereals with legumes to alleviate protein energy malnutrition in Africa. Int J Food Sci Technol 51(3):543–554. https://doi.org/10.1111/ijfs.13035

Thuita M, Pypers P, Herrmann L, Okalebo RJ, Othieno C, Muema E, Lesueur D (2012) Commercial rhizobial inoculants significantly enhance growth and nitrogen fixation of a promiscuous soybean variety in Kenyan soils. Biol Fertil Soils 48:87–96. https://doi.org/10.1007/s00374-011-0611-z

Tsion K, Steven W (2019) An overview use and impact of organic and synthetic farm inputs in developed and developing countries. Afr J Food Agric Nutr Dev 19(3):14517–14540

Vargas LK, Volpiano CG, Lisboa BB, Giongo A, Beneduzi A, Passaglia LMP (2017) Potential of rhizobia as plant growth-promoting rhizobacteria. In: Microbes for legume improvement. Springer, Cham, pp 153–174

Venance SK, Mshenga P, Birachi E (2016) Factors influencing on-farm common bean profitability : the case of smallholder bean farmers in Babati District, Tanzania. J Econ Sustain Dev 7(22):196–201

Youseif S, Abd El-Megeed F, Saleh S (2017) Improvement of faba bean yield using *Rhizobium/Agrobacterium* inoculant in low-fertility sandy soil. Agronomy 7(1):2. https://doi.org/10.3390/agronomy7010002

Zaim S, Bekkar AA, Belabid L (2017) *Rhizobium* as a crop enhancer and biofertilizer for increased non-legume production. In: Hansen AP et al (eds) *Rhizobium* biology and biotechnology. Springer, Cham, pp 25–37

Zhan Y, Yan Y, Deng Z, Chen M, Lu W, Lu C et al (2016) The novel regulatory ncRNA, NfiS, optimizes nitrogen fixation via base pairing with the nitrogenase gene nifK mRNA in *Pseudomonas stutzeri* A1501. Proc Natl Acad Sci U S A 113(30):E4348–E4356. https://doi.org/10.1073/pnas.1604514113

Chapter 9
Transformation of Agricultural Breeding Techniques Using Biotechnology as a Tool

Ekta Khare and Pallavi Singh Chauhan

Abstract New agricultural breeding techniques offer advancement for the development of quality crop trait which is used commercially in both private and public sector. The success of techniques is still challenging at scientific level, interrupted under political influences as well as affected at social level. There are few institutional and social barriers in the implication of new plant breeding technologies, reported from data obtained during survey by an international panel of experts. Major issues for the succession of breeding techniques are associated with regulatory as well environmental concerns. But there is an increasing demand of appetite with diversity in nutritional status, more productivity. Newer breeding techniques involving marker-assisted selection, gene editing and enhanced productivity are an advanced science which uses innovation at genetic and biotechnological level. The technique may also provide advances in gathering information regarding unique quality of particular plant. Developing methods to store, share, and quickly analyze these data will produce significant advances in plant breeding.

Keywords Cisgenesis · Transgenesis · Intragenesis · Conventional breeding techniques · CRISPR-Cas system · Genotyping · Genetically modified crops · Transformation · Imputation · Gene knockout

Abbreviations

CRISPR	Clustered Regularly Interspaced Short Palindromic Repeats
DNA	Deoxy ribonucleic acid
GM	Genetically modified crops
HDR	homology-directed repair
NPBT	New Plant Breeding Techniques

E. Khare (✉)
Department of Pharmacy, I.T.M. University, Gwalior, Madhya Pradesh, India

P. S. Chauhan (✉)
Department of Life Sciences, I.T.M. University, Gwalior, Madhya Pradesh, India

P. Guleria et al. (eds.), *Sustainable Agriculture Reviews 45*, Sustainable Agriculture Reviews 45, https://doi.org/10.1007/978-3-030-53017-4_9

9.1 Introduction

Agritech; agricultural biotechnology is an emerging area which involves modification of crops at genomic level by utilizing various emerging genetic manipulation techniques like genetic engineering, new tissue culture techniques, diagnostic methods at molecular level, involvement of various molecular markers etc. In recent times researchers are more focused towards crop biotechnology, which helps in production of desired trait by inbreeding two species with different genetic composition. The genetically manipulated species are known to possess advanced characteristics such as flavor, drought, disease, pests as well as temperature resistance, productivity and size of harvested products. The present study is focused on summarizing recently used new breeding techniques and their applications, because the conventional breeding techniques are known to have several disadvantages. Thus the study may provide new insight in the development as well as applications of new plant or crop breeding techniques to benefit human health and maintain cost effectiveness.

The gene transfer among organisms develops modifications in biological entities. (Shukla et al. 2014) The modification either from traditional techniques or from new molecular biology techniques may help in development of agricultural species that are important (Garcia-Sancho and Myelnikov 2019) between both the techniques, modifications are done in order to produce new varieties of agriculture with advanced traits, like development of disease resistance or high proportion of protein etc (Borrelli et al. 2018). The only thing that advances molecular methods over traditional methods of breeding is its scope and precision (Gottardo et al. 2019). When breeding was made between two genetically different species, where each parents contribute equally. It was observed that the contribution of the composition varies widely. To perform the successful recombination probability, many crosses should be made. (Han et al. 2017) Also, sometimes back cross of the progeny should be made in order to adopt new trait. Sometimes there is a probability of development of undesired traits and losses of desired traits (Cobb et al. 2019a). These disadvantages have limited the applicability of classical breeding. To overcome such disadvantages few advanced molecular methods for efficient gene transfer came into existence which at the basic level allow the manipulation of processes. For reducing failure risks, gene for specific trait needs to be inserted into an established genomic sequence in order to control the expression pattern in new varieties of plants and animals (Abdallah et al. 2015). This technique may provide improvement in varieties along with shortening the breeding time by homing desired trait in desired location.

Rather than sexual reproduction, scientists have opted newer laboratory based methods to standardize transfer of individual genes between organisms by using natural gene transfer. The transfer methods could be performed by normal cellular uptake of DNA or by viral mediated packaged genome intercellular transfer. Studies on bacterial viruses have open up new gates for understanding the mechanisms of genetic material transfer in a simplified manner. The studies thus provide a new insight on the development and exploration of research field (Figs. 9.1, 9.2, and 9.3) at remarkable rate. Scientists thus used above mechanism to transfer genes into organisms, where the organism range may vary from soybeans to sheeps. Still this area is less explored and thus needs further exploration in order to perform experiments on perfect embryo culture, plant organ as well as tissue regeneration, gene transfer etc (Ozdemir and Budak 2018).

Model systems such as fruit fly *Drosophila melanogaster* and *Echerischia coli* are advantageous for genetic manipulation because of the availability of wide scientific knowledge along with their ease of manipulation (Mirzoyan et al. 2019).The researchers are thus extending their molecular research area to various agricultural organisms that are commercially important. Research in this direction may not replace conventional agricultural breeding techniques, but this may expand the era of improvement of agricultural species with wide applications (Venken et al. 2016).

Conventionally in order to improve agricultural productivity cross and self-pollination strategies were opted (Suso et al. 2016). Whereas,the emergence of genetically modified crops (Fig. 9.2), there is lot of improvisation in the yield along with usage of pesticide and nutritional composition (Perry et al. 2016). In last three decades researchers have proved that the usage of genetically modified crops is no more risky. Few countries like Africa and Asia are in the lane of hesitating, the usage of genetically modified crops, due to risk of losses in export (Raman 2017).

Fig. 9.1 Increased productivity of legumes

Fig. 9.2 Enhanced fruiting capacity of plants

Fig. 9.3 Propagation of plants

9.2 New Breeding Techniques

New Plant Breeding Technique (NPBT) is at emergence these days, as they eradicate several fears associated with the usage of genetically modified crops (GM crops) (Oliver 2014). In order to improve the traits, this technique may allow alteration in the sequence as well as modifications in the endogenous genes without allowing any transgenic transfers across species boundaries. The new breeding techniques may include gene silencing, DNA-free CRISPR-Cas9 gene editing, homology-directed repair (HDR), and transient gene silencing or transcriptional repression (CRISPRi) (Tagliabue 2018).

As an emerging technique, particularly Clustered Regularly Interspaced Short Palindromic Repeats (CRISPR) Cas system has come into existence in order to perform the crop genome editing, where Cas stand for CRISPR-associated proteins. With rapid enhancement in these techniques, its applicability in the area of agriculture has gained attention of various researchers (Jaganathan et al. 2018). This system employs targeting mediated by site-directed nucleases to target, along with accurately modifying DNA. They have added advantages over conventional methods as they are cost effective as well as improvisation at genomic level of orphan crops, which may include genome editing of local fruits along with several vegetables, having vital role in maintenance of human health (Wang et al. 2019). The main reason for the heavy regulation of Genetically modified crops is the usage of foreign DNA.Thus in genome edited crops; the cost of the regulatory procedures can be reduced, because of the unavailability of transgenes (Hundleby and Harwood 2019). The whole procedure may ultimately results in speeding innovation, enhancement in the completion in various industries such as seed industry, along with making modified seeds available at more reasonable price for farmers (Spielman and Kennedy 2016). The lack of acceptance of such techniques by the public for employing the products in regular usage has various reasons. Some of the most potent reasons are lack of technical knowledge as well as incapability to handle transgenic techniques (Salisu et al. 2017). There is always a prerequisite to develop advanced strategies and new efforts in order to make the usage as well as adoption of Genetically modified crops, for sustainable development (Ladics et al. 2015).

Rapid generation advance and single-seed descent are two renounced techniques which has advantages in reducing crop's life cycle such as reducing breeding time and localization of suitable genes in genomic sequence. The technique is benefiting farmers for the improvisation of several grain crops (Tanaka et al. 2016). Most popular techniques regularly being used these days for the identification of phenotypically uncharacterized plants within a plant population includes genomic selection techniques such as genotyping and imputation (Bhat et al. 2016).

Still there are few limitations as well as disadvantages in these breeding methods, which could be overcome by the usage of new technique that employ effective application along with implementation of new molecular techniques by altering the genomic pattern called as clustered regularly interspaced short palindromic repeats (CRISPR) Cas systems (Saha et al. 2018). In order to improve the quality of crop

CRISPR-Cas9–mediated gene knockout is a newer technique, which is regularly being used. This technique has several applications in the improvisation of crops, some of the applications are obtaining high-yield grains, enhancing the flavor and development of disease resistant food varieties (Haque et al. 2018). Also the application area may include editing of genome sequence, replacement of genes, and improvisation of several traits by altering promoter properties, binding capabilities with transcription factors, affecting enhancer binding as well as regulatory sequence alteration for maintaining pattern of gene expression (Long et al. 2016).

Clustered Regularly Interspaced Short Palindromic Repeats genome-wide screens technique, may have advantages in the identification of several unknown traits of crop plants, improvisation of quantitative traits, development of drought as well as salinity tolerant crop varieties. In order to improve future crops, researchers nowadays are more focused towards the combination of Clustered Regularly Interspaced Short Palindromic Repeats (CRISPR)-Cas technologies with modern breeding techniques (Razzaq et al. 2019).

In the coming time newer along with interesting application area of genome editing may open new routes. By implementing advanced genome-editing technologies, various food security crops could get benefited immediately by reducing the usage of harmful pesticides. Also this may provide benefits by making plants resistant to climatic stress conditions (Zaidi et al. 2019). The only parameter to examine the advantages of using genome editing tools is to monitor the statistical records of the successful public acceptance for them odified crop ultimately benefiting them (Ishii and Araki 2016). With the increase in number of crop genome, along with development of strategies for making comparison of allelic pattern, targeting certain genes for improvisation is now very much easy for the researchers (Wambugu et al. 2018).

In order to achieve a food secured future, collaboration is mandatory for the public funded Consultative Group on International Agricultural Research to maintain unity of various regional organizations (Ryan et al. 2019). In collaboration of various regional, national as well as international research institutes, the Consultative Group on International Agricultural Research centers is now supporting maintenance of crop-specific gene banks, supply of plant genetic materials in developing countries for encouraging breeders to develop new modifications in crop varieties (Diez et al. 2018).

.The backcrossing technique is now days utilized by many crop breeders. The alternative name for backcrossing is introgression breeding. The utility of this technique is in performing a breeding. Suppose there is a case where a plant with mild resistance toward adverse environmental conditions like salinity, moisture as well as temperature is allowed to get crossed with an another plant that does not have resistance, thus in such cases the desirable backcrossing is required. After performing backcrossing technique, there must be a quality control step, which will allow one to have surety regarding that the new changes have been already made in the pure trait such as in the progeny (Yenni et al. 2019). The cross is then made between mildew resistant progenies with their high-yielding parental generation (Cobb et al. 2019a, b). The back crossing is then allowed to be done in a repeated manner, ultimately enhancing the possibility of more advanced progeny with efficacy (Li et al. 2018).

Other breeding techniques involve inbreeding, where for the development of inbred variety, plants breed by themselves and the pattern continues in future generations to preserve pure traits. They are advantageous in many ways as far as research is concerned. Hybrid breeding involves production of characteristics which are stable and also for the formation of hybrid vigor by making cross of two different inbred varieties (Fujimoto et al. 2018).

The results thus obtained clearly showed that the productivity of offspring is far more than parents. Another breeding technique involves mutation breeding, where the mutation may be random or artificially encouraged by chemicals or radiation mediation, so as to create new varieties (Sedeek et al. 2019). Molecular marker-assisted selection involves implementation of classical inbreeding and other hybridization methods, having a major difference that, instead of selecting phenotypic traits breeders selected genotypic inheritance pattern of the offspring from their parents (Hadasch et al. 2016). Other techniques involve building of insert designs and then insertion of desirable traits into plants, which ultimately results in the emergence of genetically modified organisms (Khan et al. 2016).

9.2.1 Application of New Plant Breeding Techniques

New plant breeding techniques facilitate the breeding of improved crop varieties. Crop improvement is an important endeavor if we are to meet the demands of growing population for which food production is to be increased. To respond adequately, we should apply existing tools to breed improved crops and ensure sustainable food production. Plant breeding has resulted in numerous improved food and industrial crop variety with traditional breeding based on crossing and selection while faces limitation in crop with complex genetics (polyploidy, heterozygosity or self-incompatibility).

In future development of in vitro cell selection techniques for disease resistance would be equally important. A coordination of the recent techniques of anther and microspore culture, cell suspension, irradiation of haploid cells, chromosome doubling and regeneration of doubled haploid plants could be utilized to obtain genotypes with desired traits.

(i) An induced mutagenesis technique is gaining importance in plant molecular biology as a tool to identify and isolate genes and to study their structure with function.
(ii) These new plant breeding technique have major impact on the future crop improvement.
(iii) Mutation with genetic engineering may constitute tools of plant breeder in near future.

(iv) In vitro culture and molecular methods are resulted in the creation and utilization of mutation breeding for crop improvement.
(v) With the help of in vitro culture technique, small amount of tissue can be subjected to the mutagenesis for development of crops.
(vi) In vitro mutagenesis technique has enhanced the crop yield and germ plasm innovation by the development of improved resistance traits.
(vii) Heavy ion beam irradiation has emerged as an effective and efficient way of inducing mutation in many plant varieties due to broad spectrum and high frequency.
(viii) Many seed propagated plants like wheat, rice, maize and barley etc. can be re generated from cell suspension culture.
(ix) An induced mutation technique is being used in the preparation of genetic maps that can facilitate molecular marker assisted plant breeding in the future.
(x) The direct use of mutation in the development of structural and functional genomics may lead to improvement of plant yield and quality rapidly.
(xi) In case of hypoallergenic nuts, clustered regularly interspaced short palindromic repeats will be helpful for allowing more people to enjoy protein-packed healthy snack in schools.
(xii) On average, gluten allergies affect about 1% of the global population worldwide. The manufacturing of gluten-free wheat could provide the nutritional benefits of wheat without affecting people's Celiac Disease or other gluten-related disorders.

The molecular techniques of DNA fingerprinting and mappings of molecules are contributed towards the screening and analysis of mutants such as:

(i) Random Amplified Polymorphic DNA
(ii) Amplified fragment length polymorphism
(iii) Sequence tagged Macrosatellite Sites

9.3 Conventional Breeding Techniques and Disadvantages

Agricultural practices are currently implemented with different approaches on global scale to meet sustainable environmental and economic developments. Heterozenous nature and linkage drag often influence stacking of desirable genes from wild sources in conventional breeding for crop improvement. Recombinant DNA technology and transgenesis have enabled transformation of alien gene into plants across the barriers.

Regarding about safety issues of transgenic crops, two transformation concepts was developed such as cisgenesis and intragenesis as an alternative to transgenic crop development. To meet the concern of genetically modified crops into agricultural production, cisgenesis was developed as new tools in crop modification and plant breeding (Cobb et al. 2019a, b).

Above given two concepts are based on the exclusive use of genetic material from the same species capable of sexual hybridization in contrast to transgenesis where gene or DNA can be moved between any species (Cobb et al. 2019a, b).

9.3.1 Cisgenesis

Cisgenesisis is the modification in the genetic background of recipient plant by naturally derived gene from cross compatible species. Therefore, Cisgenes shared common gene pool for traditional breeding (Schouten et al. 2006). Furthermore foreign genes such as selection marker genes and vector backbone genes should be eliminated from the primary cisgenic transforms as per their progeny (Flachowsky et al. 2011).

Genetic makeup of plant variety is maintained, this technique is particularly efficient method for cross fertilizing heterozygous plant that propagate such as potato, apple and banana (EFSA 2012).

Genetic makeup can directly improve an existing variety without disturbing the genetic make-up of the plant. Traditional introgression breeding of cross fertilizing plants does not allow the introduction of genes from wild germ plasma without mixing up the combination of alleles in the existing heterozygous elite recipient genotype.

Cisgenic and intragenic plants are produced by the same transformation techniques to develop transgenic plants. For this, Genes must be isolated, Cloned, synthesized and transferred back into a recipient where genes can be stably integrated and expressed. The development of cisgenic plants requires additional research and techniques when cisgenic plant is compared to the development of transgenic plant.

The pre requisites for developing these plants are

(i) Availability of desired genes and genetic elements within sexually compatible gene pool
(ii) Production of plant devoid of foreign DNA from marker genes and vector backbone sequences

Cisgenesis may be safer than conventional breeding. The donar sequence does not replace an allelic sequence but the donar is added to the recipient species genome. Owing to the process of gene transfer, the new sequence is inserted several times in one genome which might affect gene expression and phenotype. However, gene duplication is common natural occurrence for instance in the case of multigene or resistance gene family. Furthermore, Cisgenic plant might contain small non-coding sequence from the vector such as t-DNA where 25 base pair repeats imperfect and delimit the DNA segment which is transferred to plant cell during the case of agrobacterium mediated gene transfer.

9.3.2 Transgenesis and Intragenesis

Transgenesis is the genetic modification of recipient plant with two or more genes from any donar plant which is sexually incompatible with the recipient plant while intragenesis is the other genetic transformation which was introduced by Rommens in 2004.

Transgenesis and intragenesis involve specific genetic elements from plant belonging to the same sexually compatible gene pool. Transgenesis and intragenesis describes about the coding regions of gene with or without introns that can be combined with promoters and terminators from different genes of same sexually compatible gene pool. For example during agrobacterium mediated transformation, t-DNA border sequence should originate from the sexually compatible DNA pool (Mujjassim et al. 2019).

9.3.3 Disadvantages

(i) The genes outside the sexually compatible gene pool cannot be introduced.
(ii) The generation of intragenic or cisgenic crops are time consuming as compare to transgenic crops.
(iii) The gene fragments may not be readily available but need to be isolated from sexually compatible gene pool.
(iv) Cisgene or intragene may influence the expression of genes and phenotypic differences.

Thus, Considerable efforts are needed to produce more numbers of transformant (Lamalakshmi devi et al. 2013).

9.4 Conclusion

There is an increased desire of development of diversified as well as nutritionally rich food such as more nutrition per calorie. But this is painful to say that feeding is not just an easy task. The most efficient as well as sustainable ways we have are croplands. Croplands have been utilized by crop breeder in order to develop new and genetically varied crop varieties, having added advantages such as enhanced productivity, high nutritious value. Plants having potential to conserve water, soil, genetic diversity are in high demand. Genetic diversity provides resilience against different new disease as well as natural disaster. One of the prerequisite is to increase productivity in the limited land area so as to preserve natural resources. Also the plants need to be genetically modified to resist increased temperature and unfavorable weather conditions. Crop breeding is now considered as advancing science employing usage of biotechnological genetic manipulations in order to develop efficient and advantageous crop varieties.

Acknowledgement We would like to thank Managing Director and Vice chancellor of ITM University Gwalior, India for providing required facility & their valuable support and encouragement throughout the work.

Conflict of Interests There is no conflict of interest. Each authors share equal membership.

References

Abdallah NA, Prakash CS, Mc Hughen AG (2015) Genome editing for crop improvement: challenges and opportunities. GM Crops & Food 6(4):183–205. https://doi.org/10.1080/21645698.2015.1129937

Bhat JA, Ali S, Salgotra RK, Mir ZA, Dutta S, Jadon V, Tyagi A, Mushtaq M, Jain N, Singh PK, Singh GP (2016) Genomic selection in the era of next generation sequencing for complex traits in plant breeding. Front Genet 27(7):221

Borrelli VM, Brambilla V, Rogowsky P, Marocco A, Lanubile A (2018) The enhancement of plant disease resistance using CRISPR/Cas9 technology. Front Plant Sci 9:1245. https://doi.org/10.3389/fpls.2018.01245

Cobb JN, Biswas PS, Platten JD (2019a) Back to the future: revisiting MAS as a tool for modern plant breeding. Theor Appl Genet 132(3):647–667. https://doi.org/10.1007/s00122-018-3266-4

Cobb JN, Juma RU, Biswas PS, Arbelaez JD, Rutkoski J, Atlin G, Hagen T, Quinn M, Ng EH (2019b) Enhancing the rate of genetic gain in public-sector plant breeding programs: lessons from the breeder's equation. Theor Appl Genet 132(3):627–645. https://doi.org/10.1007/s00122-019-03317-0

Diez MJ, De la Rosa L, Martín I, Guasch Pereira LM, Cartea ME, Mallor C, Casals J, Simo Cruanyes J, Rivera A, Anastasio G, Prohens J (2018) Plant genebanks: present situation and proposals for their improvement. The case of the Spanish network. Front Plant Sci 9:1794

EFSA (2012) Scientific opinion addressing the safety assessment of plants developed through cisgenesis and intragenesis. EFSA J 10(2):1–33

Flachowsky H, Le Roux PM, Peil A, Patocchi A, Richter K, Hanke MV (2011) Application of a high-speed breeding technology to apple (Malusdomestica) based on transgenic early flowering plants and marker-assisted selection. New Phytol 192(2):364–377

Fujimoto R, Uezono K, Ishikura S, Osabe K, Peacock WJ, Dennis ES (2018) Recent research on the mechanism of heterosis is important for crop and vegetable breeding systems. Breed Sci 17(1):55

Garcia-Sancho M, Myelnikov D (2019) Between mice and sheep: biotechnology, agricultural science and animal models in late-twentieth century Edinburgh. Stud Hist Phil Sci C Stud Hist Phil Biol Biomed Sci 75:24–33

Gottardo P, Gorjanc G, Battagin M, Gaynor RC, Jenko J, Ros-Freixedes R, Whitelaw CB, Mileham AJ, Herring WO, Hickey JM (2019) A strategy to exploit surrogate sire technology in livestock breeding programs. G3: Genes, Genomes, Genetics 9(1):203–215

Hadasch S, Simko I, Hayes RJ, Ogutu JO, Piepho HP (2016) Comparing the predictive abilities of phenotypic and marker-assisted selection methods in a biparental lettuce population. Plant Genome 9(1):576–579

Han Y, Cameron JN, Wang L, Beavis WD (2017) The predicted cross value for genetic introgression of multiple alleles. Genetics 205(4):1409–1423

Haque E, Taniguchi H, Hassan MM, Bhowmik P, Karim MR, Śmiech M, Zhao K, Rahman M, Islam T (2018) Application of CRISPR/Cas9 genome editing technology for the improvement of crops cultivated in tropical climates: recent progress, prospects, and challenges. Front Plant Sci 9:889

Hundleby PA, Harwood WA (2019) Impacts of the EU GMO regulatory framework for plant genome editing. Food Energy Secur 8(2):e00161

Ishii T, Araki M (2016) Consumer acceptance of food crops developed by genome editing. Plant Cell Rep 35(7):1507–1518. https://doi.org/10.1007/s00299-016-1974-2

Jaganathan D, Ramasamy K, Sellamuthu G, Jayabalan S, Venkataraman G (2018) CRISPR for crop improvement: an update review. Front Plant Sci 9:985

Khan S, Ullah MW, Siddique R, Nabi G, Manan S, Yousaf M, Hou H (2016) Role of recombinant DNA technology to improve life. Int J Genomics 2016:2405954

Ladics GS, Bartholomaeus A, Bregitzer P, Doerrer NG, Gray A, Holzhauser T, Jordan M, Keese P, Kok E, Macdonald P, Parrott W (2015) Genetic basis and detection of unintended effects in genetically modified crop plants. Transgenic Res 24(4):587–603

Lamalakshmi Devi E, Chongtham SK, Holeyachi P, Kousar N, Singh M, Behera C, Telem RS, Singh NB, Wani SH (2013) Cisgenesis and intragenesis: twin sisters for crop improvement. Res J Agric For Sci 1(10):22–26

Li F, Li Y, Cao L, Liu P, Geng M, Zhang Q, Qiu L, Sun Q, Xie C (2018) Simultaneous transfer of leaf rust and powdery mildew resistance genes from hexaploid triticale cultivar sorento into bread wheat. Front Plant Sci 9:85

Long HK, Prescott SL, Wysocka J (2016) Ever-changing landscapes: transcriptional enhancers in development and evolution. Cell 167(5):1170–1187

Mirzoyan Z, Sollazzo M, Allocca M, Valenza AM, Grifoni D, Bellosta P (2019) Drosophila melanogaster: a model organism to study cancer. Front Genet 10:51

Mujjassim NE, Mallik M, Rathod NKK, Nitesh SD (2019) Cisgenesis and intragenesis a new tool for conventional plant breeding. J Pharmacogn Phytochem 8(1):2485–2489

Oliver MJ (2014) Why we need GMO crops in agriculture. Mo Med 111(6):492

Ozdemir BS, Budak H (2018) Application of tissue culture and transformation techniques in model species Brachypodium distachyon. Brachypodium genomics. Humana Press, New York, pp 289–310

Perry ED, Ciliberto F, Hennessy DA, Moschini G (2016) Genetically engineered crops and pesticide use in US maize and soybeans. Sci Adv 2(8):e1600850

Raman R (2017) The impact of genetically modified (GM) crops in modern agriculture: a review. GM Crops & Food 8(4):195–208. https://doi.org/10.1080/21645698.2017.1413522

Razzaq A, Saleem F, Kanwal M, Mustafa G, Yousaf S, Arshad I, Muhammad H, Hameed MK, Khan MS, Joyia FA (2019) Modern trends in plant genome editing: an inclusive review of the CRISPR/Cas9 toolbox. Int J Mol Sci 20(16):40–45

Ryan MJ, McCluskey K, Verkleij G, Robert V, Smith D (2019) Fungal biological resources to support international development: challenges and opportunities. World J Microbiol Biotechnol 35(9):139

Saha SK, Saikot FK, Rahman MS, Jamal MA, Rahman SK, Islam SR, Kim KH (2018) Programmable molecular scissors: applications of a new tool for genome editing in biotech. Mol Ther-Nucleic Acids 14:212–238

Salisu IB, Shahid AA, Yaqoob A, Ali Q, Bajwa KS, Rao AQ, Husnain T (2017) Molecular approaches for high throughput detection and quantification of genetically modified crops: a review. Front Plant Sci 8:1670

Schouten HJ, Krens FA, Jacobsen E (2006) Cisgenic plants are similar to traditionally bred plants. EMBO Rep 7:750–753

Sedeek KE, Mahas A, Mahfouz M (2019) Plant genome engineering for targeted improvement of crop traits. Front Plant Sci 10:114

Shukla SK, Mangwani N, Rao TS, Das S (2014) Biofilm-mediated bioremediation of polycyclic aromatic hydrocarbons. In: Microbial biodegradation and bioremediation. Elsevier, Oxford, pp 203–232

Spielman DJ, Kennedy A (2016) Towards better metrics and policymaking for seed system development: insights from Asia's seed industry. Agric Syst 147:111–122

Suso MJ, Bebeli PJ, Christmann S, Mateus C, Negri V, Pinheiro de Carvalho MA, Torricelli R, Veloso MM (2016) Enhancing legume ecosystem services through an understanding of plant–pollinator interplay. Front Plant Sci 7:333

Tagliabue G (2018) Scientific mistakes from the agri-food biotech critics. Life Sci Soc Policy 14(1):25

Tanaka J, Hayashi T, Iwata H (2016) A practical, rapid generation-advancement system for rice breeding using simplified biotron breeding system. Breed Sci 66:15038

Venken KJ, Sarrion-Perdigones A, Vandeventer PJ, Abel NS, Christiansen AE, Hoffman KL (2016) Genome engineering: drosophila melanogaster and beyond. Wiley Interdiscip Rev Dev Biol 5(2):233–267

Wambugu PW, Ndjiondjop MN, Henry RJ (2018) Role of genomics in promoting the utilization of plant genetic resources in genebanks. Brief Funct Genomics 17(3):198–206

Wang T, Zhang H, Zhu H (2019) CRISPR technology is revolutionizing the improvement of tomato and other fruit crops. Hortic Res 6(1):77

Yenni GM, Christensen EM, Bledsoe EK, Supp SR, Diaz RM, White EP, Ernest SM (2019) Developing a modern data workflow for regularly updated data. PLoS Biol 17(1):e3000125

Zaidi SS, Vanderschuren H, Qaim M, Mahfouz MM, Kohli A, Mansoor S, Tester M (2019) New plant breeding technologies for food security. Science 363(6434):1390–1391. https://doi.org/10.1126/science.aav6316

Chapter 10
Genetic Transformation to Confer Drought Stress Tolerance in Soybean (*Glycine max* L.)

Phetole Mangena

Abstract Plant biotechnology has evolved more than three decades ago, and it entails a wide range of technologies applied in a range of techniques intended to enhance plant growth and productivity of valuable crops, such as soybean. About one-third of the world's edible oils and two-third of protein meals are derived from such improved crop varieties. The use of genetic engineering for the improvement of soybean has promised higher yield, better seed quality, lower pesticides/fertilizers applications, affordable soy-based products and the potential to rapidly raise economic gains in developing countries. Soybean contains 35–52, 14–24, 34–41 and 7–11% of seed compositional protein, oil, carbohydrates and fibre respectively, than most of the grain legumes. However, the rise to a host of technology concerns and negative perceptions against plant transformation techniques used for soybean improvement, like the recombinant DNA technology is overwhelming. The tool is said to raise serious concerns despite the numerous benefits associated with it, such as; improved plant nutritional content, wider genetic characterization and high tolerance to pests diseases as well as abiotic constraints such as drought.

Therefore, the current review summarizes research on challenges facing soybean genetic improvement, the extent of this crop's vulnerability against drought stress, approaches used to improve growth and productivity, especially in combating food insecurity, and interrogate potential opportunities as well as scepticism surrounding the evolution of biotechnology with respect to general consumer perceptions and agribusiness.

Keywords Breeding · Consumer perceptions · Drought · Genetic improvement · Gene transfer · Growth · Transgenic plants · Plant transformation · Soybean · Yield

---

P. Mangena (✉)
Department of Biodiversity, School of Molecular and Life Sciences, Faculty of Science and Agriculture, University of Limpopo, Sovenga, South Africa
e-mail: phetole.mangena@ul.ac.za

P. Guleria et al. (eds.), *Sustainable Agriculture Reviews 45*, Sustainable Agriculture Reviews 45, https://doi.org/10.1007/978-3-030-53017-4_10

10.1 Introduction

The domestication of soybean [*Glycine max* (L.) Merrill] as a major source of proteins and oils creates the need to acquire new genetic characteristics for improved plant growth and productivity. Consequently, both conventional and modern breeding programmes are still focused on developing new methods required to achieve high yield, increased proteins, oils and to potentially use this crop to combat food insecurity in areas vulnerable to the effects of climate change. To date, the main challenge facing soybean production has been the rapidly diminishing quality of processed seeds, poor seed germplasms, and cultivation under harsh environmental conditions. The breeding of cultivars with new gene combinations and maintenance of superior genotypes is the only available solution for preservation and improvement of this crop's adaptability to conditions prone to biotic and abiotic stress factors (Rivas and Condon 2015).

Soybean is a major source of proteins (42%), oils (with saturated fatty acids ranging from 10 to 21% and about 24% of unsaturated oleic, linoleic and linolenic fatty acids), soluble carbohydrates (12%), isoflavones and minerals (Ijarotimi and Famurewa 2006; Bellaloui et al. 2011). The oil and protein constitute about 60% of seed dry matter, meanwhile the remainder of 35 and 5% are carbohydrates and hull, respectively. The cotyledons contain the highest percentage of both protein and oil compared to hull and hypocotyl axis (Lui 1997). These compositions demonstrate that, soybeans remain the basic food ingredient for human populations, in addition to their pharmaceutical benefits and use in animal feed manufacturing. Soy foods and soy derivatives are major dietary/nutraceutical solutions for countries like Argentina, Brazil, China, India, South Africa and the United States (Mangena 2018).

However, the improvement of soybean germplasm using conventional breeding remains a major challenge to breeders in selecting for better growth and yield characteristics. Tools like particle bombardment using embryogenic culture (Finer and McMullen 1991) and *Agrobacterium tumefaciens*-mediated genetic transformation using cotyledonary node culture (Hinchee et al. 1988) are widely applied in modern genetic engineering of many crops for abiotic-biotic stress resistance. Such improvements are achieved under controlled growth conditions *in-vitro,* or *in-vivo* under *in-planta* conditions. Amongst the two, *Agrobacterium*-mediated genetic transformation offers a cheap and reliable protocol for rapid introgression of novel foreign genes into hosts, positive screening and subsequent development of new transgenics with improved growth and yield characteristics (Fig. 10.1). The use of *Agrobacterium* to transfer genes into *Zea mays* L. variety Funk's G90 using shoot apex was reported by Gould et al. (1991). Nyaboga et al. (2014) also reported the successful transformation of yam (*Dioscorea rotundata*), a root and tuber crop, using axillary buds as explants. Plants developed from these cited reports indicated the expression of NPT II gene in their positively identified progenies.

Particle bombardment has also been used to develop herbicide resistant varieties to improve growth and yield characteristics. Rech et al. (2008) reported 9, 2.7 and 0.55% efficient recovery of transgenic soybean, common bean and cotton

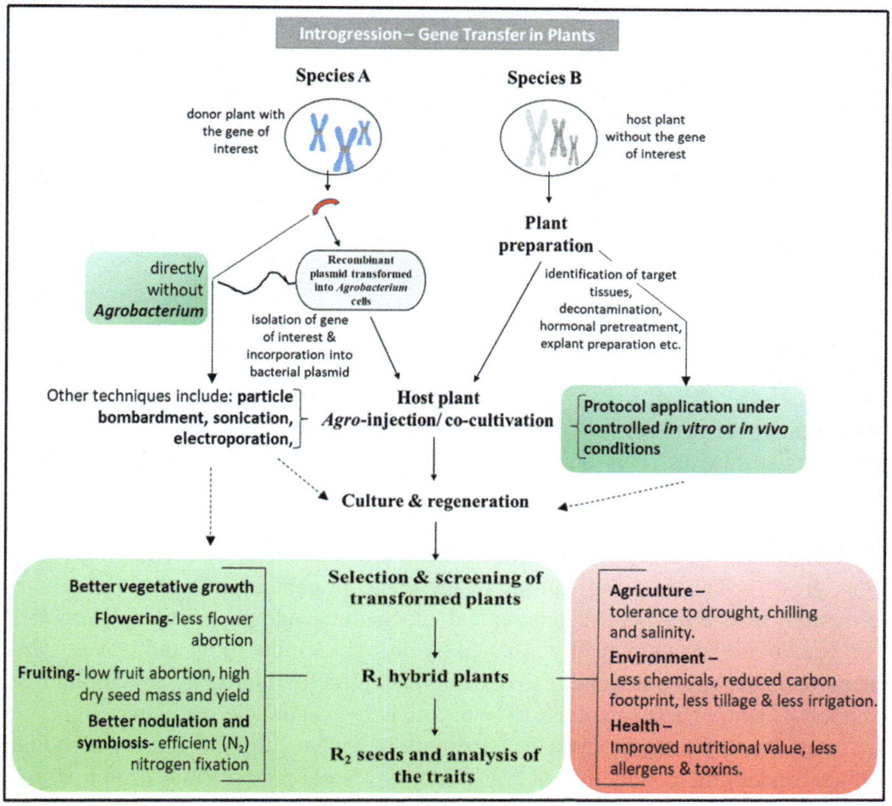

Fig. 10.1 Plant genetic improvement illustration. An example of soybean *Agrobacterium*-mediated transformation via *Agro*-injection or plant tissue culture, and associated agricultural, environmental and health benefits in the use of biotechnology

respectively using this biolistic tool for combined resistance for the herbicide imazapyr. These include previous publications by Sagi et al. (1995), Ruma et al. (2009), Sah et al. (2014), Sparks and Jones (2014) and Ghorbanzade and Ahmadabadi (2015) who genetically transformed banana, tomato, rice, wheat and African violet for this trait. Despite these, all breeding technologies still remain aloof for soybean improvement due to various setbacks.

The setbacks faced during genetic improvement of soybean through plant tissue culture-based transformation include; genotype specificity, protocol inefficiencies and the rapid loss of viability in seeds required for efficient *in-vitro* regeneration of transformed soybean plantlets (Paz et al. 2006; Mangena and Mokwala 2019). Since soybeans are also not immune to environmental stress challenges caused by climate change, particularly drought. Further research is needed to establish a procedure that combat all hurdles mentioned above. More research is therefore, needed to destigmatise and objectively assess the impacts (either positive or negative) of genetically modified organisms on the environment, human and animal health, apart from the already well-known benefits (Fig. 10.1). Therefore, the current review summarises and analyse research as follows: (1) challenges faced during soybean genetic improvement for drought tolerance, (2) highlight possible solutions to improve drought stress tolerance focusing on applicability of modern biotechnology tools used for genetic improvements, (3) influence of this biotechnology on farming practices and to combat food insecurity and (4) interrogate potential opportunities as well as the scepticism surrounding the evolution of breeding techniques with respect to general perceptions and policy.

10.2 Drought Stress

Drought stress is a severe environmental constraint and a very critical threat to food security worldwide. This abiotic stress is the greatest limiting factor, that also adversely affect growth and production of soybean (Farooq et al. 2009; Fathi and Tari 2016). Drought stress causes inhibition of both vegetative and reproductive growths from the onset, by deactivating metabolic machinery that leads to germination, growth, flowering and fruiting of the plant. In general, drought impacts crop growth by controlling several anatomical, biochemical and physiological processes. Soybean plant organs typically affected by drought are the roots and leaves, while stems remain slightly intact due to extra cellular structural integrity emanating from support by simple permanent tissues, e.g. collenchyma and sclerenchyma (Fig. 10.2).

In the vascular tissue systems, particularly xylem, drought causes embolism which renders the conduction of water by the vessels permanently or temporarily dysfunctional (Fig. 10.2c) (Minorsky 2003). It also affects the flow rates of photosynthates from the source to the sink as a result of viscosity build-up or failure to keep adequate water status and collapse of turgor pressure in the phloem (Fig. 10.2a, b) (Sevanto 2014). Drought limit photosynthesis by influencing carbon dioxide diffusion, stomal conductance and cell processes of mesophyll components (Fig. 10.2a, e, f). As a result, photosynthetic activity will be dramatically decreased due to reduced CO_2 diffusion and the formation of reactive oxygen species (ROS) (Pinheiro and Chaves 2010). These oxygen species are generated as a result of the partial reduction of atmospheric O_2, composing of characteristic half-life and an oxidizing potential. Furthermore, there are basically four forms, namely singlet oxygen, superoxide radical, hydrogen peroxide and hydroxyl radical that occur in the different cellular compartments, e.g. chloroplasts, peroxisome and mitochondria (Cruz de Carvalho 2008).

The formation of free radicals will be more prevalent in soybeans undergoing water-deficit stress (Fig. 10.2e, f), than well-watered plants (Fig. 10.2d) (Mangena 2018). All these effects indicate why water remains a major component in plant cells, including the many cellular processes such as the production of organic metabolites, hormone biosynthesis, transport of organic acids and inorganic molecules. Moreover, water shortage may still cause increased plant susceptibility to insect pests attacks, viral as well as fungal diseases. Although in contrast, Gautam

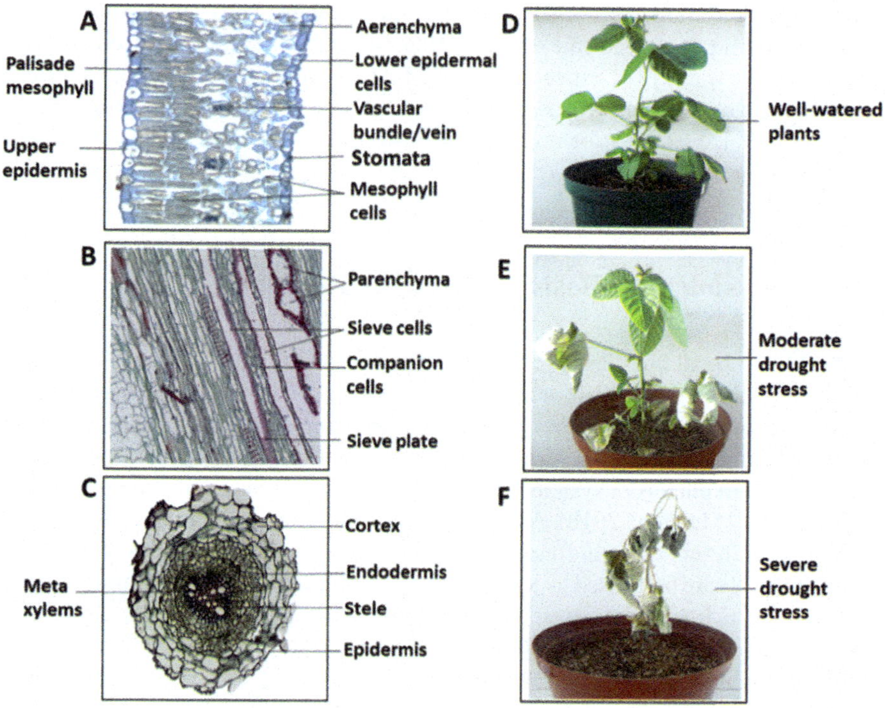

Fig. 10.2 Representation of critical plant tissue areas affected by drought. (**a**) Anatomical illustration of mesophyll cells of the leaves. (**b**) Sieve cells providing passage of sugars from the shoot to roots driven by water status of the phloem. (**c**) Root cross section. Phenotypic response of well-watered plant (**d**) and soybean plants subjected to moderate (**e**) and severe (**f**) water stress. (Mangena 2018)

et al. (2013) and Hossain et al. (2019) indicated that drought may slow down or prevent the development of plant diseases, especially of pathogens that thrive under moist conditions.

However, these limiting factors continue to cause major growth and yield reductions. To combat stress, conventional and modern breeding methods are continuously tested to achieve desired plant genetic combinations in traits of resistance. But, the high production of progenies that inherit a mixture of desirable and non-desirable genes still create problems when improvement is done using traditional breeding techniques. The many unsuccessful crosses and longer breeding cycles make conventional breeding less attractive in dealing with the current environmental challenges compared to modern biotechnology approaches.

To date, many soybean varieties were studied to evaluate morphological and physiological markers that contribute to drought tolerance for selection and use in breeding programmes. Mabulwana (2013) assessed and identified Sonop as a more drought tolerant variety amongst all six South African soybean cultivars (LS677, LS678, Mopanie, Sonop, Klap and Pan 1564) compared to the American cultivars

(R01016 and R01581) for their resistance to drought. This study showed that Sonop's petiole contained thickened sclerenchymatous tissues which functions to guard against water loss, especially under drier conditions. Such varieties may be used to enhance breeding efforts which still remain focussed on long multiple cycles of experiential selections for improved growth under dry water-limited soil conditions (Devi et al. 2014).

10.3 Physiological Basis for Drought Tolerance

Drought can be defined as a prolonged shortage of rainfall. This meteorological phenomenon has no precise definition but, it is defined by hydrologists as water deficit caused by the lack of precipitation, soil moisture, river flow and groundwater etc. (Tate and Gustard 2000). Research indicated that soil water deficit lead to shortage of water in plant root systems and the surrounding, limiting to plant growth and development (Mangena 2018). As such, more insights on how plants grow and function under different drought frequencies over time will refine how genetic improvement protocols are optimised to achieve high tolerance and yield. These physiological variations which are in response to stress continue to cause a wide range of effects on the plant's developmental and functional growth/reproductive stages.

For example, drought severely impedes physiological processes such as; ROS scavenging activity, stomatal regulations, photosynthetic responses and pigmentation (Fathi and Tari 2016). In addition to these, secondary metabolite biosynthesis (as indicated in Fig. 10.3) in the roots and leaves may also be impeded. Prolonged drought stress results in the accumulation of common toxic intermediate oxygen compounds that are highly and rapidly reactive with a wide variety of cellular constituents like proteins, DNA, RNA and lipids (Taiz et al. 2015). Such reactive forms of oxygen include hydrogen peroxide (H_2O_2), superoxide ($O_2{\bullet}^-$), singlet oxygen ($^1O_2$) and hydroxyl radicals ($OH{\bullet}$) as indicated on the previous section above.

According to Taiz et al. (2015) production of ROS can trigger membrane oxidation, complete degradation of organelles and ultimately lead to cell death. This means that, consequently, the accumulation of ROS remains one of the major toxic byproducts produced as a result of plant's exposure to abiotic stress. Drought stress frequently cause reduction in leaf water, chlorophyll content, photosynthates and increase proline content of water stressed plants compared to unstressed plants (Mannan et al. 2016). Leaf, stem and root dry matter, as well as total grain yield per hectare and other yield components such as the number of pods, number of seeds, 100-seed weight and harvest index were reported to be reduced (Mimi et al. 2017; Mangena 2018).

The extensive disruption of cellular structure, metabolism and various enzyme-catalysed reactions may further occur. Some of the altered metabolic functions include reduced synthesis of photosynthetic pigments involved in harvesting of sunlight and the generation of reducing powers (Jaleel et al. 2009). Generally, the physiological impairments that take place are attributed to the loss of turgor pressure,

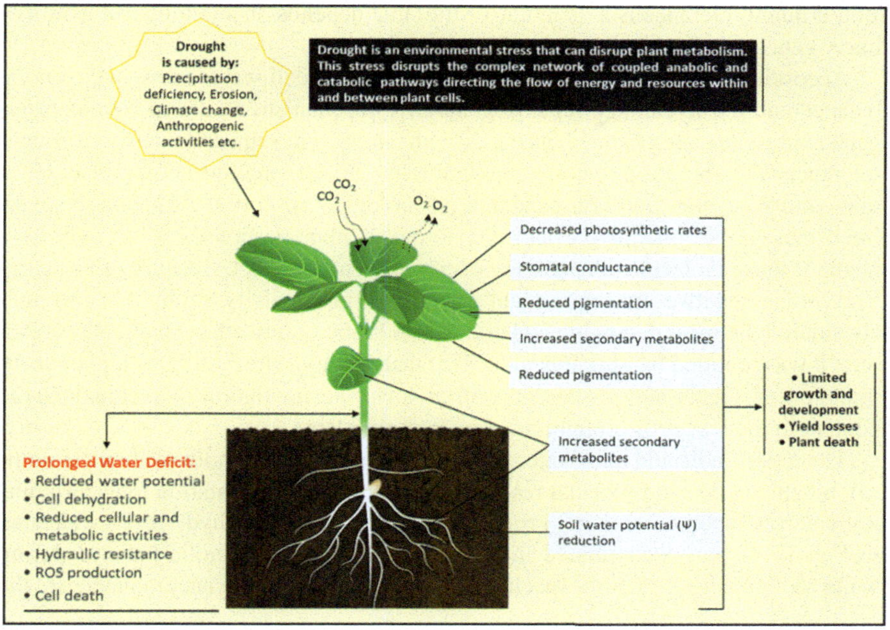

Fig. 10.3 An illustration of the effects of water deficit stress on plant metabolism, growth and yield of soybean. Reduced water (Ψ) potential in the cells may lead to the disruption of pathways involved in the flow of energy and resources within and between plant cells during growth and reproduction

disruption of metabolism, closure of stomata, and diminished plant cell water potential (Farooq et al. 2009). A better understanding of the physiological effects induced by drought may unravel fully its negative impacts on growth, especially by hampering cell enlargement, yield and ultimately the cause of deaths of plants, providing leading insights into precise means to be used for trait improvement.

10.4 Molecular Responses to Drought Stress

Technology advancement remains a key feature in the development of plants showing resistance to abiotic stress. In particular, the development of drought tolerant plants like soybean using biotechnological approaches such as genetic engineering. These techniques do not only provide opportunities to effect changes in the genetic make-up of crops but, also elucidate the complex mechanisms underlying drought resistance that is taking place in these crops. Such elucidation, when done at genetic, genomic and molecular level will assist in establishing protocols used for the improvement of new crop varieties that are stress tolerant (Hu and Xiong 2014).

But, drought tolerance is a quantitative trait that depends on a cumulative action of many genes and the surrounding environment.

According to Khan et al. (2018) oxidative stress, cell water transport, osmotic balance and tissue damage repair/regenerative mechanism regulated by numerous gene expressions are changed during drought stress. Thus, mechanisms for drought tolerance will, therefore, involve regulation by various small-effect loci and genes that control several morphological and physiological responses of plants to stress. Such measurable phenotypic and physiological plant responses to drought as a result of those molecular changes are exemplified in Fig. 10.4. Because of its nearly irreversible negative impact on plant growth and productivity, drought is currently considered the most damaging constraint than biotic or any other abiotic stress factors. It was reported by Sarkar et al. (2015) that, drought stress triggers higher accumulation of sugars and proline as well as lower accumulation of malonaldehyde content in the leaves of drought tolerant soybean plants.

The report indicated that, water deficit stress induces morphological, physiological, biochemical and molecular responses that allow crop adaptation to water limiting environmental conditions. Proline, sugars and malonaldehyde levels indicated in Fig. 10.4 were accumulated in order to keep cellular osmotic balance during water deficit stress. Those specific organic molecules are accumulated in the

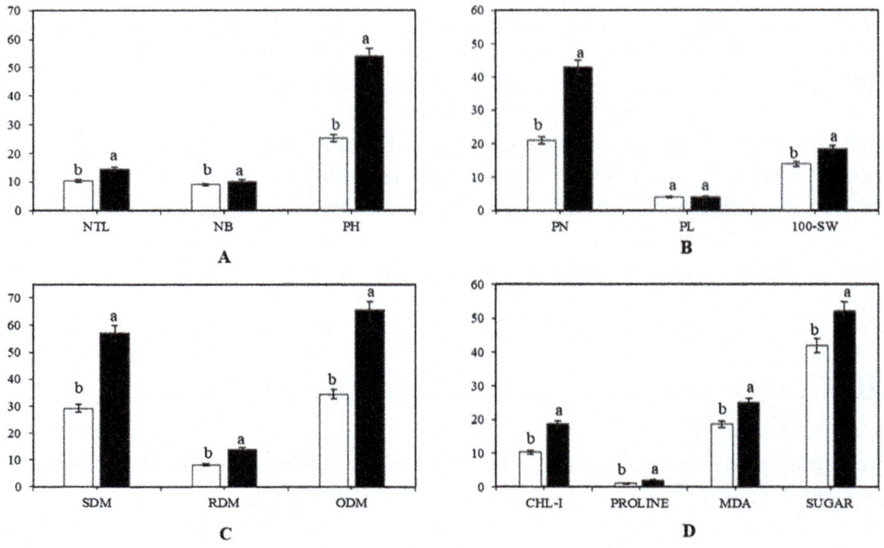

Fig. 10.4 Quantitative measurement of some of the morphological parameters; (**a**) [mean number of trifoliate leaves (NTL), number of branches (NB), plant height (PH)], (**b**) [pod number (PN), pod length (PL) and 100-seed weight (100-SW)] and physiological content; (**c**) [mean shoot dry mass (SDM), root dry mass (RDM), overall dry mass (ODM)], (**d**) [chlorophyll index (CHL-I), proline, malondialdehyde (MDA) and sugars], in plants subjected to water-deficit stress. White bar-graph refers to drought stressed plants and black bar-graph refers to plants without water stress. (Souza et al. 2013; Sarkar et al. 2015; Mangena 2018)

cytoplasm to protect the plant's membrane integrity and enzyme activity (Farooq et al. 2009; Sarkar et al. 2015). Apart from these, there are studies of genes that have been identified and well characterised to confer drought tolerance using genetic engineering in various stable crops. The altered gene expression improves drought tolerance by primarily enhancing growth and development under abiotic stress.

Soybean transgenic plants with ectopically expressed *AtABF3* gene were regenerated using *Agrobacterium tumefaciens*-mediated genetic transformation for drought and salt tolerance (Kim et al. 2018). Ghorbanzade and Ahmadabadi (2015) reported transfection of an endochitinase gene cDNA into African violet using particle bombardment technique with a gold particle coated with the pFF19G vector construct. Soybean half seed explants and African violet leaf tissue explants were targeted for both studies, respectively, and they both employed PCR and RT-PCR to confirm and analyse transgene integration and expression in the F_2 regenerated transgenic plants.

10.5 Methods of Plant Transformation

Plant transformation is now considered the most economic and highly effective method of genetic engineering that has been reported so far. The method holds the potential and promise to efficiently regenerate transgenic plants, especially for recalcitrant legume crops. Legumes like soybeans are some of the most important pulse crops and a good source of high quality proteins and oils, required for human consumption, health benefits and good industrial processing value (Wilson 2004). The vegetative and reproductive stages of soybean continuously show high sensitivity to abiotic and biotic stress constraints. Yield quality and quantity of legume crops is severely affected by high temperatures, chilling stress, waterlogging and particularly, water deficit stress (Mangena et al. 2017).

To circumvent challenges posed by all stress factors; an affordable, efficient and rapid system of transformation that develops non-chimeric transgenic plants with resistance to these stress-inducing factors must be advanced. Such gene manipulation techniques could be established from the direct and indirect methods of plant transformation, like plant tissue culture-based *Agrobacterium*-mediated transformation, particle bombardment, sonication assisted gene transfer and electroporation. Although the techniques have the ability to enable specific exogenous gene transfer into targeted plant host tissues, protocols for routine application and high transformation frequencies still need to be developed.

10.5.1 In Vitro Agrobacterium-*Mediated Genetic Transformation*

The *Agrobacterium tumefaciens*-mediated transformation involves the identification, isolation, reconstruction of the bacterial binary vector, and transfection of exogenous DNA segments into host plants (Paz et al. 2004; Liu and Wei 2005). The advantages of using this technique is that; it remains a highly efficient transformation system, provide flexibility for DNA delivery to targeted tissues, allow effective selection of the transformed from non-transformed plants and it is easy to incorporate into the *in vitro* regeneration pathways required for micropropagation purposes. This technique is still considered a cheap indirect method of genetic improvement that relies on optimised plant tissue culture conditions. The disadvantage is that, the current culture conditions remain inadequate for successful plant genetic transformation, especially in soybean. However, reasonable progress has been made thus far in conferring characteristics that include the ability to survive herbicide treatment, insecticides, disease resistance and to withstand salinity stress (Rech et al. 2008; Lee et al. 2012). Rizwan et al. (2015) reported the development of sulfonylureas herbicide resistant soybean plants, and including other crops like canola, sunflower, wheat and corn.

Regenerated transgenic plants were successfully produced through this technique, combined with the use of seed mutagenesis. Various *in vitro* cultures such as nodal culture, meristem culture, protoplast culture and cotyledonary node cultures are currently applied as the basis for the establishment of protocols used for the *in vitro* regeneration of transformed plants (Birch 1997). The transformation efficiencies were reported ranging from 2.1% in *Cicer arietinum*, 7.89% in *Pisum sativum* and over 39.4 in cultivated *Musa* sp. (Das Bhowmik et al. 2019, Aftabi et al. 2018, Shivani and Tiwari 2019). Furthermore, the development and standardisation of *Agrobacterium*-mediated soybean transformation has also showed to allow efficient establishment of transgenic soybean plants carrying genes from many different sources, including microbes, insects and animals (Korth 2008).

10.5.2 Particle Bombardment/Biolistics

As an alternative to *Agrobacterium tumefaciens*-mediated genetic transformation, biolistics was first described by Sanford et al. (1987) as the first successful method for the production of transgenic plants. This was thoroughly demonstrated in the transformation of cereals, which generally yielded between 0.4% and 1.1% transformation efficiency reported in all crops to date (Ismagul et al. 2018). Particle bombardment relies on electric discharge of gold or tungsten DNA coated particle acceleration using a gene gun device (Christou 1994). The gene constructs used in this approach may take a form of either linear or circular plasmids/ expression cassette. The method is responsible for the many genetically engineered crops currently on the market and the frequencies of transformation reported in many organisms and major agronomic crops. Such successes include, the biolistic transformation of microbial cells, mammalian cells and food crop plants such as; onion, rice, wheat, soybean, conifer and loblolly species (Bliffeld et al. 1999; Connett et al. 2002; Behrooz et al. 2008).

The technique of particle bombardment have several advantages making it a method of choice for genetic engineer of crops. Those benefits include in addition to the higher rates of stable transformation;

- the ability to engineer organised and potentially regenerable tissues (Christou 1994),
- break cultivar/genotype specificity barrier of recalcitrant species,
- and it is used in gene expression studies (Taylor and Fauquet 2002).

The disadvantages reported in this technique, however, include mixed undesirable integration patterns, relatively low throughput, high input costs and that the targeted tissues cannot be controlled. However, as indicated by Behrooz et al. (2002), many researchers avoid the use of particle bombardment due to the random nature of targeting intracellular recipients of the transgene, complexity of integration patterns and multiple copy number insertions that could cause silencing of desirable genes.

10.5.3 Liposome-Mediated Transfection

Liposome-mediated gene transfer is another tool used to inject DNA fragments coding for a foreign protein, conceivably conferring resistance to biotic or abiotic stress. The method was reported by Deshayes et al. (1985), in which plasmid-loaded liposomes were fused with mesophyll protoplasts using polyethyleneglycol treatment. The technique makes use of the positively charged lipids, e.g. cationic liposomes, for gene transfer into host cells. The foreign DNA of interest is encapsulated in the spherical lipid bilayer, taking advantage of liposome's favourable interactions with negatively charged DNA, and the cell membrane (Gad et al. 1990). Foreign DNA fragments will be free to recombine and integrate into the host genome following a successful cell invagination through endocytosis. Successful transfection based on this system was previously reported in potato and wheat (Behrooz et al. 2008), in addition to tobacco (Deshayes et al. 1985). The most recent case of high and stable transformation efficiency achieved was 3.74% in callus cells regenerated from *Elaeis guineensis* (oil palm) protoplasts (Masani et al. 2014).

Furthermore, it is also used to treat acquired diseases in mammalian cells, which entail genetic components that can be corrected through gene therapy. Cancer, cystic fibrosis, muscular dystrophy and familial hypercholesteremia are among the known mammalian genetic diseases that are treated using this gene transfer technique (Ropert 1999). Advantages associated with the liposome cell systems include;

- enhanced delivery of encapsulated DNA by membrane fusion,
- protection of nucleic acids from nuclease enzyme activity,
- specific cells could be easily targeted,
- make use of plasmodesmata in order to gain passage in reaching different cell compartments,
- and the DNA of interest is delivered in a form of intact small organelles (Gad et al. 1990).

10.5.4 Fibre-Mediated DNA Delivery

Silicon carbide fibre-mediated transformation is one of the recently developed methods of plant transformation. This method is used to deliver DNA into host cells or tissues, followed by culture and selection of transgenic tissues over non transformed cells. Plant tissues that contain identifiable transferred marker genes are proliferated and regenerated further into intact transgenic plants (Rashid and Lateef 2016). Black Mexican sweet maize and tobacco were first examples of transformed plants, in the presence of plasmid DNA encoding ß-glucuronidase (GUS) and silicon carbide fibre using a liquid suspension culture (Kaeppler et al. 1990). This report confirmed the efficiency of gene transfection of vortexed maize and tobacco cells using a scanning electron microscope. The highest transient expression of the Gus A gene was also detected in vortexed cells of mature embryo of wheat. Further GUS expression was observed in callus tissue of the same plant induced from the silicon carbide fibre of the same study (Serik et al. 1996). However, Hassan et al. (2016) reported a transformation efficiency of 6.88% using 20-day-old peanut callus cells. Comparatively, there is thus far no report of any previous study on transformation that has obtained this highest transient expression frequency, particularly of leguminous plants like peanut. The advantages of this system include; rapid transformation, cost effectiveness, simplified procedure, and has also been used to overcome barriers caused by genotype specificity. Low transformation efficiency, severe tissue damage and toxic effect of fibres are reported problems associated with the technique (Behrooz et al. 2008).

10.5.5 Laser Induced DNA Delivery

Laser induced intracellular delivery of DNA involves the use of laser beams to puncture the cell membrane through which exogenous genetic material of interest may enter into cell cytoplasm of hosts (Meacham et al. 2014). This method was first introduced by Tsukakoshi et al. (1984) in the transfer of DNA into normal rat kidney cells using an ultraviolet nanosecond laser irradiation containing 0.5 μm spot size and 1 mJ laser energy. Laser beam was reported to cause perforations on cell membrane which allowed the transfer of the *Eco-gpt* gene coding for enzyme converting xanthine to xanthine monophosphate. This technique has received extensive attention due to nontoxicity, high intensity and the duration in which cells could be perforated with DNA fragments, particularly in mammalian cell cultures.

The development of this method potentially overcomes serious limitations by subsequently being able to translate the concept of gene transfer into routinely used physical technology and nanotechnology approach, especially with high frequency of gene transfection and direct delivery of DNA fragments into targeted tissue. Laser-mediated transformation was reported to achieve high efficiency for *in vitro* and *in vivo* gene transfer if laser beam is focussed slightly within the cell near the nucleus as opposed into the cytoplasm (Meacham et al. 2014). Furthermore, this was accompanied by 100% transformation efficiency with zero tissue senescence on the transfected tissues. However, there are reported concerns over adverse effects and the destructive power of ultraviolet laser radiation on the cellular and subcellular components (Kajiyama et al. 2007; Li et al. 2017).

10.5.6 Pollen Transformation

In this method, pollen grains to be involved during pollination are soaked in a solution containing DNA fragments of interest. DNA molecules will then diffuse into the pollen as a result of this soaking. Pollen grains in this case are being targeted due to their abundance, accessibility and because they are simple to handle (Eapen 2011). Progress in introducing transgene through this technique have been limited, despite being a simple and quicker method to recover fertile transgenic plants in some species. Thus far, efforts made to improve transformation efficiency and enhanced reporter gene expression using pollen provided evidence of reasonably high transgenicity in soybean (3.0%), onion (15%), walnut (20.7%) and maize (33.3%).

Booy et al. (1989) previously highlighted difficulties in achieving transgenic maize from germinating pollen grains, incubated in a solution containing a mixture of carriers and plasmid DNA carrying a gene coding for kanamycin resistance. Further evidence indicated that, the plasmid DNA was degraded by nucleases released by the germinating pollen. Pollen transformation effectively eliminate the production of chimeras and somaclonal variation (Wang et al. 2001). Other quicker, more efficient and simpler techniques like floral dip method (Clough and Bent 1998), pistil transformation (Chumakov et al. 2006) and ovary-dip transformation (Yang et al. 2009) are related to pollen transformation. Sangwan et al. (1993) reported successful recovery of transgenic *Datura innoxia* and *Nicotiana tubacum* plants from pollen-derived embryos. These haploid plants were developed from pollen grains cocultured with *Agrobacterium tumefaciens* harbouring binary and cointegrate vectors. However, the high transformation rates (over 75%) were achieved from injured or cut cotyledonary -stage pollen embryos and not from infected pollen grains and proembryos.

10.6 Soybean Vulnerability and Adaptability to Varied Agroclimatic Conditions

Although, a wide variety of United States elite cultivars and the International Institute of Tropical Agriculture (IITA) soybean varieties have been developed via genetic engineering and classical breeding, significant knowledge on their molecular and physiological responses to the different cultivation conditions are not yet well researched (Zhao et al. 2018). A series of fundamental studies focussed on evaluating the adaptation of newly developed soybeans to a wide range of environmental conditions such as drought are required. For decades, world soybean yield per hectare has been stagnant at about 1.1 tons (Khojely et al. 2018). This low yield can be attributed to the poor performance of commercial varieties under different agroclimatic conditions, and their gauge of susceptibility to pests and drought. Vulnerability was however, reported by various researchers including; Tripp (1996), Keneni et al. (2012) and Fu (2015) as a phenomenon instigating crops' genetic defencelessness to stress.

Soybean improvement has to offer varieties with traits that give the highest returns by combining the most desirable traits when grown under different conditions, regardless of whether the conditions are suitable or harmful (Acquaah 2012). To do so, genetic engineering has been documented to be amongst the top modern technologies representing the greatest potential to end the high susceptibilities and vulnerabilities of many crops to biotic and abiotic stress. This has been reported by many researchers worldwide and often considered to be a sustainable solution to end hunger and poverty, particularly in developing countries (Falck-Zepeda et al. 2013).

Crop vulnerability to stress caused by fluctuating agroclimatic conditions reflects poor genetic homogeneity and uniformity, which must expectedly offer significant advantages on the growth and reproduction of soybean (St-Martin 1982). Although, genetic similarity may not immediately cause increased susceptibility to abiotic stress and rapid spread of infectious diseases (Bharadwaj et al. 2002), genetic stability may be required to improve adaptability and productivity since the so called "high yielding varieties" are genetically unstable, very few in number, and are often dramatically influenced by stress, as well as other harmful effects related to climate change (Al-Yozbaki et al. 2015; Mangena 2018).

10.7 Unsustainable Cultivation of Soybean

The cultivation of soybean takes place under drylands and irrigation. Therefore, high productivity also rely heavily on irrigation farming practices, and less on the identification of better performing genotypes that are rarely suitable and well adapted to a wide range of planting conditions. According to Brodt et al. (2011), changes observed in agriculture demonstrated a shift from traditional farming practices to new technologies, mechanisation and increased use of agro-chemicals as illustrated in Fig. 10.5. These developments have had many benefits like producing more food and reducing many risks associated with farming costs (Brodt et al. 2011; Emami et al. 2018). But, they were also accompanied by significant health and environmental hazards. Prominent among these, is the soil poisoning by soil fumigants, chemical pollution (air, groundwater etc.), and health safety issues for farm labourers.

To maximize production and minimise the impact of using unsustainable farming practices together with the stagnated technology, specific traits and breeding strategies require detailed knowledge, especially in relation to the environment where crops are grown. Bhatia et al. (2014) emphasized genotype x environment interactions and fine tuning of breeding techniques to produce cultivars suitable for planting in local proprietary farming systems and large scale commercialisation. This model could assist in reducing the amounts of agrochemicals used (Fig. 10.5), promote subsistent farming, and would furthermore strengthen the call for rapid expansion and adoption of modern breeding programmes to efficiently facilitates the development of cultivars that are uniformly adapted to a wide range of planting conditions. Although, advances have been made in the past decades, most commercial cultivars are still adapted to specific planting areas and only do well under certain growing conditions.

The current mechanised character of cultivation cause difficulties for small holder farmers who cannot keep-up with the high costs of agrochemicals, gasoline and heavy mechanization. These farmers rely on the use of chemicals such as fertilizers and pesticides that are toxic to the environment (Fig. 10.5). Chemicals also have direct/indirect effects on the environment accompanied by the various other

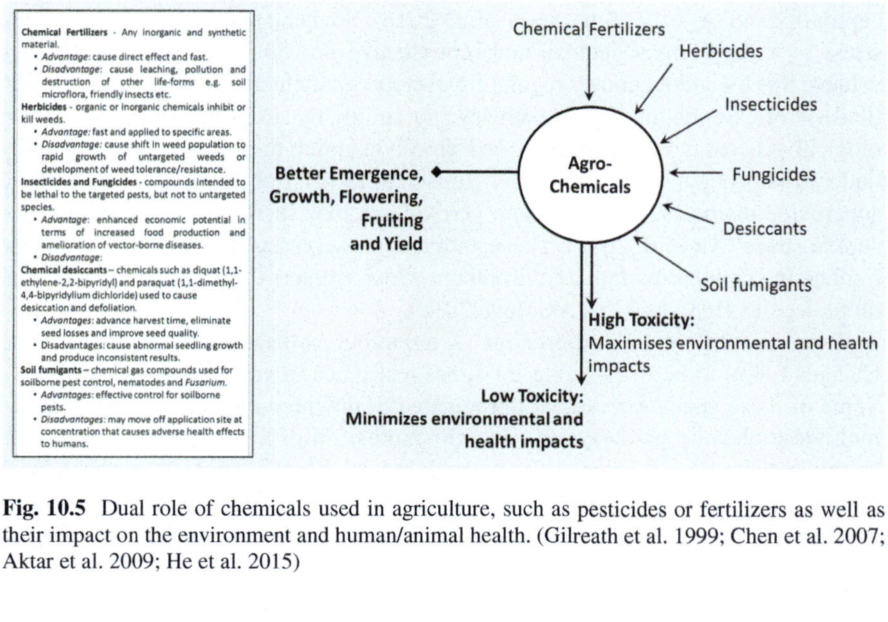

Fig. 10.5 Dual role of chemicals used in agriculture, such as pesticides or fertilizers as well as their impact on the environment and human/animal health. (Gilreath et al. 1999; Chen et al. 2007; Aktar et al. 2009; He et al. 2015)

general negative impacts on nature which continues to cause great concerns (Bickel and Dros 2003; Barona et al. 2010; Sharma and Singhvi 2017; McArthur and McCord 2017). But, to meet agricultural demands and feed increasing populations, farmers appear justified to modify farming equipment and substantially increase the use of chemical fertilizers (Sharma and Singhvi 2017). The approach certainly overcome losses in crop productions but remain unsustainable, unrenewable and do not allow for long-term environmentally friendly farming practices and conservation programmes as illustrated in Fig. 10.5 above.

10.8 Physiological and Molecular Response of Soybean to Drought Stress

Soybeans usually take about 90 days to mature, depending on the type of variety. Early maturing varieties could take 75 days while late maturing varieties can take up to 150 days to reach maturity stages (Shurtleff and Aoyagi 2009). The varied maturity periods do not directly influence productivity, but rather indicate the extent to which this crop will narrowly remain adapted to certain agro-climatic conditions. These exhibit the different rates of growth and reproduction that could be expected. However, when soybeans are confronted with severe drought stress due to unfavourable weather conditions, especially those inclined by climate change (Maisa et al. 2019), they experience disturbed metabolic functions. At most, these effects are as a result of reduced cell water potential, stomatal conductance, respiration and photosynthesis rates (Xu et al. 2018).

Such growth problems are also related to those encountered in major crops such as maize, wheat and rice in which the response always differ according to their susceptibility and severity of the stress (FAO 2016). Soybean plants cope with drought stress by controlling vegetative and reproductive growths to escape stress. They achieve this by endogenously regulating osmotic gradient and enzyme activity, particularly of essential metabolic pathways, or simply by sustaining important physiological processes already mentioned above in order to avoid stress, especially under mild drought stress conditions (Taiz et al. 2015). For example, the activity of superoxide dismutase, catalases and peroxidases may increase during plant exposure to stress (Xu et al. 2018). These antioxidants serves as the first line of defense minimising cellular oxidative environment which triggers the oxidation of essential biomolecules (Ighodaro and Akinloye 2018).

There are differential expressions of numerous annotated genes with twofold changes found to be responsible for stress resistance in soybean (Xu et al. 2018). Some of these genes were either upregulated or downregulated in order to control multiple molecular pathways that lead to expression of transcription factors, secondary metabolites, abscisic acid fatty acids etc. in order to cope with drought. High gene expression level were detected in the roots of various varieties as well in drought resistant soybean varieties compared to the drought sensitive varieties (Prince et al. 2015; Song et al. 2016).

10.9 Improvement of Soybean Transformation for Drought Tolerance

Retrospectively looking back, the introduction of genetically modified plants has simplified the exploitation of crops for many industrial applications. Thus, to greatly reap benefits from the high amount of proteins and oils stored in soybean seeds, rapid development of cultivars tolerant to drought stress must be facilitated (Mangena 2019). Numerous studies assessed the competency of soybean varieties to gauge totipotency and proliferative ability of tissue explants used for *in vitro* regeneration and transformation (Birch 1997; Paz et al. 2004; Hwang et al. 2017; Mangena et al. 2017; Raza et al. 2017). The findings showed significant variations amongst the varieties and culture conditions. In the main, although Paz et al. (2004) reported 2.0 to 6.3% transformation efficiency using cotyledonary node explants co-cultured with *Agrobacterium*, this still do not give the same results when using other varieties.

According to results obtained thus far, the already established transformation protocols remain aloof for soybean improvement due to genotype specificity, internal protocol inefficiencies and rapid loss of viability by seeds used to develop explants etc. (Paz et al. 2006; Mangena and Mokwala 2019). These clearly indicate the dual role of inefficacious culture establishment and gene transfer techniques on inability to exploit totipotence potential of targeted tissues and their transgenicity during plant transformation. Such observations have been highlighted in many transformation studies using different species. Studies in the higher leguminous plants and grasses such as *Cassia fistula*, *Leucaena leucocephala*, *Zea mays Saccharata var. rugose* have also demonstrated the importance of developing an efficient transformation protocol that could produce transgenic plants showing resistance to both biotic and abiotic stress. This remain critical because the growth and yield production potential of these crops is largely depended upon such developments (Kandil et al. 2013) and other means as described previously.

10.10 Biotechnology Applications for Abiotic Stress

The production of genetically modified plants factually involves the insertion of foreign DNA into host plants' genome. This holds potential impact of improving stable crops against various biotic and abiotic stress constraints. But, to achieve such major objectives, new holistic and integrative approaches must be established, such as the incorporation of system biology, bioinformatics, applications involving metabolomics and proteomics. System biology already deploys an interdisciplinary approach into focus when studying multiple interactions of factors within biological systems, than just providing a sum of the different parts (Karahalil 2016). This approach tackles different aspects and clarifies various biochemical and molecular constituents underlying cellular processes, and the numerous responses observed under environmental stress.

For example; proteomic technology may deal with sequencing, structural, functional, interactive and expression proteomics to fully characterise polypeptides properties, dimensional structures, functions and abundance in plants exposed to drought stress (Aizat and Hassan 2018). In soybean, advances have been made in elucidating the impact of drought on root architecture and nodule traits by analysing underpinning transcriptome, proteome and metabolome information aimed at improving its tolerance to drought stress (Kunert et al. 2016). Similarly, Goh et al. (2016) reported the effect of temperature on plant chemical composition using chromatograph-mass spectrometry and liquid chromatography time-of-flight mass spectrometry in a simulated reciprocal transplant experiments.

Metabolite fingerprinting in this study identified varied amounts of volatile organic compounds, flavonoids and terpenoids under higher temperature treatments. Transcriptional profiles of genes in seedling leaves and roots of soybean inbred line HJ-1 were also examined using transcriptome analysis and next generation sequencing in response to drought, salt and saline-alkali environmental stress (Fan et al. 2013). Simplified and interlinked mechanisms involving system biology-based approaches could help in understanding how individual factors and mechanisms (biochemical, molecular and metabolic) spatially and temporally interact, especially in improving tolerance to drought of economically important food crops, such as barley, maize, rice, soybean and wheat (Jogaiah et al. 2013). All these reports have yielded new insights into stress tolerance mechanisms of soybeans and other above mentioned stable crops.

10.10.1 Omics Tools Used in Crop Improvement and Analysis

Advances in biotechnology research over the years have raised expectations that, highly nutritious and more stress resistant crop varieties will be developed. But, due to the nature of technology challenges, some of the techniques used in biotechnological research have not yet fully translated into the development of genetically improved cultivars as reported by Ibrahim and Shawer (2014), Samaha et al. (2019) and Raza et al. (2017). There are several biotechnology successes reported to date, except the hurdles of selecting traits, increasing yield and accelerating breeding programs. Such include; revelation of fundamental information regarding the chemical compositions and value of many food crops using techniques like genomics and proteomics.

These techniques have been used in the chemical analysis and genomic characterisation of legumes such as cowpea, lentils, peas, peanuts and other pod producing plants that are cultivated commercially or privately as food sources (Bouhabidu et al. 2013; Grela et al. 2017; Gyang et al. 2017). Legume crops play a critical role in the traditional diet of many countries like Brazil, China, India and regions in the Middle East, Southern America and Africa (Messina 1999). Proteins and oils found in these crops have also received considerable attention for different health and nutritional benefits. Soybean proteins in particular, have been reported to lower blood cholesterol, reduce the risk of cancer and osteoporosis. They are very low in saturated fats, therefore, being an excellent source of dietary fibre, essential amino acids and a variety of micronutrients and phytochemicals (Messina 1999).

Although techniques like proteomics have been used in plants to purify individual components and study them in isolation, they have been rarely used for trait improvement of crops. The use of one- or -two dimensional sodium-dodecyl sulphate polyacrylamide gel electrophoresis (1D-SDS-PAGE or 2D-SDS-PAGE) is one of the popular tools utilised for protein analysis in plants. This technique has demonstrated that plant cells contain large and complex set of molecules forming chains of polypeptides (Lesk 2010). High resolution one-dimensional and two-dimensional SDS-PAGE has been used to successfully isolate and separate individual amino acids even when the protein profile is influenced by exogenous contaminating protein factors.

Zurfluh and Guilfoyle (1980) examined alterations in protein synthesis on soybean hypocotyls developed from seedlings pre-treated with 50 μM 2,4-dichlorophenoxyacetic acid (2,4-D). This technique revealed altered spectrum of polypeptide chain synthesis on the hypocotyls in response to plant hormones showing possible elucidation of changes even during the exposure of plants to abiotic or biotic stress. Other applications of this technique include isolation and detection of proteases, lipids and phosphoproteins (Karpe and Hamsten 1994; Kinoshita et al. 2009; Du Plessis 2013). Genetic modification of cotton, corn and potato expressing *Bacillus thuringiensis* (*Bt*) crystal protein conferring resistance to pests have also been analysed using these techniques (Gould 1998).

10.11 Use of Genetically Modified Soybean Plants

Numerous crop plants still encounter tremendous growth inhibition as a result of high temperatures, water deficit, high salt levels and mineral deficiencies, as well as metal toxicity, under both natural and agricultural systems (Jogaiah et al. 2013). In many cases, the crops are faced with more frequent and severe exposure to drought stress than any kind of abiotic stress (Dai 2011). However, integration of modern genetic modification techniques into traditional methods of plant improvement could allow for efficient production of stress tolerant plants. Genetic engineering have successfully achieved selection across multiple traits, reduced cost and time, resulting into the establishment of protocols efficiently used to develop varieties with improved growth and productivity (Hadiarto and Tran 2011; Jogaiah et al. 2013).

A key critical factor of the use of this technology is the potential to increase yields under both water deficit and normal well-watered growing conditions. Genetic engineering faces criticism, and the use of genetically modified soybeans is already banned in countries like Russia, Serbia and Indonesia (Table 10.1).

Table 10.1 World average yields, area and production of leading grain crops for 2018–2019 projections

	Yield (mmt)	Area (m.ha)	Production (mmt.ha[1])	GMO crops
Cotton				
Brazil	12.8	1.6	0.00175	++
China	27.8	3.5	0.00175	+
Corn				
Brazil	100.0	17.5	5.18	++
Russia	77.0	57.8	2.64	–
Algeria	4.0	19.2	1.74	–
Tunisia	1.5	0.6	1.72	+
Morocco	2.9	0.1	2.19	+
Syria	4.8	1.6	1.45	+
Iraq	4.8	2.4	2.72	+
Iran	16.8	6.7	1.81	–
India	100.0	29.8	2.28	+
Rice				
China	7.03	30.2	148.5	+
India	3.91	44.5	116.0	+
Indonesia	4.79	12.2	37.1	–
Brazil	6.29	1.75	7.48	++
Egypt	8.78	0.46	2.8	+
United States	1.18	8.62	7.12	++
Soybean				
United States	123.7	35.7	3147.9	++
Brazil	3.24	36.1	117.0	++
China	1.89	8.4	15.9	+
India	1.05	11.0	11.5	+
Canada	1.47	2.74	4.03	+
Russia	1.47	2.74	4.03	–
Indonesia	1.27	0.41	0.52	–
Serbia	2.84	0.22	0.63	–
Mexico	1.25	0.19	0.34	+
South Africa	1.75	0.73	1.28	+
Iran	2.29	0.07	0.16	–

Data sourced from Foreign Agricultural Service/United States Department of Agriculture, World Agricultural Production 5–19 LOC (2019)

Note: mmt- million metric tons, m.ha- million hectares, and mmt.ha^{-1}- million metric tons per hectare. GMO crops- countries that allow cultivation of genetically modified crops (+), expanded the area cultivated with genetically modified crops (++), restrictions and ban of import, distribution and commercialization of genetically modified crops (–)

Genetically modified organisms' bans are still lifted, although, conventional breeding is somewhat inconvenient, time consuming and labour intensive, producing unstable and un-uniform hybrids compared to genetic engineering (Jiang 2013). Serbia is one of the three largest soybean producers in Europe and at the same time the only country that is self-sufficient in the production and processing of non-genetically modified soybean for the purpose of domestic food and feed industry (Zivkov et al. 2016).

But, increased demand and price at World market signifies the need for research investment into modern technologies targeting improved growth and enhanced quality of seeds to lower soybean market price. Adoption of the law legally prohibiting trade in genetically modified soybean products by countries like Serbia, Russia and Algeria primarily increased the demand, negatively affected yield when compared to World market (Table 10.1) and influenced the increase in price. Some researchers hold the view that, benefits of genetic engineering or use of genetically modified organisms should be balanced with their public acceptance and presumable risks that they can bring (Rosculete et al. 2018).

But, it is still hard to ascertain which of the negative effects (health, nutritional or environmental) occurs solely as a consequence of the use and consumption of genetically modified soybean products. Unverified reports arguing that genetically modified foods can cause allergies of different types in humans, which may subsequently affect their immune systems caused a fierce debate globally. The importance of genetically modified organisms attached to the characteristics derived from new gene fragments inserted, giving them resistance advantages compared to non-genetically modified organisms cannot be ignored. Rosculete et al. (2018) stated that, for countries that are part of the European Union Member States, cultivation of transgenic soybean plants is prohibited, and this limit cultivation and heavily increases soybean imports from Argentina, Brazil and United States of America.

10.11.1 Consumer Perception on Genetically Modified Soybeans

Although, the cultivation of genetically modified organisms is not permitted in some countries, others like China and USA are the biggest growers and consumers of genetically modified food products. Genetically modified soybeans falls within this controversial category and seemingly, there is a divide between consumers' attitude toward genetically modified food safety and agribusiness that has economic interests. Deng et al. (2019) used regression models to reveal that agribusiness is more concerned about genetically modified food consumption than the public and opposes transgenic crop production, with those already profiting by using genetically modified crops or doing research on them supportive of this biotechnology application.

According to the National Technology Innovation Planning of China, the government vowed to strengthen research on genetically modified crops and promote industrialisation of herbicide-tolerant soybeans until the year 2021 (NTIP 2016; Deng et al. 2019). Such decision are constantly made by many countries' states officials on behalf of many consumers regardless of widespread negative reporting on genetically modified technology application worldwide. There are further notable differences in the attitudes and perceptions towards genetically modified foods amongst researchers, consumers and law makers. The controversy around Roundup Ready soybeans exacerbated the problem especially in the European Union, Japan and other countries (as indicated in Table 10.1) with several reports suggesting some health and environment risks.

Herman (2003) made exonerating findings on allergenic reactions associated with proteins expressed in genetically modified crops, including soybean. This report highlighted the lack of allergenic risks beyond intrinsic allergens naturally found in this plant. Similarly, Tsai et al. (2017) found insignificant results in the protein concentrations of *Glym4* and *GlymBd* 30 k compared between non-transgenic and transgenic soybean plants. This report further showed no increases in allergic reaction of patients following exposure to genetically modified soybean, especially persons who had pre-existing allergy to birch pollen and cow's milk casein, which triggers the expression of these aforesaid proteins.

As the commitment of organic agriculture focuses strictly on restricting the use synthetic agrochemicals, use of additives and cultivation of genetically modified organisms, consumers must, however, be given a fair choice to purchase either organic or genetically modified foods. This would mean that the increased/decreased demand for genetically modified foods should be a determining factor for either an increase or decrease in transgenic crop production. In contrary, this discourse appears fuelled by more speculations apart from the notable and acceptable consumer concerns, as well as the promotion of non-genetically modified organic foods by organic agribusiness managers (Chern and Rickertsen 2001). Unfortunately, without a fair assessment debate of transgenic crops and direct tangible benefits to the consumers, genetically modified food products, including those of soybean will remain inferior compared to their non-genetically modified counterparts (Magnusson 2004).

10.11.2 Socio-Economic Benefits of Genetically Modified Soybeans

There are several studies indicating that seed and biotechnology industries support the development and cultivation of genetically modified soybeans. Increased sales and substantial profit margins were widely recorded from the commercialisation of transgenic crops since they were first introduced in the market in 1994 starting first with a Flavr Savr tomato (Clarke et al. 2014; Deng et al. 2019). A major driving force for this growth, especially for the soybean market, has been the introduction of varieties that are herbicide tolerant, resistant to pests and a few showing tolerance to harsh environmental conditions, in addition to varieties with improved oleic acid and linoleic fatty acid profiles (Wunderlich and Gatto 2015). Such transgenic varieties had improved productivity and reduced the risks associated with cultivation costs, production and nutritional quality.

Worldwide consumption of soybean products has increased due to the nutrients and bioactive compounds contained within the seeds. Soybean also contain a wide variety of constituents (Table 10.2) and phytochemicals such as phytic acid (2.2%), sterols (0.46%) and saponins (6.16%) with potential health effects (Rizzo and Baroni 2018). The high quality plant proteins and lower levels of allergens or associated mycotoxins obtained in genetically modified soybeans influences price premia of more than 20% soy products compared to their conventional counterparts. The adoption of transgenic crops led to improved subsistence level carrying capacity of 66% for farmers, farm workers and consumers. Furthermore, increased farm income as a result of these higher yields, additional investment in family income, education, leisure and healthcare was also recorded (Brookes and Barfoot 2008).

Planting area for both non-genetically modified and genetically modified soybeans must be increased since reduced productions means less export and more import. Soybean production undoubtedly has potential economic contribution to many agricultural industries, mostly in developing countries. The demand is increasing for use in livestock feed, vegetable oil and biodiesel production (Mangena 2018). According to Brookes and Barfoot (2014) commercialisation of transgenic crops has significantly increased the net economic benefits at the farm level of about $18.8 billion in 2012 and $116.6 billion in nominal terms. This global consumption rate continued to accelerate due to the introduction of herbicide tolerant soybeans, with their industry ultimately reaching 200 million tons production levels.

Table 10.2 Popular soy products and nutritional constituents quantified from 100 g and 1 kg raw samples of homogenised soybean seeds

Soybean products	Mineral nutrients (mg)	Vitamins (mg)	Organic compounds (g)
Soymilk	Calcium (240)	Folate (0.37)	Fats (18.6)*
Soycheese	Iron (9.7)	Nicotinic acid equivalence (7.9)	Isoflavones (0.16)*
Miso	Magnesium (250)	Riboflavin (0.27)	Protein (35.9)*
Soy oil	Phosphorus (3.74)	Thiamine (0.61)	Sugars (15.8)*
Soy sauce	Potassium (18.42)	Vitamin E (2.9)	
Tofu	Sulfur (1.98)		
Soy fruit "yoghurt"	Zinc (4.3)		

Source: British Nutrition Foundation (2002), Vargas et al. (2018) and Rizzo and Baroni (2018). Only data with asterisks (*) were determined using g/kg raw seed sample.

Furthermore, Argentina and Brazil continue to generate about 2.3–10 billion US dollars per year from soybean exports to Europe (Boerema et al. 2016).

Issues like uncontrollable population growth, scarcity of fertile agricultural land, depleted irrigation/freshwater resources, unintended farming consequences and high food demands necessitate the development and implementation of crop genetic manipulation technology. These suggestions are supported by the myriad of reports demonstrating the critical roles played by biotechnology in various processing and manufacturing industries (Brookes and Barfoot 2008; Hu and Xiong 2014; Boerema et al. 2016; Zhang et al. 2018).

10.12 General Remarks and Conclusions

Despite their great diversity in form and size, all crop plants carry out similar physiological processes in order to reproduce and cope with environmental stress. All plant tissues, organs and whole organisms show a growth polarity, being derived from axial or radial polarity of cell division of the meristems, which particularly serves to improve growth under different conditions (Taiz et al. 2015). It is now clear that environmental conditions may be limiting to plant growth and reproduction. However efficient growth, differentiation and physiological processes remain essential requirements to confer stress tolerance in many varieties. As a result, the molecular approaches used to establish any form of resistance or tolerance against any attack on plant growths must be well optimised, efficient and made highly genotype independent, especially for soybean varieties.

To combat all hurdles, plant genetic engineering has now emerged as a rapid, feasible and more successful crop improvement tool. This technology involves simple introgression of desired foreign genes into hosts, subsequently altering host genome to produce newly improved genomic combinations (Lee et al. 2012; Hwang et al. 2017). The tool could be efficaciously used to genetically improve most legume species that, are an abundant source of protein and oil serving as major food crops worldwide. This is achieved with disturbing the crop's natural and favourable characteristics. Legume food crops like soybean are cultivated in both temperate and tropical environments. Therefore, the most compelling contribution of plant transformation may be anticipated for the development of water stress tolerant traits. Following this, major advantageous use of the technique expected could be to breed varieties that show high yield of nutrients and oils.

In conclusion, soybean as a recalcitrant legume crop have the potential to be improved for abiotic stress resistance against drought, salinity and extreme temperatures using the abovementioned procedure, which is central to basic and applied molecular biology. Although, it is well understood that, traits such as abiotic stress tolerance are usually complex (Telem et al. 2013). Genetic engineering is generally the most preferred method to overcome conventional breeding barriers because only specific genes of interest are isolated and inserted into hosts than a mixture of desirable and undesirable genes. It could be optimised surpassing narrow genetic base, especially of elite soybean cultivars, inefficient *in vitro* regeneration protocols and poor selection regimes devoid of chimerism (Mangena et al. 2017).

Acknowledgements Funding for this work is provided by the Department of Research Administration and Development of the University of Limpopo, and in part by both the National Research Foundation and the Department of Higher Education and Training under the New Generation of Academics Programme in South Africa. The author would like to thank Dr. PW Mokwala, Prof RV Nikolova and Colleagues in the Department of Biodiversity for their continued support.

References

Acquaah G (2012) Principles of plant genetics and breeding, 2nd edn. Wiley, Hoboken, pp 629–638

Aftabi M, Negawo AT, Hassan F (2018) Improved protocol for *Agrobacterium*-mediated transformation of pea (*Pisum sativum*). Mol Biol 7(202):1–6. https://doi.org/10.4172/2168-9547.1000202

Aizat WM, Hassan M (2018) Proteomics in system biology. Adv Exp Med Biol 1102:31–49. https://doi.org/10.1007/978-3-319-98758-3_3

Aktar MW, Sengupta D, Chowdhury A (2009) Impact of pesticides use in agriculture: their benefits and hazards. Interdiscip Toxicol 2(1):1–12. https://doi.org/10.2478/v10102-009-0001-7

AL-Yozbaki GSH, Racheed JR, Salih SM (2015) Transformation of soybean (*Glycine max* L.) via GUS- labelled *Agrobacterium rhizogenes* R1000. Int J Sci Technol 4(6):267–272. www.journalofsciences-technology.org/archive/2015/june-vol-4-no-6/815014294456.pdf

Barona E, Ramankutty N, Hyman G, Coomes OT (2010) The role of pasture and soybean in deforestation of the Brazilian Amazon. Environ Res Lett 5(2):1–9. https://doi.org/10.1088/1748-9326/5/2/024002

Behrooz D, Safar F, Mahmoud T, Saeed Z, Shahin N, Stewart NC Jr (2008) DNA-delivery methods to produce transgenic plants. Biotech 7:385–402. https://doi.org/10.3923/biotech.2008.385.402

Bellaloui N, Ebelhar MW, Gillen AM, Fisher DK, Abbas HK, Mengistu A, Reddy KN, Paris RL (2011) Soybean seed protein, oil, and fatty acids are altered by S and S + N fertilizers under irrigated or non-irrigated environments. Agric Sci 2(4):465–476. https://doi.org/10.4236/as.2011.24060

Bharadwaj CH, Satyavathi CT, Tiwari SP, Karmakar PG (2002) Genetic base of soybean (*Glycine max*) varieties released in India as revealed by coefficient percentage. Indian J Agric Sci 72(8):467–469. http://epuds.icr.org.in/ejournal/index.php/iJAgS/article/view/40986. Date accessed: 07/06/2019

Bhatia A, Yadav A, Zhu C, Gagner J, Radhakrishnan A, Steinmetz LM, Bhanot G, Sinha H (2014) Yeast growth plasticity is regulated by environment-specific multi-QTL interactions. G3 (Bethesda) 4(5):769–777. https://doi.org/10.1534/g3.113.009142

Bickel U, Dros JM (2003) The impacts of soybean cultivation on Brazilian ecosystems- there case studies commissioned by the Worldwide Fund for Nature (WWF) forest conversion initiative, pp 1–33. http://assets.panda.org/downloads/impactofsoybean.pdfGooglescholar

Birch RG (1997) Plant transformation: problems and strategies for practical application. Ann Rev Plant Biol 48:297–326. https://doi.org/10.1146/annrev.arplant.48.1.297

Bliffeld M, Mandy J, Potrykus I, Futterer J (1999) Genetic engineering of wheat for increased resistance to powdery mildew disease. Theor Appl Genet 98:1079–1086. https://doi.org/10.1007/s001220051

Boerema A, Peeters A, Swolfs S, Vandevenne F, Jacobs S, Staes J, Meire P (2016) Soybean trade: balancing environmental and socio-economic impacts of an intercontinental market. Plos ONE 11(5):e0155222. https://doi.org/10.1371/journal.pone.0155222

Booy G, Krens FA, Huizing HS (1989) Attempted pollen-mediated transformation of maize. J Plant Physiol 135(3):319–324. https://doi.org/10.1016/S0176-1617(89)80126-9
Bouhabida M, Srarfi F, Saadi I, Kharrat M (2013) Molecular characterisation of pea (*Pisum sativum* L.) using microsatellite markers 10SR. J App Chem 5(1): 57–61. www.iosrjournals.org. 26/07/2019
British Nutrition Foundation (2002) Briefing paper- soya and health. BNF Publications, pp 1–38. https://www.nutrition.org.uk. 30/06/2019
Brodt S, Six J, Feenstra G, Ingels C, Campbell D (2011) Sustainable agriculture. Nat Edu Knowl 3(10):1
Brookes G, Barfoot P (2008) Global impact of biotech crops: socio-economic and envirmnetal effects, 1996-2006. AgBioforum 11(1):21–38. https://www.agbioforum.org/v11n1/v11n1a03-brookes.pdf_31/07/2019
Brookes G, Barfoot P (2014) Economic impact of GM crops: the global income and production effects 1996-2012. GM Crop Food 5(1):65–75. https://doi.org/10.4161/gmcr.28098
Chen JH, Wu JT, Young CC (2007) The combined use of chemical, organic fertilizers and/or biofertilizer for crop growth and soil fertility. Environ Sci:1–11. https://doi.org/10.30058/SE.200706.001
Chern WS, Rickertsen K (2001) Consumer acceptance of GMO: survey results from Japan, Norway, Taiwan and the United States. Taiwan Agric Eco Rev 7(1):1–28. https://pdfs.semanticscholar.org/8837/c08febba81d04d04517e225d043ec5820439.pdf
Christou P (1994) Gene transfer to plants via particle bombardment. In: Gelvin SB, Schilperoort RA (eds) Plant molecular biology manual. Springer, Dordrecht. https://doi.org/10.1007/978-94-011-0511-8_2
Chumakov MI, Rozhok NA, Velikov VA, Tyrnov VS, Volokhina IV (2006) *Agrobacterium*-mediated *in planta* transformation of maize via pistil filaments. Russ J Genet 42:893–897. https://doi.org/10.1134/s1022795406080072
Clarke JD, Alexander DC, Ward DP, Ryals JA, Mitchell MW, Wulff JE, Guo L (2014) Assessment of genetically modified soybean in relation to natural variation in the soybean seed metabolome. Sci Rep 4(3871):1–6. https://doi.org/10.1038/srep03871
Clough SJ, Bent AF (1998) Floral dip: a simplified method for *Agrobacterium*-mediated transformation of *Arabidopsis thaliana*. Plant J 16(6):735–743. https://doi.org/10.1046/j.1365-313x.1998.00343.x
Connett MD, Gladfelter HJ, Gulledge JE, McCormack RR (2002) Enhanced transformation and regeneration of transformed embryogenic pine tissue. US Patent Published Application No. 20020100083
Cruz de Carvalho MH (2008) Drought stress and reactive oxygen species. Plant Signal Behav 3(3):156–165. https://doi.org/10.4161/psb.3.3.5536
Dai AG (2011) Drought under global warming: a review. Wires Clim Chang 2:45–39. https://doi.org/10.1002/wcc.81
Das Bhowmik SS, Chang AY, Long H, Tan GZH, Huang TML, Karbaschi MR, Williams B, Higgins TSU, Mundree SG (2019) Robust genetic transformation system to obtain non-chimeric transgenic chickpea. Front Plant Sci 10:524. https://doi.org/10.3389/fpls.2019.00524
Deng H, Hu R, Pray C, Jin Y (2019) Perception and attitude towards GM technology among agribusiness managers in China as producers and as consumers. Sustainability 11(1342):1–17. https://doi.org/10.3390/su11105342
Deshayes A, Herrera-Estrella L, Caboche M (1985) Liposome-mediated transformation of tobacco mesophyll protoplasts by an *Escherichia coli* plasmids. EMBO J 4:2731–2737. https://doi.org/10.1002/j.1460-2075.1985.tb03996.x
Devi JM, Sinclair TS, Chen P, Carter TE (2014) Evaluation of elite southern maturity soybean breeding lines for drought-tolerant traits. Agron J 106:1947–1954. https://doi.org/10.2134/agronj14.0242

Du Plessis M (2013) Cysteine protease activity and gene expression studies in soybean nodules during development and drougt stress. Masters Dissertation University of Pretoria, South Africa. http://hdl.handle.net/2263/33171. 26/07/2019

Eapen S (2011) Pollen grains as a target for introduction of foreign genes into plants: an assessment. Physiol Mol Biol Plants 17(1):1–8. https://doi.org/10.1007/S12298-010-0042-6

Emami M, Almassi M, Bolchoda H, Kalantari I (2018) Agricultural mechenization, a key to food security in developing countries: strategy formulating for Iran. Agric Food Sec 7(24):1–12. https://doi.org/10.1186/s40066-018-0176-2

Falck-Zepeda JB, Wesseler JHH, Smyth S (2013) The current status of the debate on socio-economic regulatory assessments: positions and policies in Canada, the USA, the EU and developing countries. World Rev Sci Tech Sustain Dev 10(4):203–227. https://doi.org/10.1504/WRSTSD.2013.057690

Fan XD, Wang JQ, Yang N, Dong YY, Liu L, Wang FW, Wang N, Chen H, Liu WC, Sun YP, Wu JY, Li HY (2013) Gene expression profiling of soybean leaves and roots under salt, saline-alkali and drought stress by high-throughput illumine sequencing. Gene 512(2):392–402. https://doi.org/10.1016/j.gene.2012.09.100

Farooq M, Wahid A, Kobayashi N, Fujita D, Basra SMA (2009) Plant drought stress: effects, mechanisms and management. Agron Sustain Dev 29:185–212. https://doi.org/10.1051/agro:200802

Fathi A, Tari DB (2016) Effect of drought stress and its mechanism in plants. Int J Life Sci 10(1):1–6. https://doi.org/10.3126/ijls.v10i1.14509

Finer JJ, McMullen MD (1991) Transformation of soybean via particle bombardment of embryogenic suspension culture tissue. In Vitro Cell Dev Biol Plant 27:175. https://doi.org/10.1007/BF02632213

Food and Agriculture Organisation of the United Nations (2016) Climate change and food security: risks and responses. FAO I5188E/1/11.15. http://www.fao.org/3/a-i5188e.pdf

Fu YB (2015) Understanding crop genetic diversity under modern plant breeding. Theor Appl Genet 128(11):2131–2142. https://doi.org/10.1007/s00122-015_2585-y

Gad AE, Rosenberg N, Altman A (1990) Liposome-mediated gene delivery into plant cells. Physiol Plant 79:177–183. https://doi.org/10.1111/j.1399-3054.1990.tb05883

Gautam HR, Bhardwaj MJ, Kumar R (2013) Climate change and its impact on plant disease. Curr Sci 105(12):1685–1691. https://www.jstor.org/stable124099750

Ghorbanzade Z, Ahmadabadi M (2015) Stable transformation of the *Saint paulia ionantha* by particle bombardment. Iran J Biotechnol 13(1):e1037. https://doi.org/10.1571/ijb.1037

Gilreath JP, McSorley R, McGovern RJ (1999) Soil fumigant and herbicide combinations for soilborne pest control in caladium. Proc Fla State Hortic Soc 112:285–290. https://www.journals.fcla.edu/fshs/article/download/91326/87525

Goh HH, Khairudin K, Sukiran NAB, Noor NM, Baharum SN (2016) Metabolite profiling reveals temperature effects on the VOCs and flavonoids of different plant populations. Plant Biol 1:130–139. https://doi.org/10.1111/plb.12403

Gould F (1998) Sustainability of transgenic insecticidal cultivars: integrating pest genetics and ecology. Annu Rev Entomol 43:701–726. https://doi.org/10.1146/annurev.ento.43.1.701

Gould J, Devey M, Hasegawa O, Ulian EC, Peterson G, Smith RH (1991) Transformation of *Zea mays* L. using *Agrobacterium tumefaciens* and the shoot apex. Plant Physiol 95:426–434. https://doi.org/10.1104/pp.95.2.426

Grela ER, Riczorowska B, Samolinska W, Matras J, Kickzorowski P, Rybinski W, Hanczakowska E (2017) Chemical composition of leguminous seeds: part 1-content of basic nutrients, amino acids, phytochemical compounds and antioxidant activity. Eur Food Res Technol 243(8):1385–1395. https://doi.org/10.1007/s00217-017-2849-7

Gyang PJ, Nyaboga E, Muge EK (2017) Molecular characterisation of common bean (*Phaseolus vulgaris* L.) genotypes using microsatellite markers. J Adv Biol Biotechnol 13(2):1–15. https://doi.org/10.9734/JABB/2017/33519

Hadiarto T, Tran TS (2011) Progress studies of drought-responsive genes in rice. Plant Cell Rep 30:297–310. https://doi.org/10.1007/s000299-010-0956-2

Hassan M, Akram Z, Ali S, Ali GM, Zafari Y, Shah ZH, Alghabari F (2016) Whisker-mediated transformation of peanut with chitinase gene enhances resistance to leaf spot disease. Crop Breed Appl Biotechnol 16(2):108–114. https://doi.org/10.1590/1984-70332016v16n2a17

He YQ, Cheng JP, Liu LF, Li XD, Yang B, Zhang HS, Wang ZF (2015) Effect of pre-harvest chemical application on rice dessication and seed quality. J Zhejiang Univ-Sci B (Biomed & Biotechnol) 16(10):813–823. www.zju.edu.cn/jzus;www.springerlink.com

Herman EM (2003) Genetically modified soybeans and food allergies. J Exp Bot 54(386):1317–1319. https://doi.org/10.1093/jxb/erg164

Hinchee MAW, Connor-Ward DV, Newell CA, McDonnell RE, Sato SJ, Gasser CS, Fischhoff DA, Re DB, Fraley RT, Horsch RB (1988) Production of transgenic soybean plants using *Agrobacterium*-mediated DNA transfer. Nat Biotechnol 6(8):915–922. https://doi.org/10.1038/nbt0888-915

Hossain M, Vereklaas EJ, Hardy ESJ, Poot P (2019) Tree host-pathogen interactions as influenced by drought timing: linking physiological performance, biochemical defence and disease severity. Tree Physiol 39(1):6–18. https://doi.org/10.1093/treephys/tpy113

Hu H, Xiong L (2014) Genetic engineering and breeding of drought resistance crops. Annu Rev Plant Biol 65:715–741. https://doi.org/10.1146/annrev-arplant-050213-040000

Hwang HH, Yu M, Lui EM (2017) *Agrobacterium*-mediated plant transformation: biology and applications. In: Shan L (ed) The Arabidopsis book. American Society of Plant Biologists, Rockville, pp 1–31. https://doi.org/10.1199/tab.0186

Ibrahim RA, Shawer DM (2014) Transgenic Bt-plants and the future of crop protection (an overview). Int J Agric Food Res 3(1):14–40. https://doi.org/10.24102/ijafr.v3i1412

Ighodaro OM, Akinloye AO (2018) First line defence antioxidants-superoxide dismutase (SOD), catalase (CAT) and glutathione peroxidase (GPX): their fundamental role in the entire antioxidant defence grid. Alex J Med 54(4):287–293. https://doi.org/10.1016/j.ajme.2017.09.001

Ijarotimi OS, Famurewa JAV (2006) Assessment of chemical compositions of soybean supplemented weaning foods and nutritional knowledge of nursing mothers on their utilization. Pak J Nutr 5(3):218–223. https://doi.org/10.3923/pjn.2006.218.223

Ismagul A, Yang N, Maltseva E, Iskakova G, Mazonka I, Skiba Y, Bi H, Eliby S, Jatayev S, Shavrukov Y, Bovisjuk N, Longridge P (2018) A biolistic method for high-throughput production of transgenic wheat plants with single gene insertions. BMC Plant Biol 18(135):1–8. https://doi.org/10.1186/s12870-018-1326-1

Jaleel CA, Manivannah P, Wahid A, Farooq M, Al-Juburi HJ, Somasundaram R, Panneerselvam R (2009) Drought stress in plants: a review on morphological characteristics and pigments composition. Int J Agric Biol 11: 100–105. http://fspublishers.org. 22/06/2019

Jiang GL (2013) Plant marker-assisted breeding and conventional breeding: challenges and perspectives. Adv Crop Sci Tech 1:e106. https://doi.org/10.4172/2329-8863.1000e106

Jogaiah S, Govind SR, Tran LSP (2013) System biology-based approaches towards understanding drought tolerance in food crops. Crit Rev Biotechnol 33(1):23–39. https://doi.org/10.3109/07388551.2012.659174

Kaeppler HF, Gu W, Somers DA, Rines HW, Cockburn AF (1990) Silicon carbide fiber-mediated DNA delivery into plant cells. Plant Cell Rep 9(8):415–418. https://doi.org/10.1007/BF00232262

Kajiyama S, Inoue F, Yoshikama Y, Shoji T, Fukusaki E, Kobayashi A (2007) Novel plant transformation system by gene-coated gold particle introduction into specific cell using ArF excimer laser. Plant Biotechnol 24:315–320. https://www.jcpcmbjp/

Kandil AA, Sharief AE, Sheteiwy MS (2013) Effect of seed storage periods, conditions and materials on seed quality of some soybean cultivars. Int J Agric Sci 5(1):339–346. http://www.bioinfopublication.org/jouarchive.php?opt=&jouid=BPJ0000217

Karahalil B (2016) Overview of system biology and omics technologies. Curr Med Chem 23:1–10. https://doi.org/10.2174/0929867236661609261506l7

Karpe F, Hamsten A (1994) Determination of apolipoproteins B-48 and B-100 in triglyceride—rich lipoproteins by analytical SDS-PAGE. J Lipid Res 35(7): 1311–1316. www.jir.org 23/07/2019

Keneni G, Bekele E, Imtiaz M, Dagne K (2012) Genetic vulnerability of modern crop cultivars: causes, mechanisms and remedies. J Plant Res 2(3):69–79. https://doi.org/10.5923/j.plant.2012020305

Khan A, Pan X, Nejeek V, Tan DKY, Fahad S, Zahoor R, Luo H (2018) Coping with drought stress and adaptive mechanisms, and management through cultural and molecular alternatives in cotton as vital constituents for plant stress resilience and fitness. Biol Res 51(47):1–17. https://doi.org/10.1186/s40650-018-0198-2

Khojely DM, Ibrahim SE, Sapey E, Han T (2018) History, current status and prospectus of soybean production and research in sub-Saharan Africa. Crop J:226–235. https://doi.org/10.1016/j.cj.2018.03.006

Kim HJ, Cho HS, Pak JH, Kwon T, Lee JH, Kim DH, Lee DH, Kim CG, Chung YS (2018) Confirmation of drought tolerance of ectopically expressed AtABF3 gene in soybean. Mol Cells 41(5):413–422. https://doi.org/10.14348/molcells.2018.2254

Kinoshita E, Kinoshita-Kikuta E, Koike T (2009) Separation and detection of large phosphoproteins using Phos-tag SDS-PAGE. Nat Protoc 4(10):1513–1521. https://doi.org/10.1038/nprot.2009.154

Korth K (2008) Genes and traits of interest for transgenic plants. In: Stewart CN (ed) Plant biotechnology and genetics: principles, techniques and applications. Wiley, Hoboken. https://doi.org/10.1002/9780470282014.ch8

Kunert KS, Vorster BJ, Fenta BA, Kibido T, Dionisio G, Foyer CH (2016) Drought stress responses in soybean roots and nodules. Front Plant Sci 7:1015. https://doi.org/10.3389/fpls.2016.01015

Lee H, Park SY, Zhang ZJ (2012) An overview of genetic transformation of soybean. In: Board J (ed) A comprehensive survey of international soybean research- genetics, physiology, agronomy and nitrogen relationships. InTech Open, London, pp 489–506. https://doi.org/10.5772/51076

Lesk AM (2010) Introduction to protein science. Oxford University Press, New York, pp 8–25

Li S, Xia Q, Wang F, Yu X, Ma J, Kou H, Lin X, Gao X, Liu B (2017) Laser irradiation-induced DNA methylation changes are heritable and accompanied with transpositional activation of *mPing* in rice. Front Plant Sci 8:363. https://doi.org/10.3389/fpls.2017.00363

Library of Congress (LOC) (2019) Restrictions on genetically modified organisms research and reports. http://www.loc.gov. Date accessed: 19/06/2019

Liu HC, Wei ZM (2005) Recent advances in soybean genetic transformation. J Plant Physiol Mol Biol 31(2):126–134. https://www.zju.edu.cn/jzus

Lui K (1997) Chemistry and nutritional value of soybean components. In: Boston MA (ed) Soybeans. Springer. https://doi.org/10.1007/978-1-4615

Mabulwana PT (2013) Determination of drought stress tolerance among soybean varieties using morphological and physiological markers. Masters Dissertation, University of Limpopo, Sovenga, Turfloop, South Africa. https://www.proteinresearch.net/poems/images/projects/0073/abstract/1-2-4a-1-4o-sp-mabulwana-pt-2013.pdf. 28/06/2019

Magnusson M (2004) Consumer perception of organic and genetically modified foods. Health and environmental considerations. Acta Universitatis Upsaliensis 137(71): 11–14. http://www.diva-portal.org/smash/search.jsf?dswid=6884. 28/07/2019

Maisa R, Lambert F, Ramirez-Villegas J, Challinor AJ (2019) Emergence of robust precipitation changes across crop production areas in the 21st century. Proc Nat Acad Sci 116(14):6673–6678. https://doi.org/10.1073/pnas.1811463116

Mangena P (2018) Water stress: morphological and anatomical changes in soybean (*Glycine max* L.) plants. In: Andjelkovic V (ed) Plant, abiotic stress and responses to climate change. Intech Open, London, pp 10–31. https://doi.org/10.5772/intechopen.72899

Mangena P (2019) A simplified in-planta genetic transformation in soybean. Res J Biotechnol 14(9):117–125. https://worldresearchersassociations.com/Archives/RJBT/Vol(14)2019/September2019.aspx

Mangena P, Mokwala PW (2019) The influence of seed viability on the germination and *in vitro* multiple shoot regeneration of soybean (*Glycine max* L.). Agriculture 9(2):35. https://doi.org/10.3390/agriculture9020035

Mangena P, Mokwala PW, Nikolova RV (2017) Challenges of *in vitro* and *in vivo Agrobacterium*-mediated genetic transformation in soybean. In: Kasai M (ed) Soybean- the basis of yield, biomass and productivity. InTech Open, Rijeka, pp 75–94. https://doi.org/10.5772/66708

Mannan MA, Halder E, Karim MA, Ahmed JU (2016) Alleviation of adverse effect of drought stress on soybean (*Glycine max* L.) by using poultry litter biochar. Bangladesh Agron J 19(2):61–67. https://doi.org/10.3329/baj.v19i2.31853

Masani MYA, Noll GA, Parveez GKA, Sambanthamurthi R, Prufer D (2014) Efficient transformation of oil palm protoplasts by PEG-mediated transfection and DAN microinjection. PLoS One 9(5):e96831. https://doi.org/10.1371/journal.pone.0096831

McArthur JW, McCord GC (2017) Fertilizing growth: agricultural inputs and their effects in economic development. J Dev Econ 127:133–152. https://doi.org/10.1016/j.jdeveco.2017.02.607

Meacham JM, Durvasula K, Degertekin FL, Fedorov AG (2014) Physical methods for intracellular delivery: practical aspects from laboratory use to industrial-scale processing. J Lab Autom 19(1):1–18. https://doi.org/10.1177/2211068213494388

Messina MJ (1999) Legumes and soybeans: overview of their nutritional profiles and health effects. Am J Clin Nutr 70(3):4395–4505. https://doi.org/10.1093/ajcn/70.3.4395

Mimi A, Mannan M, Khaliq Q, Mia M (2017) Yield response of soybean (*Glycine max* L.) genotypes to water deficit stress. Bangladesh Agron J 19(2):51–60. https://doi.org/10.3329/baj.v19i2.31853

Minorsky PV (2003) The hot and the classic. Plant Physiol 131:6–7. https://doi.org/10.1104/pp.900071

National Technology Innovation Planning No. 43 (2016) State Council Thirteenth Five-Year. http://www.gov.cn/zhengce/content/2016-08/08/content_5098072.htm. 29/06/2019

Nyaboga E, Tripathi JN, Manoharan R, Tripathi L (2014) *Agrobacterium*-mediated genetic transformation of yam (*Dioscorea rotundata*): an important tool for functional study of gens and crop improvement. Front Plant Sci. https://doi.org/10.3389/fpls.2014.00463

Paz MM, Shou H, Guo Z, Zhang Z, Benerjee AK, Wang K (2004) Assessment of conditions affecting *Agrobacterium*-mediated soybean transformation using the cotyledonary node explant. Euphytica 136:167–179. https://doi.org/10.1023/B:EUPH.0000030669.75809.dc

Paz MM, Martinez JC, Kalvig AB, Fonger TM, Wang K (2006) Improved cotyledonary node method using an alternative explant derived from mature seed for efficient *Agrobacterium*-mediated soybean transformation. Plant Cell Rep 25(3):206–213. https://doi.org/10.1007/s00299-005-0048-7

Pinheiro C, Chaves MM (2010) Photosynthesis and drought: Can we make metabolic connections from available data? J Exp Bot 62(3):869–882. https://doi.org/10.1093/jxb/erg340

Prince SJ, Joshi T, Mutava RN, Syed N, Vitor MDSS, Patil G, Nguyen H (2015) Comparative analysis of the drought-responsive transcriptome in soybean lines contrasting for canopy wilting. Plant Sci 240:65–78. https://doi.org/10.1016/j.plantsci.2015.08.077

Rashid AHA, Lateef DD (2016) Novel technique for gene delivery into plants and its applications for disease resistance in crops. Am J Plant Sci 7:181–193. https://doi.org/10.4236/ajps.2016.71019

Raza G, Singh MB, Bhalla PL (2017) *In vitro* plant regeneration from commercial cultivars of soybean. Biomed Res Int 1(1):1–9. https://doi.org/10.1155/2017/7379693

Rech EL, Vianna GR, Aragao FJL (2008) High-frequency transformation by biolistic of soybean, common bean and cotton transgenic plants. Nat Protoc 3:410–418. https://doi.org/10.1038/nprot.2008.9

Rivas CF, Condon F (2015) Plant domestication and utilization: the case of the Pampa biome. In: Al-Khayri JM, Jain SM, Johnson DV (eds) Advances in plant breeding strategies: breeding, biotechnology and molecular tools. Springer, Cham, pp 3–24. https://doi.org/10.1007/978-3-319-22521-0_1

Rizwan M, Akhtar S, Aslam M, Asghar MJ (2015) Development of herbicide resistant crops through induced mutations. Adv Life Sci 3(1):34–38

Rizzo G, Baroni L (2018) Soy, soy foods and their role in vegetarian diets. Nutrients 10(1):43. https://doi.org/10.3390/nu10010043

Ropert C (1999) Liposomes as a gene delivery system. Braz J Med Biol Res 32(2):163–169. https://doi.org/10.1590/S0100-879X1999000200004

Rosculete E, Bonciu E, Rosculete CA, Teleanu E (2018) Detection and quantification of genetically modified soybean in some food and feed products. A case study on product available on Romanian market. Sustainability 10(1325):2–13. https://doi.org/10.3390/su10051325

Ruma D, Dhaliwal MS, Kaur A, Gassal SS (2009) Transformation of tomato using biolistic gun for transient expression of the *ß-glucuronidase* gene. Indian J Biotechnol 8:363–369. http://nopr.niscair.res.in/handle/123456789/45280

Sagi L, Panis B, Remy S, Schoofs H, Smet KD, Swennen R, Cammue BPA (1995) Genetic transformation of banana and plantain (*Musa spp.*) via particle bombardment. Bio/Technology 13:481–485. https://doi.org/10.1038/nbt0595_481

Sah SK, Kaur A, Kaur G, Cheema GS (2014) Genetic transformation of rice: problems, progress and prospects. J Rice Res 3:132. https://doi.org/10.4172/2375-4338.1000132

Samaha GA, Ahmed MA, Abd El-Hameid AR (2019) Assessment of growth and productivity of five peanut cultivars and genetic diversity using RAPD markers. Bull Natl Res Cent 43(1):168. https://doi.org/10.1186/s42269-019-0201-z

Sandford JC, Klein TM, Wolf ED, Allen N (1987) Delivery of substances into cells and tissues using a particle bombardment process. Part Sci Technol 5:27–37. https://doi.org/10.1080/02726358708904533

Sangwan RS, Ducrocq C, Sangwan-Norreel B (1993) Agrobacterium-mediated transformation of pollen embryos in *Datura innoxia* and *Nicotiana tubacum*: production of transgenic haploid and fertile homozygous diploid plants. Plant Sci 95(1):99–115. https://doi.org/10.1016/0168-9452(93)90083-C

Sarkar KK, Mannan MA, Haque MM, Ahmed JU (2015) Physiological basis of water stress tolerance in soybean. Bangladesh Agron J 18(2):71–78

Serik O, Ainur I, Murat K, Tetsno M, Masaki I (1996) Silicon carbide fiber-mediated DNA delivery into cells of wheat (*Triticum aestivum* L.) mature embryos. Plant Cell Rep 16(3–4):133–136. https://doi.org/10.1007/13F01890853

Sevanto S (2014) Phloem transport and drought. J Exp Bot 65(7):1751–1759. https://doi.org/10.1093/jxb/ert467

Sharma N, Singhvi R (2017) Effect of chemical fertilizers and pesticides on human health and environment: A review. Int J Agric Environ Biotechnol 10(6):676–679. https://doi.org/10.5958/2230-372x.2017.00083.3

Shivani, Tiwari S (2019) Enhanced *Agrobacterium*-mediated transformation efficiency of banana cultivar Grand Naine by reducing oxidative stress. Sci Hortic 246:675–685. https://doi.org/10.1016/j.scienta.2018.11.024

Shurtleff W, Aoyagi A (2009) History of soybeans and soyfoods in Africa. Soyinfo Centre, Lafayette, pp 15–26. http://www.soyinfocenter.com/pdf/134/Afri.pdf

Song L, Prince S, Valligodan B, Joshi T, dos Santos JVM, Wang J, Nguyen HT (2016) Genome-wide transcriptome analysis of soybean primary root under varying water-deficit conditions. BMC Genomics 17(57):1–17. https://doi.org/10.1186/s12864-016-2378-y

Souza GM, Catuchi TA, Bertolli SC, Soratto RP (2013) Soybean under water deficit: physiological and yield responses. In: Board J (ed) A comprehensive survey of international soybean research: genetics, physiology, agronomy and nitrogen relationships. Intech Open, London, pp 273–298. https://doi.org/10.5772/54269

Sparks CA, Jones HD (2014) Genetic transformation of wheat via particle bombardment. Methods Mol Biol 1099:201–218. https://doi.org/10.1007/978-1-62703-715-0_17

St Martin SK (1982) Effective population size for the soybean improvement programme in the maturity groups 00 to IV. Crop Sci 22(1):151–152. https://doi.org/10.2135/cropsci1982.0011183X002200010035x

Taiz L, Zeiger E, Moller IM, Murphy M (2015) Plant physiology and development, 6th edn. Sinauer Associates, London, pp 756–760

Tate EL, Gustard A (2000) Drought definition: a hydrological perspective. In: Vogt JV, Somma F, advances in natural and technological hazards research. Springer, Dordrecht, pp 461–431. https://doi.org/10.1007/978-94-015-9472-1_3

Taylor NJ, Fauquet CM (2002) Microparticle bombardment as a tool in plant science and agriculture biotechnology. DNA Cell Biol 21(12):963–977. https://doi.org/10.1089/104454902762053891

Telem RS, Wani SH, Singh NB, Nandini R, Sadhukhan R, Bhattacharya S, Mandal N (2013) Cisgenics- A sustainable approach for crop improvement. Curr Genomics 14:468–476. https://doi.org/10.2174/13892029113146660013

Tripp R (1996) Biodiversity and modern crop varieties: sharpening the debate. Agric Hum Values 13:48–63. https://doi.org/10.1007/BF01530523

Tsai JJ, Chang CY, Liao EC (2017) Comparison of allergenicity at Gly m 4 and Gly mBd 30k of soybean after genetic modification. J Agric Food Chem 65(6):1255–1262. https://doi.org/10.1021/acs.j9fc.6b05135

Tsukakoshi M, Kurata S, Nomiya Y, Ikowa Y, Kasuya T (1984) A novel method of DNA transfection by laser microbeam cell surgery. Appl Phys B Lasers Opt 35:135. https://doi.org/10.1007/BF00697702

Vargas RL, Schuch LOB, Barros WS, Rigo GA, Szareski VJ, Carvalho IR, Primentel JR, Troyjack C, Jacques LBA, deSouza VQ, do Rosa TC, Aumonde TZ, Pedo T (2018) Macronutrients and micronutrients variability in soybean seeds. J Agric Sci 10(4):1–14. https://doi.org/10.5539/jas.v10n4p209

Wang JX, Sun Y, Cui G, Hu J (2001) Transgenic maize plants obtained by pollen-mediated transformation. Acta Bot Sin 43:275–279

Wilson R (2004) Seed composition. In: Boerma H, Specht JE (eds) Soybean: improvement, production and uses. ASA, CSSA and SSSA Agronomy Monograph, Madison, pp 621–677

Wunderlich S, Gatto KA (2015) Consumer perception of genetically modified organisms and sources of infection. Adv Nutr 6(6):842–851. https://doi.org/10.3945/an.115.008870

Xu C, Xia C, Xia Z, Zhou X, Huang J, Huang Z, Liu Y, Jiang Y, Casteel S, Zhang C (2018) Physiological and transcriptomic responses of reproductive stage soybean to drought stress. Plant Cell Rep 37(12):1611–1624. https://doi.org/10.1007/s00299-018-2332-3

Yang A, Su Q, An L (2009) Ovary-drip transformation: a simple method for directly generating vector- and marker-free transgenic maize (*Zea mays* L.) with a linear GFP cassette transformation. Planta 229(4):794–801. https://doi.org/10.1007/s00425-008_0871-5

Zhang L, Yan C, Guo Q, Zhang J, Ruiz-Menjivar J (2018) The impact of agricultural chemical inputs on environment: global evidence from informetrics analysis and visualisation. Int J Low-Carbon Technol 13(4):338–352. https://doi.org/10.1093/ijlct/cty039

Zhao T, Aleem M, Sharmin RA (2018) Adaptation to water stress in soybean: morphology to genetics. In: Andjelkovic V (ed) Plant, abiotic stress and responses to climate change. Intech Open, London, pp 33–68. https://doi.org/10.5772/intechopen.72229

Zivkov G, Teofilovic N, Tar DC (2016) Effects of liberalization of law on GMO on the soybean market in the Republic of Serbia. Danube Soya- Regional Centre Nove Sadi, pp 1–84. https://www.donausoja.org. 28/06/2019

Zurfluh LL, Guilfoyle TJ (1980) Auxin-induced changes in the patterns of protein synthesis in soybean hypocotyl. Proc Natl Acad Sci USA 77(1):357–361. https://doi.org/10.1073/pnas.77.1.357

Index

P. Guleria et al. (eds.), *Sustainable Agriculture Reviews 45*, Sustainable Agriculture Reviews 45, https://doi.org/10.1007/978-3-030-53017-4

U

V

W

X

Y

Z

The manufacturer's authorised representative in the EU is Springer Nature Customer Service Centre GmbH, Europaplatz 3, 69115 Heidelberg, Germany. If you have any concerns regarding our products, please contact ProductSafety@springernature.com

Printed and bound by CPI Group (UK) Ltd, Croydon, CR0 4YY
30/09/2026
02225491-0004